OPTICAL BISTABILITY, DYNAMICAL NONLINEARITY AND PHOTONIC LOGIC

OPTICAL BISTABILITY, DYNAMICAL NONLINEARITY AND PHOTONIC LOGIC

PROCEEDINGS OF
A ROYAL SOCIETY DISCUSSION MEETING
HELD ON 21 AND 22 MARCH 1984

ORGANIZED BY
S. D. SMITH, F.R.S., A. MILLER
AND B. S. WHERRETT
AND EDITED BY
B. S. WHERRETT AND S. D. SMITH, F.R.S.

LONDON
THE ROYAL SOCIETY
1985

Printed in Great Britain for the Royal Society
at the
University Press, Cambridge

ISBN 0 85403 239 8

First published in *Philosophical Transactions of the Royal Society of London*,
series A, volume 313 (no. 1525), pages 187–451

Published by the Royal Society
6 Carlton House Terrace, London SW1Y 5AG

PREFACE

This Discussion Meeting on optical bistability, dynamical nonlinearity and photonic logic has brought together, from a widespread international base, reports on the status, in 1984, of a novel set of signal processing devices that use light beams instead of electric currents to carry information. Just as nonlinear electronic circuit elements are the building blocks of today's digital electronics, so the successors of the devices reported here will constitute the components of digital optical circuits. The meeting is timely since the first optical logic gates capable of addressing succeeding elements in a purely optical manner are reported.

The basic macroscopic physical theory of optical bistability, the microscopic physics of optical nonlinearity and optical computer architecture are the concepts that will mix together to produce a new technology which at present lacks a suitable generic title. 'Optical computing' is at this time too pretentious and too restrictive, 'photonics' has the merit of direct analogy with 'electronics' while 'optonics', suggested by someone in the Commission of the European Communities in Brussels, would be a new word for a new technology. It seems certain that, whatever it is termed, the subjects discussed here will grow in importance as optics steadily invades information technology from its established bases in fibre optic communication and displays.

We take this opportunity to thank Dr Keay and the Royal Society staff, in particular Miss Christine Johnson as Meetings Organizer and Mr Richard Barrett as copy editor of these Proceedings, for their courteous and efficient contributions to the success of this meeting.

September 1984
S. D. SMITH
B. S. WHERRETT
A. MILLER

CONTENTS

CONTENTS

CONTENTS

Phil. Trans. R. Soc. Lond. A **313**, 191–193 (1984)
Printed in Great Britain

Opening address

By Sir Peter Swinnerton-Dyer, F.R.S.
Chairman, University Grants Committee

It is an honour to have been invited to open this important conference. Every good physics conference has a few mathematicians hanging around on the outskirts, like scavenger fish, hoping to pick up an interesting problem. They are tolerated because occasionally they do something useful, although most of the time mathematicians modify the problems to make them more elegant or more tractable, and thereby accidentally destroy their relevance to the real world. But it is seldom that such a person is invited into the limelight.

But in fact I am here in quite a different role. Today, when the headlines once again suggest that the Common Market is collapsing under the intolerable burden of the Common Agricultural Policy, it is right to remind you that the E.E.C. serves other and better purposes too. It needs the whole of Europe to provide a market large enough to allow high-technology industry to prosper. Just as national governments do within their own boundaries, the Commission subsidizes across Europe the strategic research from which the next generation of industries will spring. Its latest device for doing so is CODEST, the Committee for the Development of European Science and Technology, which was set up to encourage research across national boundaries. CODEST, of which I am a member, was set up last summer and given a pilot budget of 7 million European currency units for its first two years; if it proves itself successful, it can hope to get very much larger sums in the future. Most of that money will go on small grants, to help cooperative research between pairs of laboratories in different countries; but we were instructed to fund if possible one major project. In two months over the summer, Desmond Smith and Paul Mandel put together just such a project, with collaborators from all across Europe, in a key area of the technology of the future (see table 1). We were not merely happy to fund it, but amazed that such a team could be put together in so short a time. One of the side effects is this conference.

What this conference is about is the technology that will be needed for the computers and control systems of the future. Charles Babbage is always described as the father of the computer. But even if someone had assembled the vast collection of cogwheels, powered by a team of horses, which he designed, the result would not have been a true computer but only a calculating engine. What it lacked was the power to control its own procedures: the ability to do logical operations. In that sense, the first step towards the computer age was taken 80 years ago, when Lee de Forest invented the triode valve. By that device, for the first time, electricity could control itself (see table 2). It is less than 40 years ago that the first stored-program computer worked: EDSAC 1 in Cambridge, which (though wholly electronic) was not all that much smaller than Babbage's monstrous engine would have been. Progress since then has been amazing, but it seems that we are now close to the limits of what electronics can do unaided, and it is time to call in optics as a partner. That that should be possible has only become clear within the last 10 years. First the invention of optically bistable devices, then the implementation of the basic logical operations at what one hopes will become picosecond speeds, have made available the building blocks of an optical computer. Putting them together is still to come; but it is bound to come. I hope this conference will be a substantial step forward towards that goal.

TABLE 1. SUMMARY OF THE EUROPEAN JOINT OPTICAL BISTABILITY (E.J.O.B.) PROJECT

(Total funds: 1.8 million European currency units. Period: 1984 and 1985.)

institution		subject area	leaders
Brussels (Université Libre)	theory:	dynamics computer architecture	P. Mandel
Edinburgh (Heriot-Watt University)	theory:	macroscopic device modelling transverse effects	W. J. Firth
		microscopic nonlinearity	B. S. Wherrett
	experiment:	logic gates transphasors optical processor new materials	S. D. Smith
Frankfurt (University of Frankfurt)	theory:	nonlinearity in solids many-body theory	H. Haug
	experiment:	cadmium sulphide	K. Klingshirn
Freiburg (Fraunhofer Institute)	experiment:	infrared diode lasers lead salts	H. Preier
Milan (University of Milan)	theory:	two-dimensional dynamics	L. Lugiato
Munich (Max-Planck-Institute for Quantum Optics)	theory:	device modelling radiation pressure	P. Meystre H. Walther
	experiment:	radiation pressure	
Pisa (University of Pisa)	experiment:	clocks	E. Arimondo
Strasbourg (C.N.R.S.)	experiment:	biexciton processes picosecond speeds new materials	B. Grun, R. Levy B. Honerlage

subcontracts and associated laboratories

Florence (University of Florence)	theory:	noise limits	F. T. Arecchi
Dublin (University College) (Trinity College, from October 1984)	theory:	propagation	D. L. Weaire
Berlin (Technical University)	experiment:	silicon	H. Eichler
Royal Signals and Radar Establishment (Malvern)	experiment:	materials	A. Miller

TABLE 2. HISTORY OF THE COMPUTER AGE

electronics

1908   Lee de Forest's triode valve patent

'Electricity controls electricity'

1908–47   development of vacuum tube electronics
1947   transistor – solid state electronics
1960   integrated solid state circuits
1960–75   digital electronics: computing, *ca.* 1 ns switching, memories
1970s   optical fibres (larger bandwidth for communications) replaced cables by 1983: *opto-electronics*
1975   VLSI – smaller circuits – interconnections controlled by capacitance – 1 nanosecond? – Von Neumann
            bottleneck

photonics

1976   first optical bistability observed
1979   first semiconductor bistability in small devices (first 'optical transistor') with gain

'Light controls light'

1982   optically bistable devices as externally addressed logic gates: AND, OR, NAND, NOR; picosecond
            switching?
1983   transphasor gain 10^4; incoherent–coherent conversion; one gate switches another

Phil. Trans. R. Soc. Lond. A **313**, 195–204 (1984)

Printed in Great Britain

An introduction to optically bistable devices and photonic logic

By S. D. Smith, F.R.S.

Department of Physics, Heriot-Watt University, Riccarton, Currie, Edinburgh EH14 4AS, U.K.

The physical principles of the phenomenon of optical bistability are described and it is shown that a family of devices or 'optical circuit elements' can be realized including memories, logic gates and amplifiers. The current range of materials employed and of properties demonstrated is reviewed: it is concluded that with the present state of knowledge a primitive optical computer could be demonstrated and that other signal-processessing devices are likely to emerge.

Introduction

The earliest ideas of 'optical bistability' relate to laser systems themselves and can be traced back nearly 20 years; as devices, the size and power consumption of such systems appear impracticable and this Discussion Meeting concentrates on small 'passive' devices made possible by recent discoveries of giant optical nonlinearities. An exception is made by consideration of such effects in semiconductor diode lasers for comparison as optical processing elements.

Szöke *et al.* (1969) proposed that a saturable absorber inside a Fabry–Perot optical resonator could exhibit two bistable states of transmission for the same input intensity. The simple idea is that at high intensities the induced transparency allows constructive interference at resonant wavelengths and, owing to the large internal field in this situation, the system can be held 'on' to lower incident intensities than those required to induce the transparency. The experiments quoted were, however, inconclusive and it was not until 1976 that Gibbs *et al.* (1976) demonstrated optical bistability with sodium vapour in an interferometer. They deduced that the effect was caused by nonlinear refraction rather than saturable absorption. The physics and mathematics underlying optical bistability has attracted much theoretical interest; a review was given by Abraham & Smith (1982), citing 250 papers, around 80% of which are theoretical.

All observations of bistability reported so far have involved a combination of *nonlinearity* and *feedback*. We shall find that both these concepts can have quite varied manifestations.

Nonlinearities

If we begin with intensity-dependent refraction and absorption we can simply express linear effects in intensity I by:

$$n(I) = n_0 + n_2 I, \tag{1}$$

where the nonlinear refractive index, n_2, can conveniently be measured in square centimetres per kilowatt, for refraction and to relate the absorption coefficient $\alpha(I)$ to linear absorption α_0,

$$\alpha(I) = \alpha_0 - \alpha_2 I. \tag{2}$$

Both n_2 and α_2 can be described by the conventional expansion of polarization P_i in powers of the electric field:

$$P_i = \chi_{ij}^{(1)} E_j(\omega_1) + \chi_{ijk}^{(2)} E_j(\omega_1) E_k(\omega_2) + \chi_{ijkl}^{(3)} E_j(\omega_1) E_k(\omega_2) E_l(\omega_3). \tag{3}$$

In the early nonlinear optical experiments it was assumed, consistent with experiment, that high laser powers (in the megawatt region) and intensities (of order gigawatts per square centimetre) were necessary to give electric fields (*ca.* 10^7 V cm^{-1}) comparable with atomic fields in order for second- and third-order polarization to be significant. We are concerned here with third-order polarization for intensity-dependent effects proportional to $\langle E(\omega)\rangle^2$, with $\omega_1 = \omega_2 = \omega_3 = \omega$. Values of $\chi^{(3)}$ reported before 1976 varied from 10^{-8} to 10^{-11} e.s.u.† as reviewed by Wherrett (1983, and this symposium).

In 1976, nonlinear refraction was explicitly observed at milliwatt powers at wavelengths near the absorption edge of a narrow gap semiconductor in our laboratory (Miller *et al.* 1978). Retrospectively, the effect could be recognized in earlier work on the spatial distortion of modes of the spin-flip Raman laser (see, for example, Scragg & Smith 1975; Ironside 1977).

The explanation of this enormous (10^9) decrease in required power lies in the near resonance between the triply degenerate frequency (ω) of the three field components $E_j(\omega), E_k(-\omega), E_l(\omega)$ and the frequency difference between initial and various intermediate states, corresponding (say) to a semiconductor energy gap or a two-level resonance such as an exciton. All energy denominators in a quantum mechanical calculation then resonate together. In a real system some damping is present to broaden the resonance. Away from resonance, $\chi^{(3)}$ can be separated into real and imaginary parts so that $\mathrm{Re}\,\chi^{(3)}$ leads to a description of nonlinear refraction n_2, and $\mathrm{Im}\,\chi^{(3)}$ gives the intensity dependence of absorption $\alpha(I)$; near resonance the influence of both effects will be present simultaneously. The presence of absorption allows *real excitation* of the system: redistribution of the electron population will temporarily change the properties of the material. The nonlinearity is thus said to be 'active', persisting for a characteristic population lifetime τ_r (varying from microseconds to picoseconds), and can also be said to be 'dynamic'. Just as linear refraction and absorption are related by the Kramers–Kronig relation, the same integral can be used to calculate the nonlinear refraction from the change in absorption induced by population redistribution if the relaxation processes are fast compared with the measurement time. This method has been successfully used to predict the magnitude and sign of 'active' effects in semiconductors (Miller *et al.* 1981). For band edge effects in small-gap materials it is sometimes known as the 'dynamic Moss–Burstein effect' and explains both the resonance behaviour and magnitude of n_2 near (a few tens of reciprocal centimetres) the band edge with values of n_2 of order 0.1–1 cm^2 kW^{-1}. Since a device can be switched with an increment $\Delta n \approx 10^{-3}$, power densities of order watts per square centimetre can be used in practice. This nine orders of magnitude improvement in power requirement clearly opens the way to micrometre dimension devices based on nonlinear optics in which light controls light.

Although the theory outlined above gives a simple explanation of the new effects, it is as yet imperfectly understood, particularly in the absorption mechanisms responsible for carrier

† 1 e.s.u. = 1 cm^3 erg^{-1} $\equiv$ 1.4×10^{-8} m^2 V^{-2}.

excitation. Excitation from band tail or line wing is assumed with a photon energy deficit to be supplied by further carrier interaction. In the absence of theory, a measured α_{eff} has been used (Miller *et al.* 1981), giving the results shown in figure 1, showing good agreement with experiment.

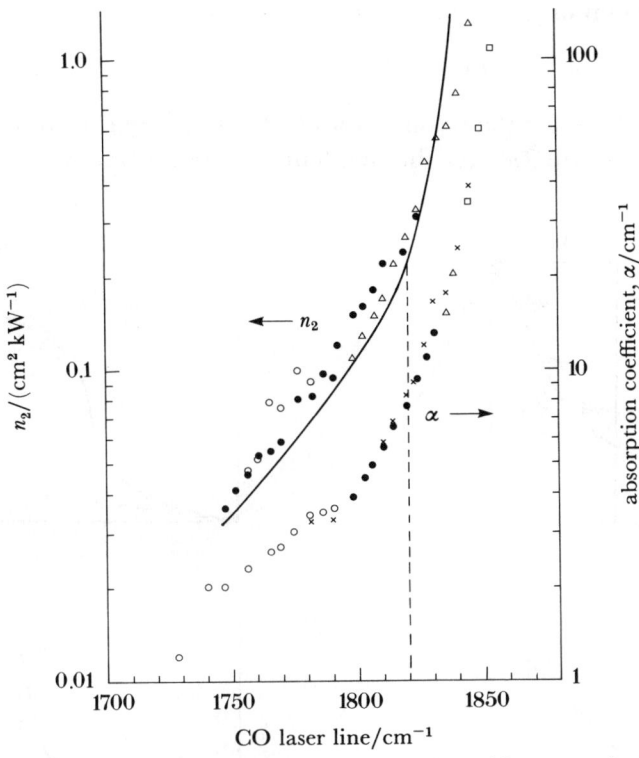

FIGURE 1. Experimentally measured absorption and n_2 as a function of frequency for InSb near 77 K. The n_2 curve is theoretical. (From Miller *et al.* (1981), with additional absorption data.)

In all cases the nonlinear refractive index, n_2, is proportional to the equilibrium carrier density generated by the incident intensity, I, from

$$n_2 \propto N = (I/\hbar\omega)\,\alpha_{\mathrm{eff}}\,\tau_{\mathrm{r}}.$$

However, the carrier density, N, can be generated arbitrarily quickly by rapidly applying a sufficiently high intensity I, i.e. more quickly than the carrier life time, τ_{r}. Thus the nonlinearity, and hence a required Δn, can be 'switched on' as rapidly as desired by the application of sufficient intensity. This proves to be useful in activating all-optical logic gates and memories. However, to 'switch off' the nonlinearity and hence the device, removal of the field leaves the material to relax within its characteristic lifetime, τ_{r}. At this level of argument there is a direct linear trade-off between magnitude of nonlinearity and switching speed. The smaller 'passive' nonlinearities are of course fast, probably substantially sub-picosecond, and therefore unnecessarily fast for devices limited by transit times dependent on their physical size. There exists considerable scope for 'engineering' τ_{r} to an optimum value by using well-known semiconductor procedures such as doping, use of surface recombination and sweep out.

[7]

Feedback

The simplest way to understand the effect of feedback is to consider the internal field in a nonlinear Fabry–Perot resonator. An interferometer of optical thickness nL shows peaks of transmission as a function of wavelength when

$$\tfrac{1}{2}M\lambda = n(I)\,L = (n_0 + n_2\,I_{\text{int}})\,L, \tag{3}$$

where I_{int} is the intensity inside the resonator and M is an integer. Consider an initial detuning from resonance of $\delta\lambda$ (figure 2a). As the incident intensity I_{i} increases, the optical thickness

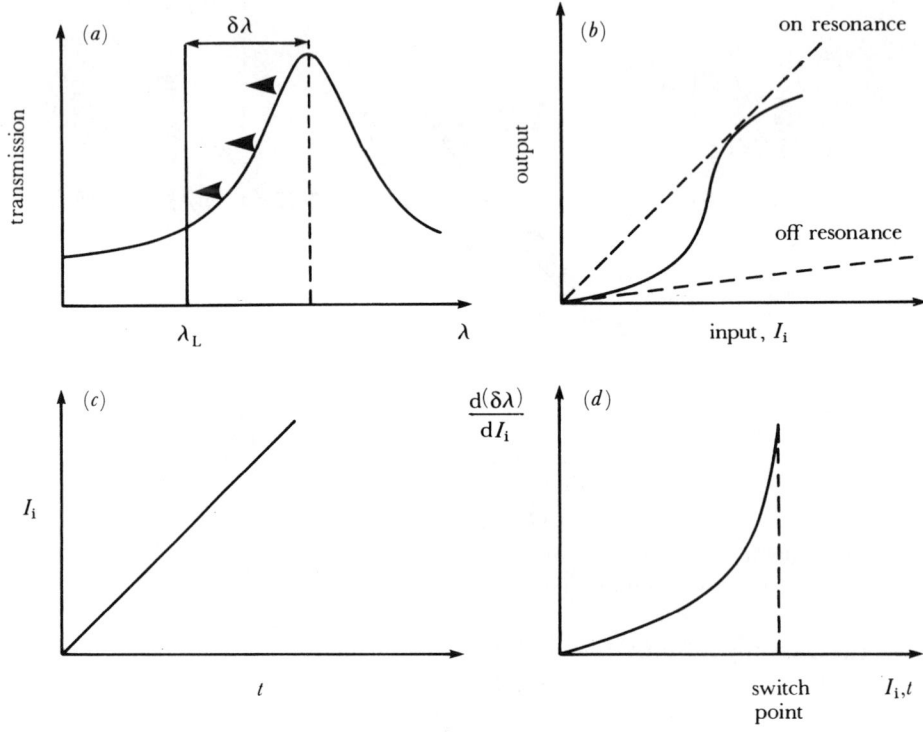

FIGURE 2. A typical Fabry–Perot characteristic (a) and the evolution of nonlinear transmission to switching behaviour (b–d). (See text.)

changes as the *internal* intensity I_{int} rises, through the term $n_2\,I_{\text{int}}\,L$ in (3). The transmission peak (figure 2a) thus moves towards resonance, increasing the transmission $T(\lambda)$ and thus giving a nonlinear characteristic (figure 2b). However, the internal intensity I_{int} is related to the 'incident' intensity I_{i} by

$$I_{\text{int}} = T(\lambda)\,\frac{(1+R)}{(1-R)}\,I_{\text{i}}, \tag{4}$$

so that I_{int} increases as $T(\lambda)$ rises, giving positive feedback to the nonlinear shift in optical thickness. In fact, then, the peak in figure 2a moves increasingly quickly toward the laser wavelength λ_{L}. If a linear intensity ramp (with time) is used to address the device, the rate can be expressed either as a function of incident intensity or time (figure 2c) and is given by

$$\frac{\mathrm{d}(\delta\lambda)}{\mathrm{d}I_{\text{i}}} = T(\lambda)\left/\left(\frac{T_{\max}}{2n_2 L} - I_{\text{i}}\frac{\mathrm{d}T}{\mathrm{d}\lambda}\right)\right. \tag{5}$$

[8]

The two terms in the denominator of the right-hand-side of (5) represent the nonlinear shift and feedback respectively. When the feedback, changing with I_i and $dT/d\lambda$, becomes large enough this denominator approaches zero and the rate of approach of the moving resonant wavelength to the pump wavelength diverges and becomes infinite. At this point the device switches suddenly on to resonance. The large internal field being now established, reduction in I_i leaves the resonator in its upper state and hysteretic, i.e. memory behaviour is seen. The simplest plane-wave theory is obtained by simultaneous solution of (4) for $T(\lambda)$ with the standard Airy formula describing figure 2a,

$$T(\lambda) = 1/(1+F\sin^2\theta), \tag{6}$$

where $F = 4R/(1-R)^2$ and $\theta = 2\pi nL/\lambda$.

The input–output characteristic may appear as in figure 2b, but varies according to choice of initial detuning, $\delta\lambda$. An experimental example is given in figure 3 for an InSb resonator activated by a CO laser at 1819 cm⁻¹. It can be seen that either a memory function, a region of differential gain or simply a nonlinear modulator can be obtained according to the device desired. The control of the characteristic depends upon choice of feedback characteristic and initial detuning. This leads to the family of optical circuit elements.

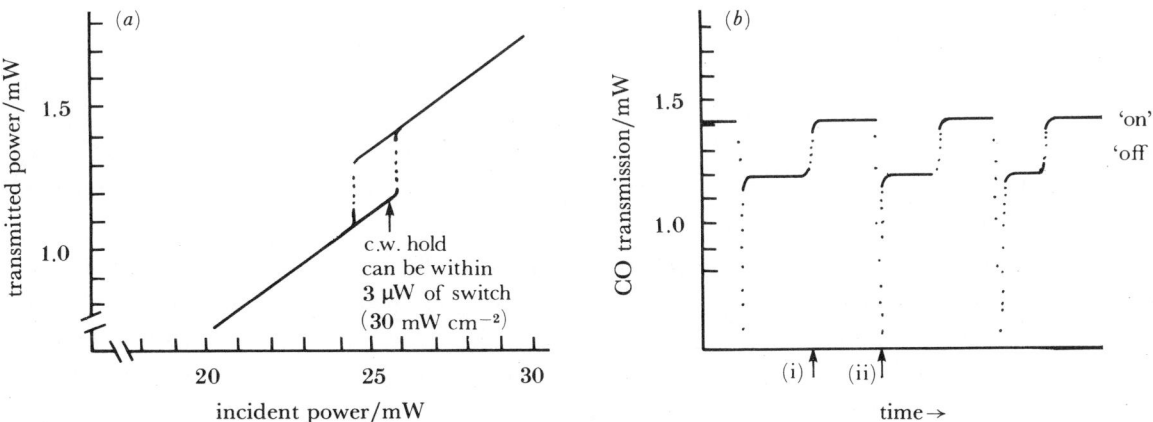

FIGURE 3. (a) Input–output characteristic for a 210 μm thick InSb etalon near 77 K, with 1819 cm⁻¹ radiation. (Surface reflectivities 36%, spot size ca. 0.2 mm.) (b) External switching of the above device with the use of 35 ps, 1.06 μm wavelength pulses: (i) 5 nJ Nd:YAG pulse external address; (ii) interrupt.

Other nonlinearities and feedback mechanisms

Nonlinearities other than refraction have been employed for optically bistable devices. These include absorption, electronically simulated nonlinearity and radiation pressure.

Similarly, *feedback* can be contrived externally via electronic circuits, by using various geometries such as focusing or defocusing, by using guided wave configurations with and without distributed gratings, thermal effects via absorption and even, it is suggested, by microscopic internal field modification. So far, however, the 'system function' appears to resemble the refractive example cited here in all reported systems.

One case of note leads to a permanent as opposed to volatile memory and was reported by Hajto & Janossy (1983a, b). The optical behaviour of self-supporting α-GeSe₂ films showed nonlinearity at intensity levels around $50\ \mathrm{W\ cm^{-2}}$ and exhibited hysteresis and bistability without placing the sample in an optical resonator. Whereas above I have considered only electronic effects, Hajto & Janossy invoke thermal effects and photostructural effects – the latter possibly laser-induced reversible microcrystallization of the amorphous material – to explain a variety of intensity-dependent effects.

The thermal effects on absorption are very instructive. The temperature-dependent band tail is assumed described by Urbach's expressions

$$\alpha = \alpha_0 \exp\left[-\{(E_s - h\nu)/kT\}\right]. \tag{7}$$

The temperature rise of an illuminated spot is determined by the heat loss proportional to the difference between 'spot temperature' T, and film temperature, T_f. Thus

$$\tau_0\,\delta T/\delta t = \eta J D(T) - (T - T_f), \tag{8}$$

where τ_0 and η are constants, J is incident laser flux, D is a dissipation coefficient (dissipation/incident flux) that depends on α and thus on T. The stationary value of spot temperature is then given by

$$\eta J D(T) = T - T_f, \tag{9}$$

together with the stability condition

$$\frac{\partial}{\partial T}\left\{D(T) - \frac{T - T_f}{\eta J}\right\} < 0. \tag{10}$$

At certain critical intensities, the increasing absorption causes 'thermal run-away' and rapid switching occurs. A feedback occurs between the value of the absorption coefficient, α, and the temperature rise, which can give 'induced absorption bistability', there being stable and unstable regions in a plot of α_0 against $T - T_f$.

The additional existence of photostructural changes also affecting α gives phenomena showing memory functions that can exist in the absence of the laser beam. In comparison with the electronic effects the photostructural effects are slow (seconds) with thermal effects of intermediate size (milliseconds to microseconds). Corresponding effects in intensity-dependent refractive index will also occur and will be important in resonator-type devices.

Although this survey is incomplete we may conclude that there already exists a rich variety of nonlinear optical and feedback effects requiring fundamental study and giving excellent device prospects.

OPTICAL CIRCUIT ELEMENTS

The principles of nonlinearity and feedback inherent in a bistable optical switch have been shown to have wider application to a whole family of both analogue and digital optical devices equivalent to those currently used in electronic signal processing.

So far, devices able to perform all fundamental logic operations have been demonstrated, for example figure 3 b. I illustrate in figure 4: (a) optical memory, (b) AND gates and OR gates that are self-resetting or if configured by initial detuning as in (a) can remain latched, (c) a transphasor amplifier, enabling the cascading of elements, fan-out and cross-modulation of differing frequencies to be achievable, and (d) a power limiter. Using these devices already demonstrated, we may deduce that a further series of devices in figure 5 can be predicted to be practical: (a) optically addressable directional optical switches, (b) two-dimensional

[10]

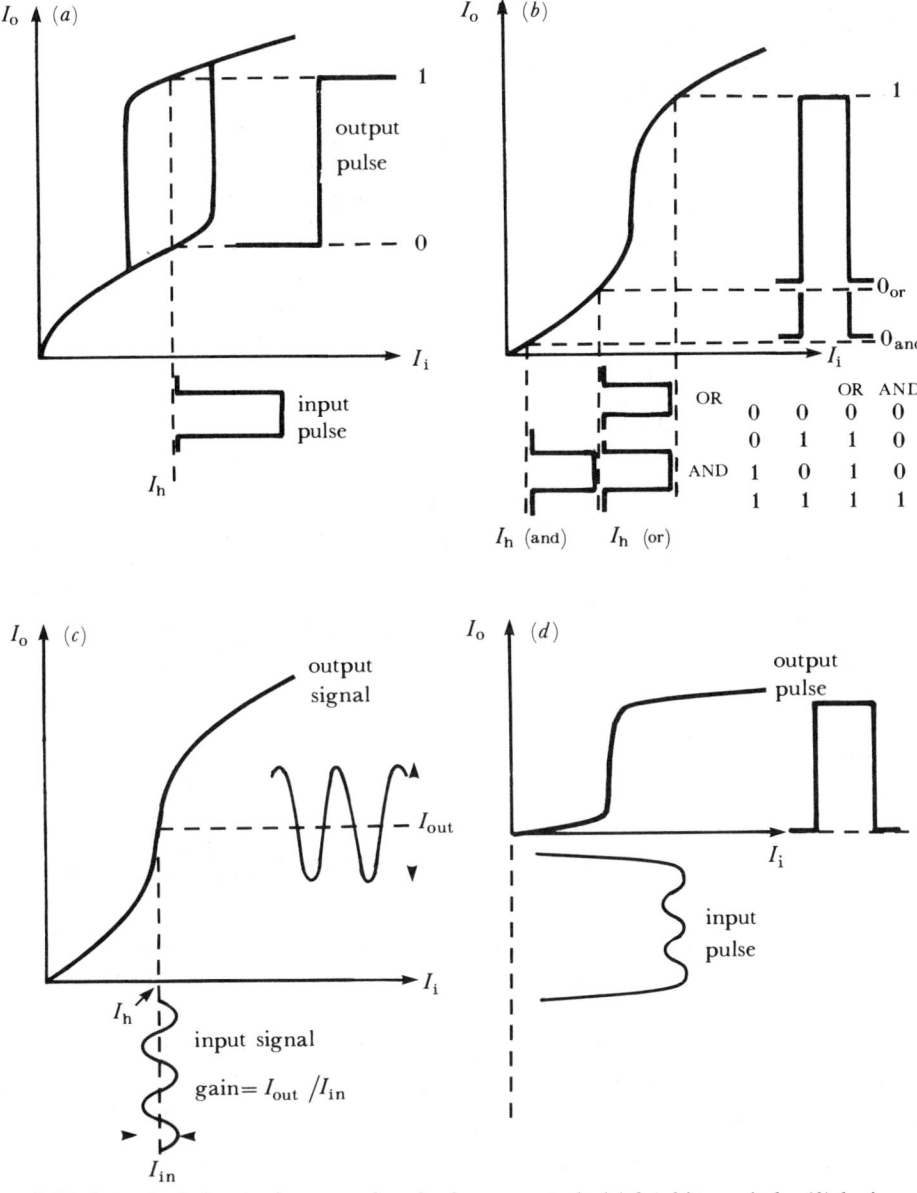

FIGURE 4. Basic optical circuit elements already demonstrated: (a) latching switch; (b) logic gate; (c) transphasor; (d) power limiter. I_h is the holding intensity.

parallel-to-series and series-to-parallel convertors, (c) array processors that can be externally programmed and can address each other in sequences, and (d) the possibility of optically addressable display devices.

MATERIALS, WAVELENGTHS AND CONDITIONS OF EXPERIMENTS IN OPTICAL BISTABILITY AND RELATED DEVICES

Table 1 lists chronologically some of the experimental systems that have been used to demonstrate optical bistability. The first two systems are included for historical purposes but it has been assumed that atomic vapour systems are not of practical interest, nor are those

[11]

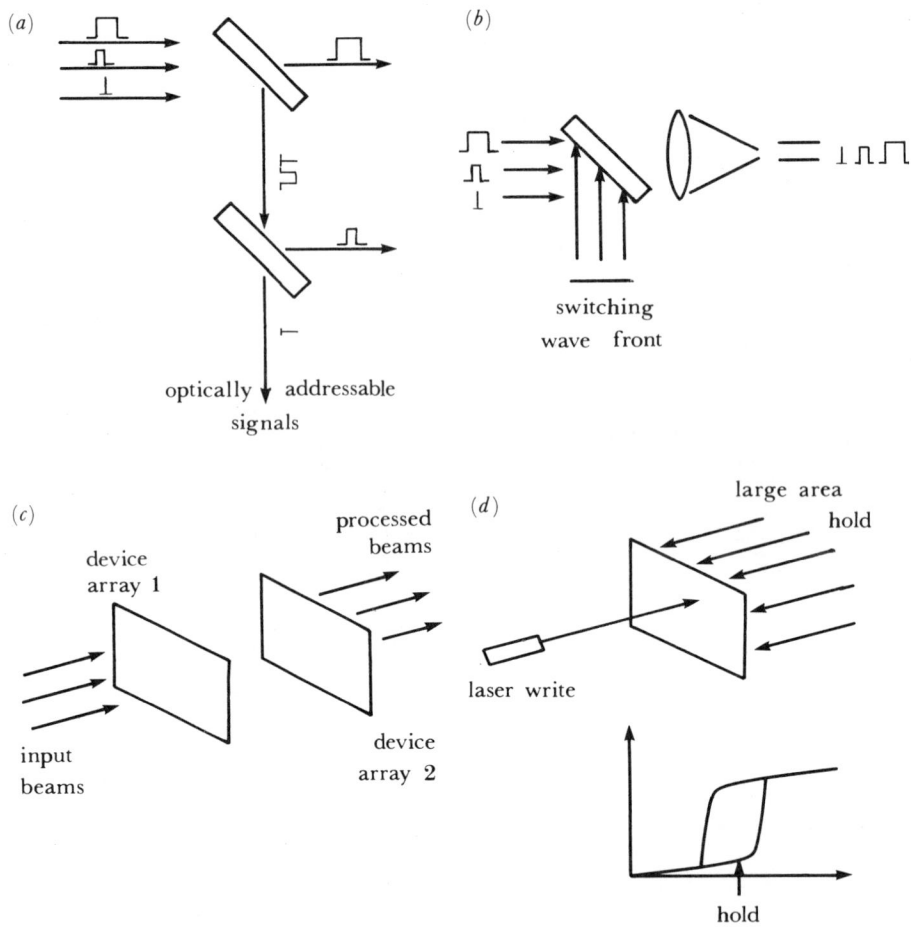

FIGURE 5. Potential devices based on optically bistable elements: (a) optically addressing signal; (b) parallel-to-series array processing; (c) serial array processing; (d) display.

systems that require very large laser powers. There is by now a substantial number of examples in solid materials, usually semiconductors. At the time of writing, no one material satisfies such desirable practical specifications as operation in the visible or 'optical fibre' wavelength region with milliwatt powers and at room temperature. Taking the best examples, however, switching on picosecond timescales, operation with milliwatt powers and a variety of wavelengths have been demonstrated experimentally.

The ability to 'hold' the system on a given part of its characteristic with a continuous wave (c.w.) beam is important for optical logic gates and transphasor amplifiers. In this respect InSb at 77 K is currently unique, enabling, for example, a device to be held within 3 μW of switch point and then externally addressed with short pulses. Figure 3b illustrates this system in use as a pulse-addressable optical memory and an AND gate. Use of the device in reflection produces a NOR gate. Two-beam signal amplification or transphasor action was observed in 1979, and recently gain as high as 10^4 has been demonstrated (Smith & Tooley 1984).

The other successful semiconductor so far, GaAs, has a nonlinearity approximately 10^{-3} as large, implying power densities around kilowatts per square centimetre compared with watts per square centimetre for InSb. However, the absorption levels are in general high in GaAs ($\alpha \approx 10^4$ cm^{-1}) so that devices have not been operated or 'held' with c.w. beams. Future

[12]

TABLE 1. OPTICALLY BISTABLE SYSTEMS AND DEVICES

year	material	typical power	observation	comment	reference
1976	Na atoms	milliwatts	optical bistability	large and slow	1
1978	CS_2, etc.	gigawatts, pulsed	o.b.	large and fast	2
1979	InSb	milliwatts, c.w.	o.b.	small, fast	3
			transphasor with gain	c.w. + external address	4
			logic gates	5 μm	5
			incoherent switching	77 K	6
1984	InSb	kilowatts, pulsed	optical bistability limiter	10 μm, 300 K 2-photon excitation	7
1979	GaAs	milliwatts, modulated	o.b.	0.8 μm, not very small, 300 K	8
1983	GaAlAs m.q.w.	watts, modulated	o.b.	not c.w.	9
1981	Te	megawatts, pulsed	o.b.	300 K, 10 μm	10
1982	$GeSe_2$	milliwatts, c.w.	o.b., etc.	slow, visible photo-refraction	11
1982	Si	megawatts, pulsed watts	tuning o.b.	300 K, 1.06 μm slow, thermal	12
1983	CdS	megawatts, c.w.	o.b.	5 K, visible	13
1983	CuCl	megawatts, pulsed	o.b.	5 K, visible biexciton	14
1984	InAs	milliwatts	o.b.	3 μm	15
1984	CdHgTe	kilowatts, pulsed	o.b.	12 μm, 2-photon	16

References: 1, Gibbs *et al.* (1976); 2, Bischofberger & Shen (1978); 3, Miller *et al.* (1979); 4, Miller & Smith (1979); 5, Seaton *et al.* (1983); 6, Smith & Tooley (1984); 7, Kar *et al.* (1983); 8, Gibbs *et al.* (1979); 9, Gibbs *et al.* (1982); 10, Staupendahl & Schindler (1982); 11, Hajto & Janossy (1983 a, b); 12, Eichler (1982, 1983); 13, Dagenais & Winful (1984); 14, Grun *et al.* (1984); 15, Poole & Garmire (1984); 16, Miller & Parry (this symposium).

investigation of materials will no doubt be directed to obtaining such operation, since this will be necessary for optical circuits to be built. Other materials of interest include liquid crystals and the molecular gases NH_3 and SF_6, which may find use in 'optical clock' driving devices.

CONCLUSIONS

In recent years the physics of giant refractive and absorptive nonlinearities has progressed sufficiently rapidly to allow the demonstration of a family of optical circuit elements. Although there remains much to be done to make the systems sufficiently practical to constitute a new technology, the capability to demonstrate simple optical signal processing now exists, and a demonstration of a primitive optical computer should be possible.

REFERENCES

Abraham, E. & Smith, S. D. 1982 *Rep. Prog. Phys.* **45**, 815–885.
Bischofberger, T. & Shen, Y. R. 1978 *Appl. Phys. Lett.* **32**, 156–158.
Dagenais, M. & Winful, H. H. 1984 *Appl. Phys. Lett.* **44**, 574–576.
Eichler, H. J. 1982 *Optics Commun.* **40**, 302.
Eichler, H. J. 1983 *Optics Commun.* **45**, 62.
Gibbs, H. M., McCall, S. L. & Venkatesan, T. N. C. 1976 *Phys. Rev. Lett.* **36**, 1135.

Gibbs, H. M., McCall, S. L., Venkatesan, T. N. C., Gossard, A. C., Passner, A. & Wiegman, W. 1979 *Appl. Phys. Lett.* **35**, 451–453.

Gibbs, H. M., Tarng, S. S., Jewell, J. L., Weinberger, D. A., Tai, K., Gossard, A. C., McCall, S. L., Passner, A. & Weigmann, W. 1982 *Appl. Phys. Lett.* **41**, 221–222.

Grun, B., Hönerlage, B. & Levy, R. 1984 In *Proc. Topical Meeting on Optical Bistability, Rochester, New York.* (In the press.)

Hajto, J. & Janossy, J. 1983*a* *Phil. Mag.* B **47**, 347.

Hajto, J. & Janossy, J. 1983*b* *Phil. Mag.* B **48**, 321.

Ironside, C. N. 1977 Ph.D. thesis, Heriot-Watt University.

Kar, A. K., Mathew, J. G. H., Smith, S. D., Davis, B. & Prettl, W. 1983 *Appl. Phys. Lett.* **42**, 334.

Miller, D. A. B. & Smith, S. D. 1979 *Optics Commun.* **31**, 331–3.

Miller, D. A. B., Mozolowski, M. H., Miller, A. & Smith, S. D. 1978 *Optics Commun.* **27**, 133.

Miller, D. A. B., Seaton, C. T., Prise, M. E. & Smith, S. D. 1981 *Phys. Rev. Lett.* **47**, 197.

Miller, D. A. B., Smith, S. D. & Johnston, A. 1979 *Appl. Phys. Lett.* **35**, 658–660.

Poole, C. D. & Garmire, E. 1984 *Appl. Phys. Lett.* **44**, 363–365.

Scragg, T. & Smith, S. D. 1975 *Optics Commun.* **15**, 166–168.

Seaton, C. T., Smith, S. D., Tooley, F. A. P., Prise, M. E. & Taghizadeh, M. R. 1983 *Appl. Phys. Lett.* **42**, 131–133.

Smith, S. D. & Tooley, F. A. P. 1984 In *Proc. Topical Meeting on Optical Bistability, Rochester, New York.* (In the press.)

Staupendahl, G. & Schindler, K. A. 1982 *Opt. Quantum Electron.* **14**, 157–167.

Szöke, A., Daneu, V., Goldhar, J. & Kumit, N. A. 1969 *Appl. Phys. Lett.* **15**, 376.

Wherrett, B. S. 1983 *Proc. R. Soc. Lond.* A **390**, 373–396.

Phil Trans. R. Soc. Lond. A **313**, 205–211 (1984)

Printed in Great Britain

205

Impact of technological advances and architectural insights on the design of optical computers

By A. Huang

AT&T Bell Laboratories, Crawfords Corner Road, Holmdel, New Jersey 07733, U.S.A.

Communication problems such as interconnection bandwidth, clock skew, and connectivity are restricting computational throughput. Bandwidth and clock skew problems limit the speed and add to the design complexity of a processor. Constrained connectivity forces much of the speed of a processor to be used to compensate for the limited number of interconnections.

Philosophically, the large bandwidth, innate parallelism and non-interfering propagation of optics offer mechanisms for overcoming these communication problems. The difficulty in exploiting these capabilities has been the absence of suitable optical logic and memory devices. Advances in optical nonlinearities offer the possibility of cascadable optical logic gates that are competitive with electronics. Advances in computer architecture can be used to simplify the optical memory requirements and utilize the large bandwidth, parallel, non-interfering communications of optics.

People have been interested in the possibility of an optical digital computer for some time. Previous efforts were limited by the absence of suitable logic and memory devices as well as the lack of a clear understanding of what optics has to offer and what the computer community needs. The technological spin-offs associated with the development of optical communications now offer the possibility of optical logic with speeds and power comparable with conventional electronic logic. What is missing is a better understanding of the capabilities of optics and the problems facing future computer systems.

PHYSICAL CONSTRAINTS ON COMPUTATIONAL THROUGHPUT

One problem confronting future computer systems is interconnection bandwidth. As system cycle times and pulse widths shrink, the bandwidth needed to preserve the rising and falling edges of these signals increases. This forces the need for bulky, expensive, terminated coaxial interconnections.

Another problem is clock skew. This problem occurs when signals from different parts of a circuit arrive at a gate at different times. This skew of the inputs can cause a gate to generate an erroneous output. The desire for shorter cycle times limits the amount of clock skew that can be tolerated. This in turn limits the maximum difference in interconnection length. A processor such as the Cray-1 has a maximum interconnection length of six inches (*ca.* 15 cm). All interconnections less than this length must be padded with gate delays to make the propagation time equivalent to that of the maximum interconnection length. This restriction complicates both the physical and electrical designs of a processor and hints at some of the difficulties associated with the next generation of processors.

14-2

The fear of clock skew precludes the use of logic in a pulsed mode. The accepted approach when using digital logic is to wait for the inputs to settle before using the output of a gate. This input settling time is dependent on the amount of time it takes to fully charge the connection. In most circuits, the RC time constant-dominated settling time is longer than the transistor switching time. This makes it difficult to utilize ultrafast logic gates.

The difficulties associated with settling time are not solved by very large scale integration. As the length of a wire shrinks by a factor of α and the cross-sectional area of the wire is reduced by a factor of α^2, the capitance of the wire decreases by a factor of α while the resistance increases by the same amount. The RC time constant remains the same and thus the input charging time remains unaltered, independent of scaling (Mead & Conway 1980). Given the parameters of very large scale integration it is estimated that the signals will be communicated at approximately 0.5 % the speed of light (Wilkes 1983).

The 'Von Neumann bottleneck': the use of time to reduce interconnections

Conventional computers suffer from a 'bottleneck' that is caused by the limited number of interconnections that can be supported in a practical manner by electronics. This problem is referred to as the 'Von Neumann bottleneck' and involves the performance limitations imposed by the sequential and address-oriented communications between the *central processing unit* and the *memory* in a conventional computer (Backus 1982).

The source of this bottlneck can be found by examining the 'classical finite state machine', an ancestor of modern computers. This processor, shown in figure 1, consists of storage elements,

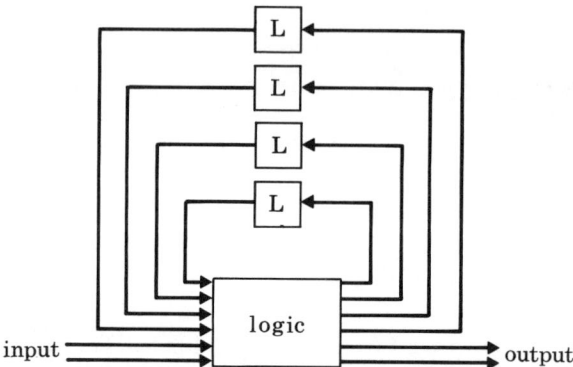

FIGURE 1. A 'classical finite state machine' does not suffer from the Von Neumann bottleneck since it can update all of its memory in parallel without the need for addresses.

a combinatoric logic unit, inputs, outputs and various interconnections. What is unusual about this processor is that it does not suffer from a Von Neumann bottleneck. All storage elements are updated in parallel without the need for addresses. The bottleneck emerges when more storage variables are added. It becomes impractical to support N interconnections between the storage and logic units with wires. As a result, the classical finite state machine is modified to the structure shown in figure 2, in which a binary encoding scheme is used to reduce the interconnections from the output of the logic unit to the memory and a common return line

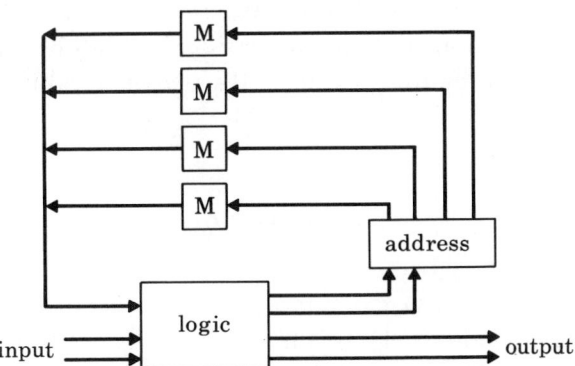

FIGURE 2. A 'modified finite state machine' suffers from the Von Neumann bottleneck since it can only update one memory element at a time and it needs an address to do so.

is used to reduce the interconnections between the memory elements and the input of the logic unit. This reduces the number of interconnections, but it also seriously degrades the performance since the 'modified finite state machine' can only address one storage element at a time and an address is needed to specify which element. This inability to support N interconnections in parallel is the origin of the Von Neumann bottleneck.

OTHER BOTTLENECKS THAT USE TIME TO REDUCE INTERCONNECTIONS

Computers suffer from other constrictions similar to the Von Neumann bottleneck. For ease of design and construction a processor is partitioned into modules. Inevitably, each module must communicate with the other modules. The impracticality of fully interconnecting M modules with $M \times (M-1)$ bus-wire interconnections leads to the use of a broadcast bus structure that uses time to reduce the number of interconnections. This results in sequential, address-oriented communications identical to those of the Von Neumann bottleneck.

A similar bottleneck occurs at an even lower level. Ideally, it would be nice if each bit of a memory chip could be independently read or written. To accomplish this a $P \times 1$ bit memory chip would have to have P input and P output lines. Since this is impractical, a binary encoding scheme is used to reduce the P inputs to $\log_2 P$. This approach is similar to that used in the modified finite state machine and results in another sequential, address-oriented communications bottleneck.

These communication bottlenecks at the architectural, bus and chip level all stem from the use of time multiplexing to compensate for an inability to communicate N channels of information in parallel. These bottlenecks, along with the previously mentioned bandwidth and clock skew problems, are interrelated. Increasing the switching speed might ease the connectivity problem, but it also aggravates the bandwith and clock skew problems.

HOW CAN OPTICS HELP?

Optics is capable of communicating many high bandwidth channels in parallel. Lenses, prisms and mirrors can convey images consisting of millions of resolvable spots. Each spot is capable of supporting a very large bandwidth channel. Optics also has the benefit of

[17]

non-interfering propagation. Optical beams can cross without interaction. These attributes have been exploited in analog optical signal processing, but they have yet to be applied to digital processing. The main reason for this has been absence of suitable optical logic and memory devices.

Optical logic

Many optical logic gates have been demonstrated and discussed in the literature. The main problem is that these gates are not cascadable (Basov 1972). The inputs of these gates are represented by a phenomenon such as phase while the output is expressed in another phenomenon such as intensity. As a result, the output of one gate cannot be used as the input of another gate.

The prospects for a cascadable optical logic have recently changed (Abraham *et al.* 1983). Optical nonlinearities that have previously only been observed at high power levels have now been demonstrated at switching energy levels comparable to transistors (Smith 1981). Nonlinearities requiring 10 s of watts per bit but reacting in 100 s of femtoseconds, as well as nonlinearities requiring 10^{-8} W bit^{-1} and reacting in 10 s of nanoseconds, have been observed (Gibbs *et al.* 1982; Miller 1982, 1983). Primitive cascadable AND, OR, NOR and NAND gates have also been demonstrated (Abraham *et al.* 1983).

Optical memory devices

The use of optics in a computer has also been hindered by the lack of a suitable optical memory. Several optical memories were developed but they were slow, awkward and expensive. In retrospect, the optical memories that were developed were for a modified finite state machine rather than a classical finite state machine. This seemingly minor difference sacrificed the parallelism of optics and forced optical memories to incorporate addressing mechanisms consisting of beam deflectors, page composers and detector arrays. Another important distinction between these architectures is the type of storage required. The storage elements of a modified finite state machine must be capable of preserving information indefinitely since the elements are addressed in random order. The search for an optical material capable of storing information in this manner has proven quite difficult. It would have been considerably simpler for optics to implement the memory as required by a classical finite state machine, since the storage elements need only preserve their information for one cycle. This architecture would also have used the parallel communications capabilities of optics and avoided the need for an addressing mechanism.

A direct descendant of the classical finite state machine is a 'parallel pipelined processor' as shown in figure 3. The alternating levels of latches and logic of the parallel pipelined processor can be viewed as an unravelled version of the classical finite state machine. The latches of the parallel pipelined processor provide the necessary storage, do not require an addressing mechanism, and are simpler to implement for optics since they need only hold information for one cycle.

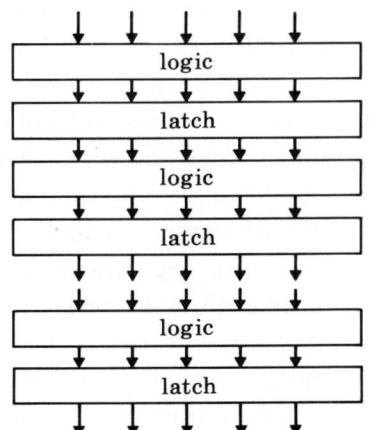

FIGURE 3. A 'parallel pipelined processor' with its alternating stages of latches and logic can be viewed as an unravelled version of a classical finite state machine.

BENEFITS OF A PIPELINED ARCHITECTURE

Power considerations have been used to dismiss optical logic and play down the potential femtosecond switching rates of optics (Basov 1972; Keyes 1975). A pipelined architecture can be used to reduce this problem. Only a small fraction of a conventional central processor is being used during any given cycle. A considerable amount of power can be saved by turning off the unused circuits. One way of accomplishing this is by casting the processor into a pipelined architecture and only partly filling the pipeline.

The memory of modern computers also consumes considerable amounts of power. Given an ultrafast logic it becomes practical to consider the use of the non-dissipative propagation delays to provide the latching as required by a pipelined architecture. This might seem a bit unusual, but it should be remembered that mercury delay lines were used to provide storage in early computers.

A pipelined architecture can also be used to structure the interconnections (Huang 1984). This simplifies the task for optics and also opens the possibility of using femtosecond optical nonlinearities. The difficulties associated with trying to communicate femtosecond signals without clock skew has discouraged the investigation of logic devices based on these phenomena. While it is difficult to route and fabricate equidistant waveguides, it should be remembered that mirrors and lenses can provide equidistant optical paths by using free-space propagation. This oversight illustrates some of the difficulties in extrapolating between technologies. One of the hidden assumptions was that since electronics must use wires to communicate, optics must use waveguides. A second assumption was that a digital circuit always involves an awkward mix of interconnection lengths.

OPTICS COMPARED WITH ELECTRONICS

The question as to which technology switches the fastest is only part of the overall problem. An analysis of the switching speed of various technologies must be coupled with a measure of the communications ability of that technology. The previous discussion pertaining to the Von Neumann bottleneck demonstrates how an inability to communicate can quickly compromise any advantage a technology might have in terms of speed. Optics has a definite advantage over electronics in terms of both the bandwidth and number of free propagating channels that can be supported per given volume.

At first glance, the relative merits of electronics and optics seem to follow from the basic characteristics of electrons and photons. Electrons affect each other even at a distance. As a result, it is easy for one electrical signal to affect itself as well as another signal. This makes it easy to perform switching, but this interaction complicates the task of communication. Interaction in the form of inductance and capacitance limits the propagation speed and bandwidth of signals. Interaction in the form of electrical or magnetic coupling threatens the integrity of a signal. Photons behave quite differently. It is very difficult to get two photons to interact. This makes it difficult for optics to perform switching, but this lack of interaction simplifies the task of communication. The symmetry of this conjecture is very appealing, but advances in optical nonlinearities show that optics can be used to perform switching at energies comparable to electronics. This puts electronics with its communication difficulties at a disadvantage.

ECONOMICS

Economics has always played a part in determining the viability of optical digital computation. It is important to consider how optics can compete against the billions of dollars already invested in electronics. These concerns have prompted the search for problems that would need and could pay for the capabilities of optics. Trying to sell a problem, a solution and a new technology at the same time is difficult. This situation is being changed by the optical communications revolution. Long haul communications are being dramatically changed by light-wave technology. The large data rates involved will force similar changes in the communications between processors. These changes will encourage changes in the communications between modules within a processor and will eventually force the communications between chips and even devices on the same chip to rely on optics. Optical local area networks and optical computer buses are already being developed. Ways of using optics to increase the communications on and off of chips are being vigorously pursued. The insatiable demand for increased data rates makes this trend inevitable. This evolutionary path is self-perpetuating. Optics is generating problems that only optics can solve. This trend will pay for and generate a new optics technology.

SUMMARY

Current computers suffer from several communication problems that limit their computational throughput, such as bandwidth, clock skew and the Von Neuman bottleneck. The bandwidth and clock skew problems limit the speed and add to the design complexity of a processor. The Von Neumann bottleneck forces much of the speed of a processor to be used to compensate for the limited number of interconnections. Philosophically, the large bandwidth, innate parallelsim and non-interfering propagation of optics offer mechanisms for overcoming these communication problems.

Historically, the development of optical digital processors has suffered from the lack of suitable logical and memory devices. Recent advances in optical bistability offer the possibility of cascadable optical logic gates with speeds and power consumption competitive with electronic gates. A parallel pipelined architecture can be shown to simplify the optical memory requirements; utilize the large bandwidth, parallel, non-interfering communications of optics; and solve some of the expected problems associated with using optical technology that operates on a femtosecond timescale.

References

Abraham E., Seaton, C. T. & Smith, S. D. 1983 *Scient. Am.* **248**, 85–93.

Backus, J. 1982 *IEEE Spectrum,* **18**(8), 22–37.

Basov, N. G. 1972 In *Laser handbook* (ed. F. T. Arecchi & E. O. Schulz-Dubois), pp. 1649–1693. Amsterdam: North Holland.

Gibbs, H., Tarng, S. S., Jewell, J. L., Weinberger, D. A., Tai, K., Gossard, A. C., McCall, S. L., Passner, A. & Wiegmann, W. 1982 *Appl. Phys. Lett.* **41**, 221–222.

Huang, A. 1984 *Proc. Inst. elect. Electron. Engrs* **72**, 780–786.

Keyes, R. 1975 *Proc. Inst. elect. Electron. Engrs.* **63**, 740–767.

Mead, C. & Conway, L. 1980 *Introduction to VLSI systems.* Reading, Massachusetts: Addison-Wesley.

Miller, D. A. B. 1982 *Laser Focus* **18** (4), 79–84.

Miller, D. A. B. 1983 *Laser Focus* **18** (7), 61–68.

Smith, P. W. & Tomlinson, W. J. 1981 *IEEE Spectrum* **19** (6), 26–33.

Wilkes, M. V. 1983 *Conf. Proc. 10th Ann Int. Symp. on Computer Architecture,* IEEE cat. no. 83CH1889-5, pp. 2–4

Phil. Trans. R. Soc. Lond. A **313**, 213–220 (1984)

Printed in Great Britain

One-electron theory of nonlinear refraction

By B. S. Wherrett

Department of Physics, Heriot-Watt University, Riccarton, Currie, Edinburgh EH14 4AS, U.K.

The design of optically bistable cavities for low-power switching requires an understanding of the variation of nonlinearities in refractive index with material, temperature, wavelength and cavity parameters. Analytic expressions, suitable for scaling, are discussed. The connections between the giant nonlinearities under current investigation and the nonlinear optical mixing processes studied since the 1960s are reviewed and the self-consistent semiconductor optical bistability problem is summarized.

1. Introduction

The key to recent advances in the application of optical bistability has been the realization that semiconductor materials exist in which the intensity dependence of the refractive index is large enough to produce bistable characteristics at low power levels and in small samples.

The demands upon the nonlinearity for bistable operation are determined by reference to the transmission of a Fabry–Perot cavity (Miller 1981; Wherrett 1984). One can define a critical switching intensity, I_c, below which bistability and hysteresis cannot be achieved and above which it can occur for suitable initial conditions.

Given an internal irradiance I and associated contribution to the refractive index, $\Delta n(I)$, then I_c can be determined in terms of a material factor and a cavity factor. Optimizing the cavity reflectivities and length to minimize I_c one concludes that a Δn of order 10^{-3} is necessary to achieve bistability in semiconductors. Concentrating on the case $\Delta n(I) = n_2 I$, one requires for switching at frequency ω in a medium of linear refractive index n:

$$\Delta n(I) = n_2 I \approx nc\alpha/3\omega. \tag{1}$$

The coefficient n_2 is directly proportional to the real part of the third-order optical susceptibility, $\chi^{(3)}(\omega, -\omega, \omega)$, such that for switching at, say, 1 kW cm^{-2} one requires:

$$\mathrm{Re}\,\chi^{(3)} = (n^2 c/4\pi\omega^2)\,n_2 \approx 10^{-3}\ \text{e.s.u.}\dagger \tag{2}$$

2. First estimates

A quasi-dimensional analysis provides the simplest, first assessment of $\chi^{(3)}$. Interest lies in near-resonance situations where specific optical transitions dominate the nonlinearity. Consider N discrete atoms per unit volume, characterized by two-level systems of energy difference E_{10} and a transition dipole-moment er_{10}. Introducing the electromagnetic field interaction three times to attain a polarization, $\langle Ner \rangle$, third-order in the field, one must obtain

$$\chi^{(3)}_{\mathrm{atom}} \propto Ne^4 r_{10}^4 E_{10}^{-3}\,F(\hbar\omega/E_{10}). \tag{3}$$

† 1 e.s.u. = 1 cm^3 erg^{-1} $\equiv 1.4 \times 10^{-8}$ m^2 V^{-2}.

The dimensionless function F is as yet unspecified; if we were to set F equal to unity then a characteristic gaseous $\chi^{(3)}$ of a mere 10^{-18} e.s.u. may be obtained.

Considerable improvement occurs for the semiconductor case. Now the density of atoms is replaced by a sum over semiconductor k-states per unit volume and E_{10} is replaced by the k-dependent energy gap, $E(k) = E_g + \hbar^2 k^2/2m_r$. The reduced effective mass, m_r, is determined from the band interaction Hamiltonian ($H' = (\hbar/m)\, k \cdot p$), which is formally very similar to the radiative interaction ($H' = \mathrm{i}(e/m\omega)\, \mathscr{E} \cdot p \exp(-\mathrm{i}\omega t) + \text{c.c.}$). Here $\mathscr{E}$ is the electric field vector and p the momentum operator. If we introduce the Kane P-parameter, which is directly proportional to the momentum matrix element for interband transitions ($P = (\hbar/m)\, |p_{cv}|$), then E_g, P and $\hbar$ form a suitable set of parameters for dimensional analysis, leading to

$$\chi^{(3)}_{\text{semiconductor}} \propto e^4 P E_g^{-4} F(\hbar\omega/E_g). \tag{4}$$

For a small-gap semiconductor $\chi^{(3)} \approx F \times 10^{-9}$ e.s.u. Over the range of semiconductors P is essentially constant and the linear refractive index varies by a factor of less than two over most, whereas the band gap can vary by a factor of ten or more. Thus the E_g^{-4} dependence is expected to dominate any scaling. The factor F must be considered if magnitudes such as required for bistability are to be achieved. To do so one must recognize that F will contain resonance enhancement factors. The likelihood of a large F is seen by considering figure 1. This shows

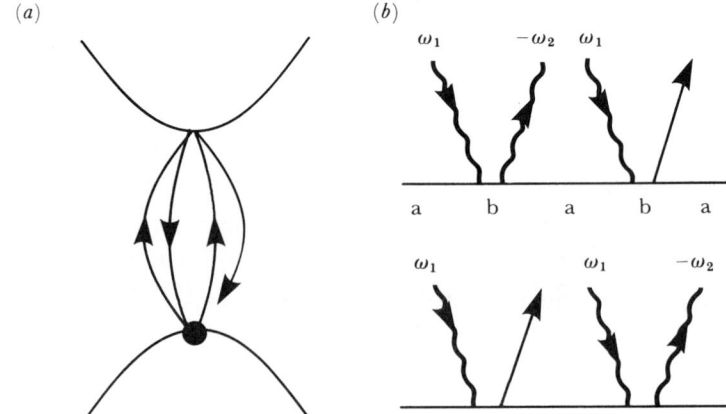

FIGURE 1. The most resonant contribution to nonlinear refraction (a) Four-stage transition scheme;
(b) time-ordering diagrams ($\omega_2 = \omega_1$ for refraction).

a four-stage transition scheme and two associated time-ordering diagrams, describing the three electromagnetic field interactions and the emission of a photon via the generated polarization. The corresponding mathematical contributions to $\chi^{(3)}$ will be proportional to the times for which each intermediate state can exist (Δt). In turn each Δt is inversely proportional to the energy mismatch between the intermediate (virtual) state of the system, and the initial state ($\Delta t \approx \hbar/\Delta E$). But for the chosen schemes the second intermediate state has precisely the same energy as the initial state. Thus $\Delta E = 0$ and the contributions to $\chi^{(3)}$ diverge to infinity. As we would expect, in practice the divergence is prohibited; nevertheless some nine orders of magnitude increase in $\chi^{(3)}$ will be derived from F (§4).

3. OBSERVATIONS AND INTERPRETATIONS

Franken & Ward recognized the existence of an intensity dependence of refraction in 1963, commenting in their theoretical paper that the phenomenon had 'not yet been observed', although in 1962 frequency-mixing processes in calcite had been observed by Terhune *et al.*, with $\chi^{(3)}$ magnitudes around 10^{-15} e.s.u. The first refractive nonlinearities (Mayer & Gires 1964; Maker *et al.* 1964), observed in liquids, gave values up to 10^{-12} e.s.u.

One must turn to $(2\omega_1 - \omega_2)$ four-photon mixing experiments for the earliest relevant semiconductor studies. This mixing is close to being a nonlinear refraction process and would be so if ω_2 were adjusted to be equal to ω_1. In the large gap material CdS, Maker & Terhune (1965) observed a correspondingly small $\chi^{(3)}$ of 5×10^{-12} e.s.u. With the invention of the infrared CO_2 laser, however, Patel *et al.* (1966) was able to study small gap semiconductors, achieving 8×10^{-10} e.s.u. in InSb with ω_1, $\omega_2 \approx 1000$, 1100 cm^{-1} respectively (the InSb band edge corresponds to approximately 1900 cm^{-1}).

Between 1965 and 1975 interest in nonlinear refraction centred on self-focusing studies, primarily in liquids. Here one is concerned with the spatial refractive index change brought about by an intense Gaussian beam and the resulting beam trapping and filamentation. The physical origin of self-focusing nonlinearities has been predominantly the molecular-orientational Kerr effect, inappropriate in solids, although as long ago as 1966 Javan & Kelley suggested that the saturation of anomalous dispersion in gases should lead to high Δn values.

Renewed interest in semiconductor nonlinear refraction came about as a culmination of several independent investigations. The first was the prediction of Szöke *et al.* (1969) that optical bistability should be possible for a saturating medium with feedback, and Gibbs's demonstration (1976) that optical bistability could be observed (in sodium vapour) but that the origin was dispersive. The second was the growing interest, since the transient holography work of Woerdman & Bolger (1969), in the dynamic refractive index gratings established when non-collinear laser beams interfere across a semiconductor sample. The diffraction of a third beam allows one to study the dynamics of the excited carriers that produced the index grating (Jarasiunas & Vaitkus 1974) or to achieve phase conjugation (Bergmann *et al.* 1978). Thirdly the 1970s were the era of the spin-flip Raman laser. In this instrument one is pumping a semiconductor (albeit in a high magnetic field) at a frequency just below the band gap (Dennis *et al.* 1972). At the time $\chi^{(3)}$ refractive index effects associated with the Raman gain were known to be present, causing mode-pulling, and the observed break-up of the transmitted pump-beam spatial profile was likely to have been due to nonlinear refraction.

It was not until the late 1970s, however, that the giant nonlinearities required for bistability were recognized. Thus, at continuous-wave (c.w.) CO-laser power levels, transmitted beam profile studies and analysis for InSb (Miller *et al.* 1978; Weaire *et al.* 1979) produced $\chi^{(3)}$ values up to 10^{-2} e.s.u. for frequencies just below the band edge, as have studies of phase conjugation and degenerate four-wave mixing in cadmium mercury telluride (Jain 1982). Finally, in 1979 reports were made of the first semiconductor optical bistability experiments (Gibbs *et al.* 1979; Miller *et al.* 1979) with $\chi^{(3)}$ values of 10^{-5} to 10^{-4} e.s.u. in GaAs and 1 e.s.u. in InSb.

A number of theoretical models have been used to interpret the various observations discussed above. Early mixing work was explained in terms of (i) free-carrier non-parabolicity (Wolff & Pearson 1966) and (ii) virtual interband transitions (Jha & Bloembergen 1968; Wynne 1969). Giant nonlinear refraction has been discussed (iii) using a direct saturation model (Miller *et al.* 1980), (iv) in terms of a dynamic Burstein–Moss effect (Wherrett & Higgins 1982; Moss 1980), (v) as a free carrier plasma effect (Jain & Klein 1980) and (vi) as an enhanced non-parabolicity effect (Khan *et al.* 1981). In addition a many-body treatment has been made by Haug and coworkers (Koch & Haug 1981; Haug & Schmitt-Rink, this symposium).

These models, and the range of experimental data, lead one to pose the following questions: How do the models relate? What resonance factors should one use? Can the eleven orders of magnitude between the mixing $\chi^{(3)}$, 10^{-11} e.s.u. in CdS and the 1 e.s.u. refractive $\chi^{(3)}$ in InSb be bridged smoothly? And how should one best calculate the giant nonlinear indices now being studied?

4. UNIFICATION OF THEORETICAL MODELS

In the non-parabolicity model one considers the momentum dependence of the effective mass of free carriers. As a result, in an applied electric field $\mathscr{E}$ the carrier velocities contain terms in $\mathscr{E}^3$ so that the conductivity (and susceptibility) contain third-order terms. In essence one obtains

$$\chi^{(3)} \approx + N_0 \, e^4 p^4 \, E_g^{-7} \, F(E_F/E_g), \tag{5}$$

where E_F is the carrier Fermi energy. N_0 is the initial density of carriers and there is no need to mention interband transitions in the theoretical approach.

By contrast, starting with four-stage virtual interband transitions for mixing at $(2\omega_1 - \omega_2)$ one obtains a non-diverging expression for $\chi^{(3)}$ dominated by the processes shown in figure 1*b*. Summing over the two schemes there is a partial cancellation of terms, following which ω_2 can be set equal to ω_1 without producing the divergence problem discussed earlier (Wherrett 1983). As a result, for N_0 electrons at the bottom of the conduction band,

$$\chi_{N_0}^{(3)} \approx N_0 \, e^4 \frac{P^4}{E_g^7} \left[\frac{E_g}{E_g - \hbar\omega} \right]^3. \tag{6}$$

Expression (5) is regained in the limit $\hbar\omega \ll E_g$ showing that the non-parabolicity model is based on interband processes. For a full valence band and empty conduction band this virtual transition scheme model gives

$$\chi^{(3)} \approx - e^4 \frac{P}{E_g^4} \left[\frac{E_g}{E_g - \hbar\omega} \right]^{\frac{3}{2}}. \tag{7}$$

Note that this has precisely the scaling form predicted by dimensional analysis; also $\chi^{(3)}$ displays a minus three-halves resonance behaviour and is negative. The latter is significant experimentally because a negative nonlinear refraction leads to beam defocusing, which is stable by comparison with the catastrophic self-focusing.

The efficiency of resonance enhancement of virtual processes is demonstrated by recent $(2\omega_1 - \omega_2)$-mixing observations in InSb (MacKenzie *et al.* 1984). As one approaches the band edge $\chi^{(3)}$ values up to 10^{-6} e.s.u. have been observed, some three orders of magnitude greater than are observed with mid-gap frequencies. These values are, however, still six orders of magnitude smaller than the refractive $\chi^{(3)}$ near resonance. To bridge this gap one must reduce the 4 cm^{-1} difference in ω_1 and ω_2 used for the above mixing experiment, thereby detecting the effect of real (rather than virtual) excitation of carriers.

By introducing two effective scattering times one can make the link between the virtual transition picture and the plasma or Burstein–Moss pictures. By analogy with a two-level system consider the optically coupled valence and conduction states, lying at the same $\boldsymbol{k}$. Describe intraband scattering, due for example to electron–electron and electron–phonon interactions, by a damping or dephasing lifetime T_2. Describe interband recombination by a time T_1. Real (long-term) excitation becomes important for $(E_g/\hbar - \omega) \lesssim 1/T_2$ and, in mixing, for $|\omega_1 - \omega_2| \lesssim 1/T_1$. The same time-ordered diagrams (figure 1b) dominate $\chi^{(3)}$, which can be calculated by using density matrix theory with the essential result

$$\chi^{(3)} \approx \chi^{(3)}_{\text{virtual}} \, T_1/T_2. \tag{8}$$

Introduction of T_2 and T_1 is the basis of the direct saturation model of the nonlinearity. The T_2 lifetime can also be used, phenomenologically, to describe the band-tail absoption that must be present for real excitation:

$$\alpha_{\text{tail}} \approx \frac{e^2}{n\hbar c P} \frac{1}{T_2} \left[\frac{E_g}{E_g - \hbar\omega} \right]^{\frac{1}{2}}. \tag{9}$$

By fitting to the observed tail one obtains an effective T_2 (Higgins 1983). Alternatively one can re-express $\chi^{(3)}$ directly in terms of α, or Δn in terms of the equilibrium excited-carrier concentration ($\Delta N = \alpha I T_1/\hbar\omega$):

$$\chi^{(3)} \approx -\frac{e^2 P^2}{E_g^4} nc\alpha \, T_1 \left[\frac{E_g}{E_g - \hbar\omega} \right] \tag{10a}$$

or

$$\Delta n \approx \frac{-e^2 P^2}{n E_g^3} \Delta N \left[\frac{E_g}{E_g - \hbar\omega} \right]. \tag{10b}$$

Within small factors precisely these results are also achieved in the Burstein–Moss and plasma models. The differences lie only in that a different distribution of excited carriers is described in each model. Figure 2 shows such conduction electron distributions schematically. Under direct saturation the carriers partly populate the optically coupled states and remain in these states, those nearest resonance being the most strongly saturated (figure 2a). In the Burstein–Moss model it is recognized that T_2 scattering will thermalize the carriers (figure 2b). The plasma model is the low-temperature limit to the Burstein–Moss description (figure 2c). In the

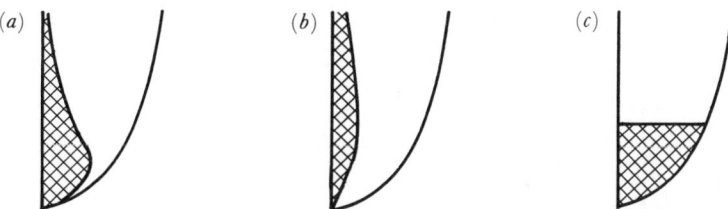

FIGURE 2. Excited conduction electron distribution for (a) direct saturation (b) the Burstein–Moss model and (c) Burstein–Moss at zero temperature or the plasma model.

latter two models the excitation and refraction may be treated separately. One calculates N at equilibrium, allows a thermal distribution to establish, then calculates the refractive index change due to this change in electron distribution among the bands. One obtains a nonlinear index only through the intensity dependence of N. The conventional Burstein–Moss effect is a shift of the effective band edge for a doped material. In the dynamic Burstein–Moss picture one thinks of the radiative generation of carriers causing a dynamic edge-shift and of the nonlinear refraction as a result of the absorption change. In the plasma picture the N carriers have a plasma frequency ω_p and near the semiconductor band edge one has a dielectric response

$$\epsilon \approx \epsilon_\infty \left(1 - \frac{\omega_p^2}{\omega^2} \frac{E_g^2}{E_g^2 - \hbar^2 \omega^2} \right). \tag{11}$$

Equation $(10\,b)$ is regained on analysis of this result, demonstrating once again that the band-edge enhancement in (11) is merely a recognition of the blocking of *interband* transitions due to the carrier plasma.

The Burstein–Moss and plasma pictures are more flexible than the direct saturation model as they can accommodate any interband absorption source (direct, two-photon, etc.) and any recombination process (via traps as described by T_1, or via radiative or Auger recombination). Furthermore the use of a Kramers–Kronig analysis on the absorption change due to the thermal carrier distribution enables any temperature to be considered. Thus for example in the Boltzmann high-temperature limit (Miller 1981),

$$\chi^{(3)} \approx \frac{e^2 P^2}{E_g^3 k_B T} nc\alpha T_1 J\left(\frac{k_B T}{E_g - \hbar\omega} \right). \tag{12}$$

These advantages lead us to the self-consistent approach to the Δn problem.

5. The self-consistent semiconductor optical bistability problem

Figure 3 demonstrates the calculations necessary for c.w., plane-wave, dispersive optical bistability in a semiconductor cavity. Starting at the left hand side one has a semiconductor characterized by a refractive index n, interband adsorption α_i and a parasitic absorption (due for example to the created free-carriers) α_p. For an incident irradiance I_1 at frequency ω one determines the internal irradiance $I(\omega)$ and the transmitted or reflected (or both) irradiances: this is the cavity feedback problem.

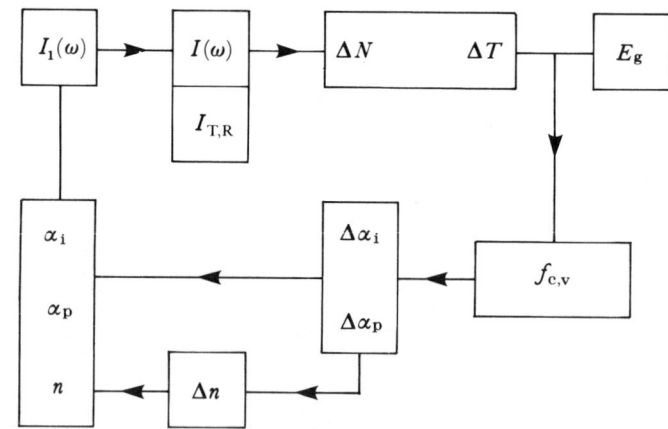

FIGURE 3. The semiconductor optical bistability self-consistency problem.

The internal irradiance generates ΔN carriers: the recombination problem; simultaneously there may be a change of sample temperature, ΔT. As a consequence the band edge and structure will alter, directly through ΔT and via many-body effects (gap renormalization). Assuming separate thermalization within the conduction band and the valence band one calculates the new carrier distributions f_c and f_v and, given an understanding of all absorption processes, both absorption changes $\Delta\alpha_i$ and $\Delta\alpha_p$ at all frequencies may be calculated. A Kramers–Kronig-type transformation will then produce the refractive index change Δn at the frequency of interest (ω). All three of $\Delta\alpha_i$, $\Delta\alpha_p$ and Δn must then be cycled around this loop self-consistently.

The linear proportionality of Δn to I will be true only under special conditions (low intensities, small carrier densities). Further, one must in practice consider time-dependent effects (switching, transients, pulsed operation) and spatially dependent effects (beam diffraction and lensing, and carrier diffusion) as discussed in following articles in this symposium. Nevertheless the $\chi^{(3)}$ or n_2 approximation has been demonstrated to be valid for certain experiments (Miller et al. 1981; Garmire et al., this symposium). Figure 4 shows just such a case, and shows that a fit to n_2 (solid line) can be achieved within a description in which the absorption (modelled by a T_2-tail) is allowed to saturate self-consistently due to a thermal carrier distribution. This technique allows us to generate *effective* T_2 and T_1 times of around 10 ps and 1 μs respectively, a ratio $T_1/T_2 \approx 10^5$.

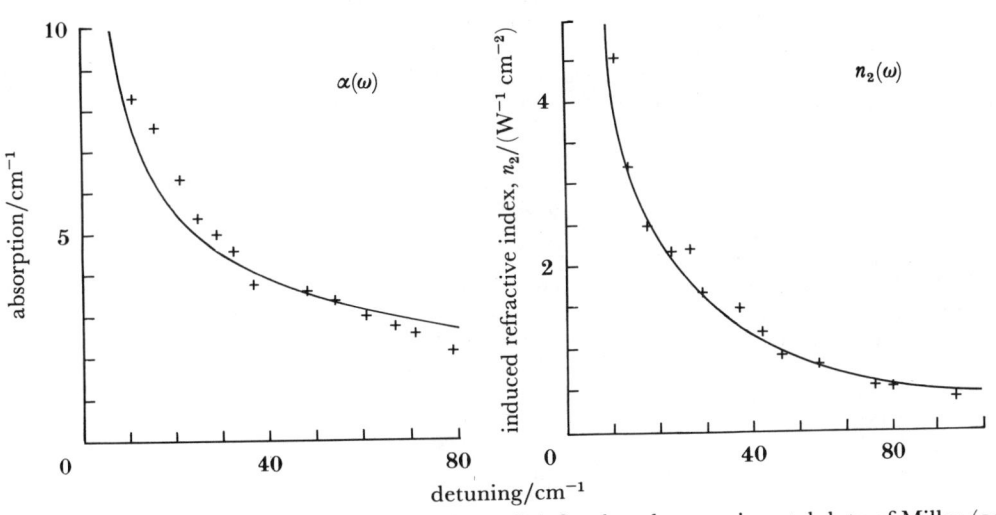

FIGURE 4. Self-consistent calculations for n_2 (Higgins 1983) fitted to the experimental data of Miller (1979) indicating an effective T_1/T_2 of 10^5 (see text).

6. CONCLUSION

In conclusion it can be demonstrated that the various models for $\chi^{(3)}$ and Δn discussed above are all based on interband carrier excitation (virtual or long term) and that one can scale n_2 essentially as

$$n_2 \approx \frac{-e^4}{c} \frac{P}{n^2 E_g^4} \left[\frac{E_g}{E_g - \hbar\omega}\right]^{\frac{3}{2}} \frac{T_1}{T_2}. \tag{13}$$

The eleven orders of magnitude increase from $\chi^{(3)}(\omega_1, -\omega_2, \omega_1)$ in CdS to $\chi^{(3)}(\omega, -\omega, \omega)$ in InSb are bridged by moving to smaller gap material ($\times 10^2$), approaching resonance ($\times 10^3$), and equating the frequencies ω_2 and ω_1($\times 10^5 - 10^6$). Finally I give as an example in the high temperature limit an expression for the critical bistability switching intensity,

$$I_c \approx \frac{\hbar c}{e^2} \frac{n^2 E_g^2}{P^2 T_1} k_B \, T J^{-1}\left(\frac{k_B \, T}{E_g - \hbar\omega}\right) f(\text{cavity}). \tag{14}$$

This expression indicates the considerations of material, temperature, frequency and cavity parameters that must be taken into account to optimize for low-intensity switching.

220 B. S. WHERRETT

This work was done in the framework of an operation launched by the Commission of the European Community under the experimental phase of the European Community Stimulation Action (1983–85).

REFERENCES

Bergmann, E. E., Bigio, I. J., Feldman, B. J. & Fisher, R. A. 1978 *Optics Lett.* **3**, 82.
Dennis, R. B., Pidgeon, C. R., Smith, S. D., Wherrett, B. S. & Wood, R. A. 1972 *Proc. R. Soc. Lond.* A **331**, 203.
Franken, P. A. & Ward, J. F. 1963 *Rev. mod. Phys.* **35**, 23.
Gibbs, H. M., McCall, S. L. & Venkatesan, T. N. C. 1976 *Phys. Rev. Lett.* **36**, 1135.
Gibbs, H. M., McCall, S. L., Venkatesan, T. N. C., Passner, A., Gossard, A. C. & Weigmann, W. 1979 *Solid St. Commun.* **30**, 271.
Higgins, N. A. 1983 Ph.D. thesis, Heriot-Watt University.
Jain, R. K. 1982 *Opt. Engng* **21**, 199.
Jain, R K. & Klein, M. B. 1980 *Appl. Phys. Lett.* **37**, 1.
Jarasuinas, K. & Vaitkus, J. 1974 *Physica Status Solidi* A **23**, K19.
Javan, A. & Kelley, P. L. 1966 *IEEE Jl Quantum Electron.* **QE-2**, 470.
Jha, S. S. & Bloembergen, N. 1968 *Phys. Rev.* **171**, 891.
Khan, M. A., Bogant, T. J., Kruse, P. W. & Ready, J. F. 1981 *Optics Lett.* **5**, 469.
Koch, S. W. & Haug, H. 1981 *Phys. Rev. Lett.* **46**, 450.
MacKenzie, H. A., Al-Attar, H. & Wherrett, B. S. 1984 *J. Phys.* B **17**, 2141.
Maker, P. D. & Terhune, R. W. 1965 *Phys. Rev.* A **137**, 801.
Maker, P. D., Terhune, R. W. & Savage, C. M. 1964 *Phys. Rev. Lett.* **12**, 507.
Mayer, G. & Gires, F. 1964 *C.r. hebd. Seanc. Acad. Sci., Paris* **258**, 2039.
Miller, D. A. B. 1979 Ph.D. thesis, Heriot-Watt University.
Miller, D. A. B. 1981 *IEEE Jl Quantum Electron.* **QE-17**, 306.
Miller, D. A. B., Mozolowski, M. H., Miller, A. & Smith, S. D. 1978 *Optics Commun.* **27**, 133.
Miller, D. A. B., Seaton, C. T., Prise, M. E. & Smith, S. D. 1981 *Phys. Rev. Lett.* **47**, 197.
Miller, D. A. B., Smith, S. D. & Johnston, A. 1979 *Appl. Phys. Lett.* **35**, 658.
Miller, D. A. B., Smith, S. D. & Wherrett, B. S. 1980 *Optics Commun.* **35**, 221.
Moss, T. S. 1980 *Physica Status Solidi* B **101**, 555.
Patel, C. K. N., Slusher, R. E. & Fleury, P. A. 1966 *Phys. Rev. Lett.* **17**, 1010.
Szöke, A., Damen, V., Goldhar, J. & Kurnit, N. A. 1969 *Appl. Phys. Lett.* **15**, 376.
Terhune, R. W., Maker, P. D. & Savage, C. M. 1962 *Phys. Rev. Lett.* **8**, 401.
Weaire, D. L., Wherrett, B. S., Miller, D. A. B. & Smith, S. D. 1979 *Optics Lett.* **4**, 831.
Wherrett, B. S. 1983 *Proc. R. Soc. Lond.* A **390**, 373.
Wherrett, B. S. 1984 *IEEE Jl Quantum Electron.* **QE-20**, 646.
Wherrett, B. S. & Higgins, N. A. 1982 *Proc. R. Soc. Lond.* A **379**, 67.
Woerdman, J. P. & Bolger, B. 1969 *Phys. Lett.* A **30**, 164.
Wolff, P. A. & Pearson, G. A. 1966 *Phys. Rev. Lett.* **17**, 1015.
Wynne, J. J. 1969 *Phys. Rev.* **138**, 1296.

Phil. Trans. R. Soc. Lond. A **313**, 221–227 (1984)

Printed in Great Britain

Non-perturbative many-body theory of the optical nonlinearities in semiconductors

By H. Haug and S. Schmitt-Rink

*Institutut für Theoretische Physik der Universität Frankfurt, Robert-Mayer-Str. 8,
D-6000 Frankfurt am Main, F.R.G.*

We discuss the various many-body effects that contribute to the large optical nonlinearities of semiconductors close to the band edge. Such effects are: band filling, screening due to the excited electron–hole pairs, band-gap shrinkage, energy broadening and excitonic enhancement. Compared to the Coulomb self-energy corrections, the radiative self-energies are normally small. However, they determine the transition rates in the rate equations of the photons and electrons.

I. Single-particle properties

In the region close to but below the band edge, semiconductors show large optical nonlinearities that can be exploited to get optical bistability in transmission or reflection experiments. These nonlinearities are due to the generation of electron–hole pairs in the exciton states and band-tail states. The long-range Coulomb interaction between the mobile electronic excitations requires a genuine non-perturbative many-body treatment. For recent reviews see Klingshirn & Haug (1981), Haug (1982) and Haug & Schmitt-Rink (1984). At low pair concentrations the attractive electron–hole interaction dominates. With increasing excitation the Coulomb forces are increasingly screened until an electron–hole plasma is formed. In the plasma state the band gap is reduced due to exchange and correlation effects below the original exciton energy level. Furthermore, the single-particle energies are smeared out by collision broadening. All these effects of the Coulomb interactions can be described as *intra-band* scattering processes due to the exchange of *longitudinal photons* ($\{i, j\} = \{e, h\}$):

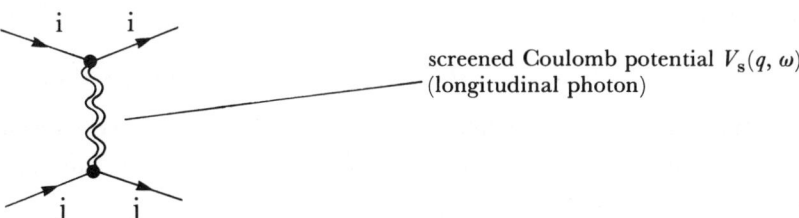

screened Coulomb potential $V_s(q, \omega)$
(longitudinal photon)

On the other hand, the interaction with the light field causes *inter-band* transitions due to the exchange of *transverse photons*:

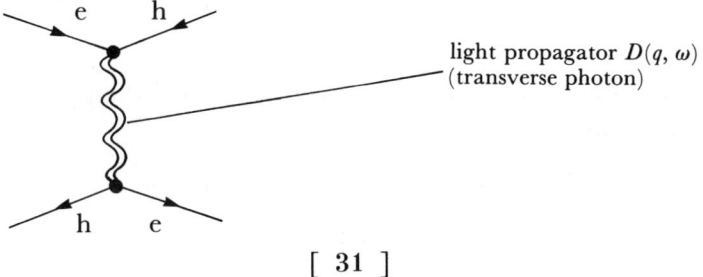

light propagator $D(q, \omega)$
(transverse photon)

15-2

With these two types of interactions the Dyson equation of the electron, for example, is in random phase approximation (r.p.a.):

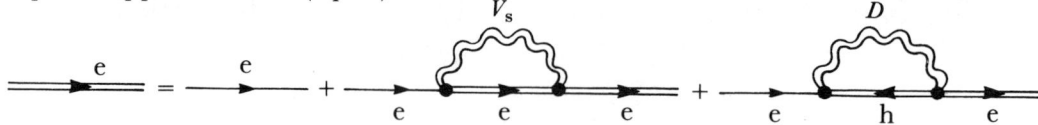

Normally, the Coulomb self-energy is larger than the radiative self-energy correction. Only for extremely intensive laser fields, where a complete bleaching occurs, the radiative self-energy modifies the single-particle spectrum appreciably by causing a gap around the laser frequency (Galitskii *et al.* 1970). Otherwise, the radiative self-energy can be neglected for the calculation of the single-particle spectrum. As shown by Kadanoff & Baym (1962) and systematically developed by Keldysh (1965) in terms of non-equilibrium Green function, one gets the density of particles from the time-dependent Green function (with $\hbar = 1$ throughout the article)

$$G_e(r, t; r, t^+) = -i\langle T\psi_e(r, t)\,\psi_e^\dagger(r, t^+)\rangle = i\langle \psi_e^\dagger(r, t)\,\psi_e(r, t)\rangle = in_e(r, t).$$

The total number of electrons is changed solely by the radiative interaction. Therefore, if one calculates the rate equation for the total number of electrons (in the conduction band) from the electron Dyson equation, only the radiative self-energy contributes and yields the radiative transition rates (Haug & Schmitt-Rink 1984)

$$\partial N_e/\partial t = \alpha(\omega)\,I/\omega - R_{\text{spon}},$$

where $\alpha(\omega)$ is the absorption coefficient, I is the laser intensity and R_{spon} is the rate of spontaneous transitions:

$$R_{\text{spon}} = \int_0^\infty d\omega(\omega n(\omega)/\pi c)^2\alpha(\omega)/[\exp\{\beta(\omega-\mu)\}-1].$$

Here, $n(\omega)$ is the index of refraction and μ the quasi-chemical potential.

II. Polarization in a dielectric medium

The optical properties of the medium can be determined by calculating the (transverse) photon Green function. Classically, the retarded Green function of the Maxwell wave equation for the transverse vector potential is given by

$$D(q, \omega) = 4\pi c/(\omega^2\epsilon(q, \omega) - q^2c^2),$$

where $\epsilon(q, \omega)$ is the transverse or optical dielectric function. The whole of one-beam spectroscopy is contained in the complex function $\epsilon(q, \omega)$; for example absorption, refraction, reflection and luminescence spectra can be calculated if $\epsilon(q, \omega)$ is known.

Quantum mechanically, the photon Green function is determined by its Dyson equation

$$D(q, \omega) = 4\pi c/(\omega^2 - q^2c^2 - 4\pi\Pi(q, \omega)),$$

where $\Pi(q, \omega)$ is the irreducible photon self-energy, also called the polarization function. The solution of the Dyson equation is

$$D(q, \omega) = 4\pi c/(\omega^2 - q^2c^2 - 4\pi\Pi(q, \omega)),$$

which yields the relation
$$\epsilon(q, \omega) = 1 - 4\pi \Pi(q, \omega)/\omega^2.$$

The inter-band part of the polarization function $\Pi(q, \omega)$ is in lowest approximation just the electron–hole pair Green function

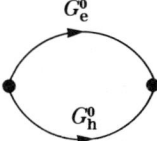

so that in the long-wavelength limit

$$\epsilon(\omega) = \epsilon_\infty\{1 - \omega_{\mathrm{pl}}^2/(\omega + \mathrm{i}\delta)^2\} - 4\pi e^2 \sum_{k} |r_{cv}(k)|^2 (1 - f_{ek} - f_{hk})\, G_{\mathrm{eh}}(k, \omega),$$

where ω_{pl} is the plasma frequency

$$\omega_{\mathrm{pl}}^2 = \frac{4\pi e^2}{\epsilon_\infty} \sum_{k} \left(\frac{f_{ek}}{m_{ek}} + \frac{f_{hk}}{m_{hk}}\right)$$

and f_{ek}, f_{hk} and m_{ek}, m_{hk} are the Fermi distribution functions and masses of the electron and holes, respectively; r_{cv} is the dipole matrix element and $G_{\mathrm{eh}}(k, \omega)$ is the pair Green function

$$G_{\mathrm{eh}}(k, \omega) = 2(\epsilon_{ek} + \epsilon_{hk})/\{(\omega + \mathrm{i}\delta)^2 - (\epsilon_{ek} + \epsilon_{hk})^2\}.$$

We illustrate this result by calculating the intensity-dependent changes of the index of refraction for the narrow-gap semiconductor $\mathrm{Pb}_{1-x}\mathrm{Sn}_x\mathrm{Se}$ by using the rate equation of the last section together with the linear response polarization function.

Figure 1 shows the low-frequency, intra-band contributions and the inter-band contributions,

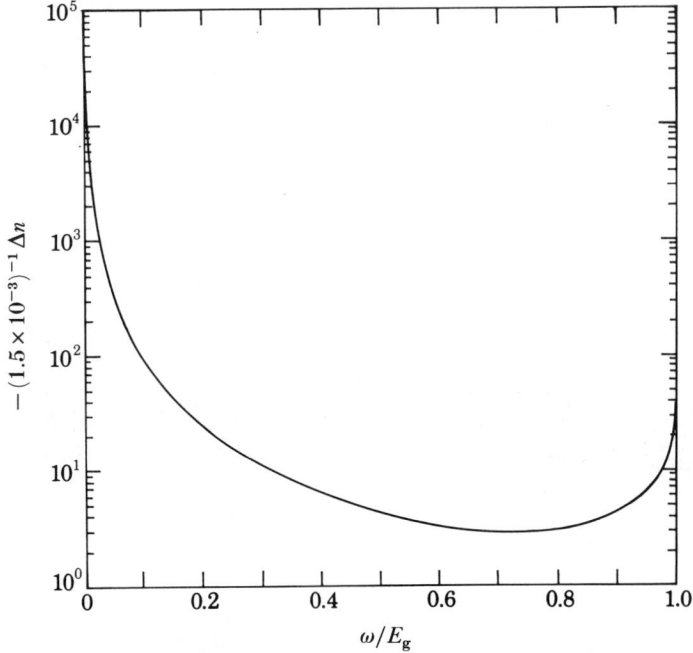

FIGURE 1. Changes of the index of refraction of $\mathrm{Pb}_{1-x}\mathrm{Sn}_x\mathrm{Se}$ at 77 K under laser excitation at $\omega_0 = 0.96\, E_{\mathrm{g}}$ with an intensity $I = 1.5 \times 10^3\ \mathrm{W\ cm^{-2}}$; $x = 0.051$, $E_{\mathrm{g}} = 0.117$ eV.

[33]

which become resonant at the band gap. Often, this simple free-particle approximation is not sufficient. The Coulomb interaction has to be taken into account by replacing the free electron–hole pair Green function by

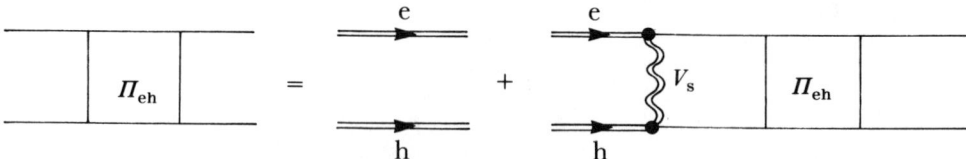

Radiative vertex corrections are completely non-resonant and can be neglected. The integral equation for Π_{eh} is the Bethe–Salpeter equation in the screened ladder approximation. In the low-density limit, where screening and single-particle renormalization are unimportant, one obtains analytically the exciton spectrum with an infinite number of bound states and an ionization continuum (i.e. the so-called Elliot spectrum). For arbitrary densities the integral equation for the polarization function has to be solved numerically. For static free-carrier

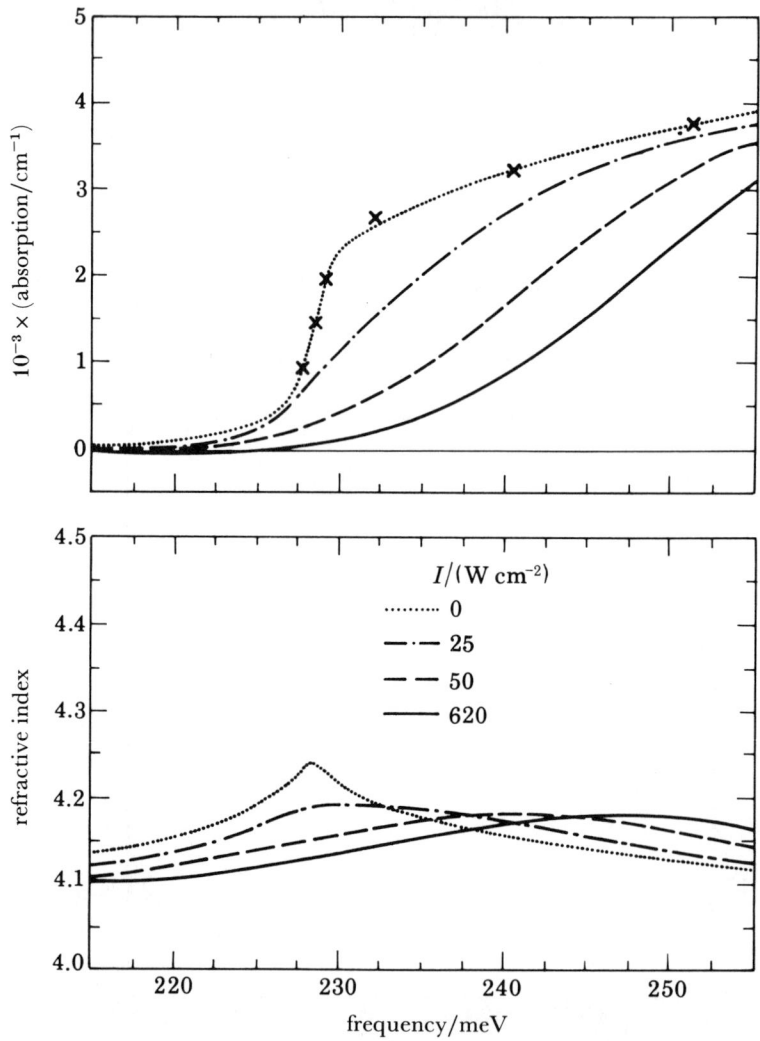

FIGURE 2. Absorption and refraction spectra of InSb at 77 K for various excitation intensities (from Löwenau *et al.* 1982). The frequency of the exciting laser beam is $\omega_0 = 0.225$ eV. The experimental points of the low-intensity absorption spectrum are taken from Gobeli *et al.* (1960).

screening accurate solutions can be obtained by matrix inversion (Löwenau *et al.* 1982). A finite single-particle energy broadening γ is incorporated in the treatment. For InSb this broadening is larger than the exciton Rydberg E_0, so that no well-resolved bound state exists. The absorption and refraction spectra for InSb are shown in figure 2 for various light intensities. The curve for vanishing light intensity shows that the exciton effects tend to produce a 'squared' absorption continuum. For increasing light intensity there is a pronounced compensation between the effects of the band-gap shrinkage and the reduction of excitonic enhancement, so that the dominating effect is band filling. The decrease of the index of refraction has been exploited (Miller *et al.* 1979) to obtain optical bistability. In bulk GaAs the broadening γ is smaller than the exciton Rydberg, at least at low temperatures ($T \lesssim 77$ K), in superlattice structures even at room temperature. Therefore, the calculated spectra (figure 3) exhibit in the low-density limit sharp exciton lines. For higher light intensities, and therefore higher free-carrier concentrations, the exciton vanishes due to the increased screening. This bleaching of the exciton resonance causes again a decrease of the index of refraction for $\omega < E_g - E_0$, which is large enough to allow the realization of optical bistability (Gibbs *et al.* 1979). In wide-gap semiconductors, where the (longitudinal) dielectric functions are smaller, large band-gap

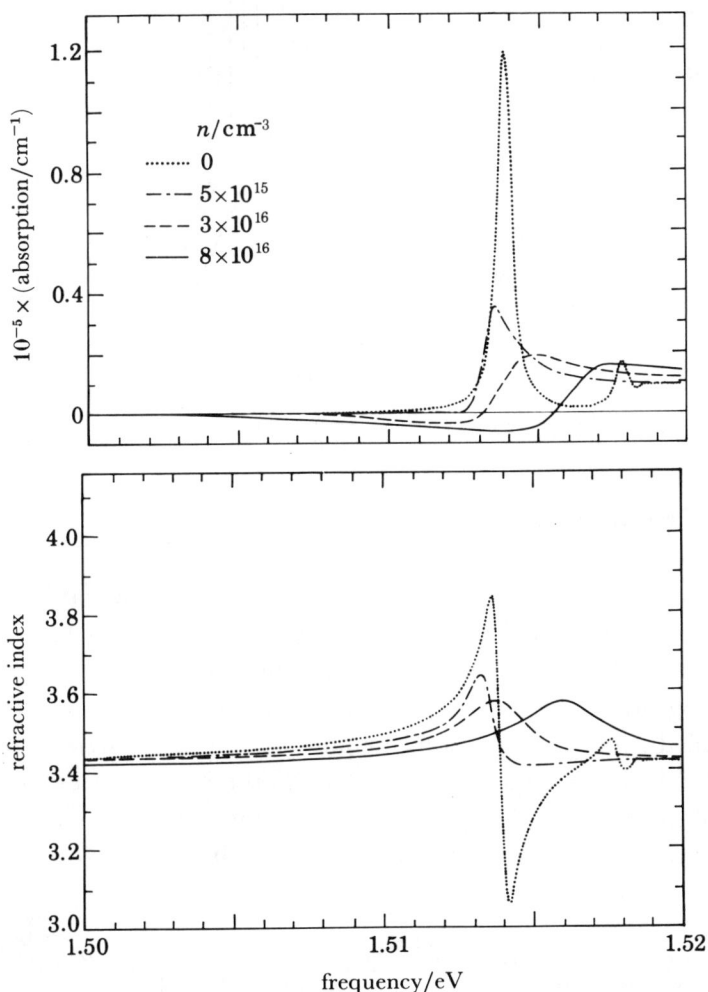

FIGURE 3. Absorption and refraction spectrum for GaAs at 10 K for various free-carrier densities n (from Löwenau *et al.* 1982).

[35]

reductions are measured with increasing concentration of electronic excitations (Bohnert *et al.* 1981), as shown in figure 4. The calculated exchange and correlation energy shifts are in good agreement with the measured ones. For laser frequencies below the lowest exciton resonance the absorption is initially weak, but increases strongly at higher intensities when the band gap passes the laser frequency.

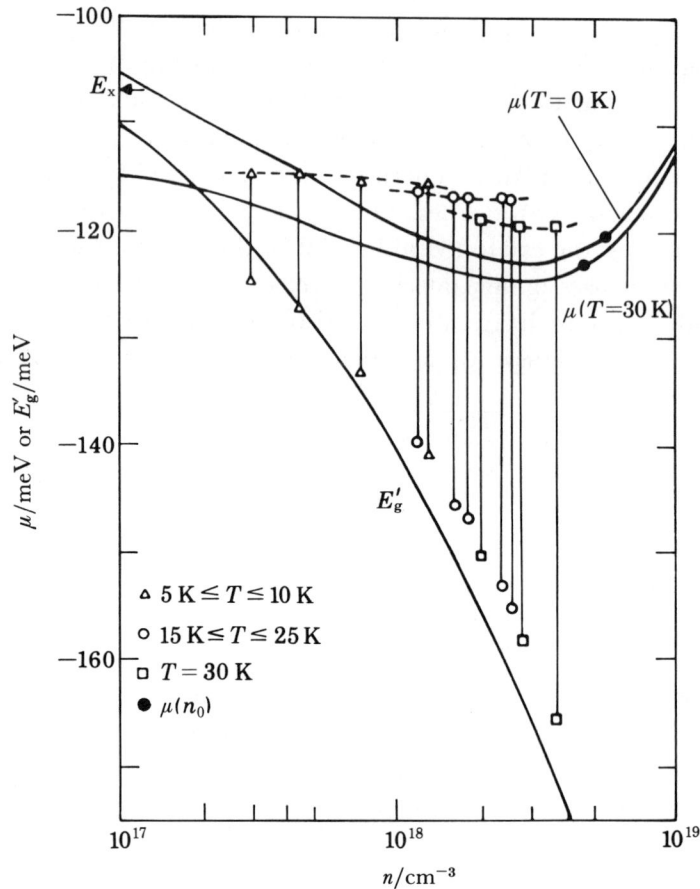

FIGURE 4. Renormalized energy-gap E'_g and quasi-chemical potential μ for CdS (from Bohnert *et al.* 1981). The solid line is for theory and the broken line is for experiment.

This induced absorption can be used to obtain an absorptive optical bistability in CdS (for the experiment see Bohnert *et al.* (1984): for the theory, see Schmidt *et al.* (1984)) with a clockwise hysteresis loop. Owing to the nonlinear absorption, density kinks develop in the medium, which give rise to interesting time dependences of the transmitted light. These effects can be studied by solving the coupled transport equations for photons and free carriers (Koch *et al.* 1984).

Finally, in CuCl, which has a gap of about 3.5 eV, the biexcitons determine the nonlinear optical properties of the medium. The polarization function Π is now explicitly nonlinear. The two-photon absorption is described by the following irreducible diagrams:

$$\Pi = \underset{G^0_x}{\bullet \!\!\!\longrightarrow\!\!\! \bullet} + \cdots$$

G_x and G_m are the exciton and biexciton Green functions, $\tilde{G}_x$ is the exciton Green function without polariton renormalization. An evaluation of these diagrams yields, if damping is neglected (Schmitt-Rink, unpublished work)

$$\epsilon(\omega) = \epsilon_\infty - 4\pi\mu_{gx}^2 \, \tilde{G}_x^2/G_x^0,$$

where
$$\tilde{G}_x = (G_x^{0-1} - \tilde{\Pi}_x)^{-1},$$

and $\tilde{\Pi}_x$ is the exciton self-energy contained in the second polarization diagram. This result was first derived by Abram & Maruani (1982). In earlier treatments we omitted the third polarization diagram, which is, however, also important close to the biexciton resonance. Optical bistability due to the nonlinearity calculated here has been observed recently (Levy *et al.* 1983; Peyghambarian *et al.* 1983).

This work has been supported by the Commission of the European Communities (under the experimental phase of the European Stimulation Action).

REFERENCES

Abram, I. & Maruani, A. 1982 *Phys. Rev.* B **26**, 4759–4761.

Bohnert, K., Kalt, H. & Klingshirn, C. 1983 *Appl. Phys. Lett.* **43**, 1088–1090.

Bohnert, K., Anselment, M., Kobbe, G., Klingshirn, C., Haug, H., Koch, S. W., Schmitt-Rink, S. & Abraham, F. F. 1981 *Z. Phys.* B **42**, 1–11.

Galitskii, V. M., Goreslavskii, S. P. & Elesin, V. P. 1970 *Soviet Phys. JETP* **30**, 117–122.

Gibbs, H. M., McCall, S. L., Venkatesan, T. N. C., Gossard, A. C., Passner, A. & Wiegmann, W. 1979 *Appl. Phys. Lett.* **35**, 451–453.

Gobeli, G. W. & Fan, H. Y. 1960 *Phys. Rev.* **119**, 613–620.

Haug, H. 1982 In *Festkörperprobleme* XXII (*Advances in solid state physics*, vol. 22) (ed. P. Grosse), pp. 149–171. Vieweg: Braunschweig.

Haug, H. & Schmitt-Rink, S. 1984 Electron theory of the optical properties of laser-excited semiconductors. *Progress in quantum electronics*. Oxford: Pergamon Press. (In the press.)

Kadanoff, L. P. & Baym, G. 1962 *Quantum statistical mechanics*. New York: W. A. Benjamin.

Keldysh, L. V. 1965 *Soviet Phys. JETP* **20**, 1018–1026.

Klingshirn, C. & Haug, H. 1981 *Phys. Rep.* **70**, 315–398.

Koch, S. W., Schmidt, H. E. & Haug, H. 1984 Spatial effects in absorptive optical bistability. *Appl. Phys. Lett.* (In the press.)

Levy, R., Bigot, J. Y., Hönerlage, B., Tomasini, F. & Grun, J. B. 1983 *Solid State Commun.* **48**, 705–708.

Löwenau, J. P., Schmitt-Rink, S. & Haug, H. 1982 *Phys. Rev. Lett.* **49**, 1511–1514.

Miller, D. A. B., Smith, S. D. & Johnston, A. 1979 *Appl. Phys. Lett.* **35**, 658–660.

Peyghambarian, N., Gibbs, H. M., Rushford, M. C. & Weinberger, D. A. 1983 *Phys. Rev. Lett.* **51**, 1692–1695.

Schmidt, H. E., Haug, H. & Koch, S. W. 1984 *Appl. Phys. Lett.* **44**, 787–789.

Phil. Trans. R. Soc. Lond. A **313**, 229–237 (1984)

Printed in Great Britain

Optical bistability in CuCl

By R. Levy, B. Hönerlage and J. B. Grun

*Laboratoire de Spectroscopie et d'Optique du Corps Solide Associé au C.N.R.S. no. 232,
Université Louis Pasteur, 5 rue de l'Université, 67000 Strasbourg, France*

Optical bistability due to biexcitons in CuCl is discussed with respect to its physical origin. We show that switching times and switching intensities are functions of the photon energy of the exciting light beam.

I. Introduction

Optical bistability was first observed by Gibbs *et al.* (1976) using a Fabry–Perot etalon filled with sodium vapour. Although this material shows saturable absorption, it turned out that the intensity dependence of the refractive index was extremely important for the observation of optical bistability. Since the use of vapour requires Fabry–Perot resonators with a long 'round-trip' time, research switched to more dense materials like liquids and solids. In semiconductors, optical nonlinearities may be very important. Near resonances, optical bistable behaviour was first observed in the infrared and visible spectral region in InSb and GaAs by Miller *et al.* (1979) and Gibbs *et al.* (1979), respectively. In both materials, bistability is due to the saturation of an absorption (band to band or exciton transitions) and the resulting intensity dependence of the refractive index. In both cases, switch-on times of the devices are limited in principle by the cavity round-trip time, which is in the picosecond range. Since the index change depends on the density of the quasi-particles excited throughout the process, the switch-off time is limited by the lifetime of the particles, which is in the nanosecond range. Although this time constant may be diminished by doping, it would be particularly important to look for nonlinear optical devices with switch-off times in the picosecond timescale.

Recently, Hanamura (1981) has studied the transient behaviour of the dielectric function in semiconductors like CuCl. As will be discussed in more detail, it can be described by a three-level system, in which virtual transitions are responsible for some nonlinearities. Therefore, the lifetimes of the excited quasi-particles do not influence the switching times, which are then only limited by the round-trip time in the cavity. In addition, in CuCl, the band gap is about 3.38 eV, which leads to an optical bistability in a new spectral region around 3.186 eV (at about 390 nm). To understand the physical origin of this bistability, we will first give some insights into the band structure of quasi-particles in CuCl.

II. Absorption and dispersion anomalies in CuCl

Copper chloride is a zinc-blende-type material (T_d) with a direct band gap at the centre of the Brillouin zone. It has a lower conduction band of point group symmetry Γ_6 and two uppermost valence bands of Γ_7 and Γ_8.

This band structure gives rise to two exciton series called Z_{12} and Z_3, respectively, having the symmetry properties

$$Z_3 : \Gamma_6 \otimes \Gamma_7 = \Gamma_2 \oplus \Gamma_5,$$
$$Z_{12} : \Gamma_6 \otimes \Gamma_8 = \Gamma_3 \oplus \Gamma_4 \oplus \Gamma_5.$$

For CuCl, the Z_3 series is the exciton series of the lowest energy and only its ground state will be considered here.

Two excitons may couple together to give rise to a bound state: the biexciton or excitonic molecule. The symmetry of the biexciton ground state with energy E_{bi} is given by the antisymmetric product of the Bloch functions of two electrons and two holes:

$$\Gamma_{env} \otimes (\Gamma_6 \times \Gamma_6)^- \otimes (\Gamma_7 \times \Gamma_7)^- = \Gamma_1.$$

Because of these symmetries, optical dipole transitions are allowed between the crystal ground state and the transverse Γ_5 exciton state, and between Γ_5 exciton states and the Γ_1 biexciton ground state. All other one-photon transitions are forbidden. However, biexcitons may be excited by two-photon absorption. More details of this system can be seen in a recent review article by Grun et al. (1982).

Owing to the strong coupling between the Γ_5 transverse exciton and the electromagnetic radiation field, eigenstates of the interacting system are polaritons. Their dispersion $E_t(Q_t)$ is given at low intensities of excitation by Hopfield's one-oscillator model, which depends on four parameters: the transverse and longitudinal exciton energy E_T and E_L, the exciton effective mass m_{ex} and the background dielectric constant ϵ_b. The polariton dispersion is directly related to the refractive index by the relation $n = \hbar c Q_t / E_t(Q_t)$.

We have determined the excitonic polariton dispersion, using hyper-Raman scattering (Phach et al. 1978). By a self-consistent analysis of the spectral positions of the emission lines, we could obtain the different parameters that describe the polariton dispersion (Hönerlage et al. 1978).

What happens when such a crystal is excited by an intense laser beam with a photon energy $\hbar \omega_p$? When the laser is tuned close to the exciton resonance, an exciton population is created and, since the transition between excitons and biexcitons is dipole allowed, a test beam with frequency ω_t may be absorbed if $\hbar \omega_t = E_{bi} - E_T$. This corresponds to an induced oscillator strength at this energy and gives rise to an anomaly in the polariton dispersion curve. In the absence of any damping, a gap is induced, the magnitude of which is roughly proportional to the number of excitons created by the pump beam. If this pump beam is detuned from the exciton resonance, we obtain qualitatively a similar behaviour: an anomaly of the dispersion, due to the two-photon ($\hbar \omega_t$, $\hbar \omega_p$) absorption to the biexciton ground state, is created at $E_t = E_{bi} - \hbar \omega_p$. This new resonance shifts when the photon energy of the pump beam is tuned. In this case, the magnitude of the gap is proportional to the number n_p of polaritons in the mode of frequency ω_p (May et al. 1979; Märtz et al. 1980).

To describe these phenomena quantitatively, our group have developed a theoretical model to calculate the dispersion of a three-level system, by using the density matrix formalism (Bigot et al. 1984; Hönerlage et al. 1984). Typical results are shown in figure 1. The pump beam is fixed at an energy of 3.184 eV and has an intensity of 10^{15} photons per cubic centimetre. The dispersion of a test beam is considerably changed at the energy 3.1695 eV, which corresponds to the induced absorption from the exciton to the biexciton state, as discussed above. A second anomaly at 3.188 eV is due to two-photon absorption to the biexciton state that uses one photon of the pump beam and one of the test beam. At very high intensities, a quasi-resonance appears

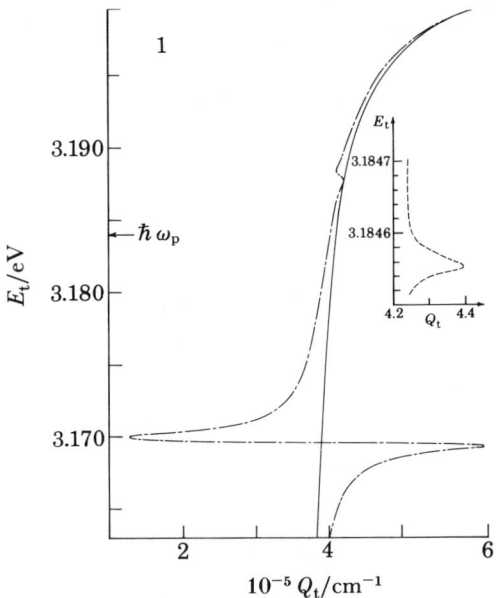

FIGURE 1. The chained curve represents the polariton dispersion of CuCl, when excited at $\hbar\omega_{\mathrm{p}} = 3.184$ eV by a pump beam with a photon density $n_{\mathrm{p}} = 10^{15}$ cm^{-3}, the damping constant being $\hbar\Gamma = 0.2$ meV. The full line gives the result for a one-oscillator model without damping and with $n_{\mathrm{p}} = 0$. The inset shows the pseudoresonance around the energy of the pump beam for $n_{\mathrm{p}} = 10^{16}$ cm^{-3}

also near the energy of the pump beam. It is due to a periodic variation of the exciton and biexciton populations.

We have discussed till now the change of the dispersion of a test beam in the presence of a pump beam; however, the pump beam can also change its own dispersion, especially when the crystal is excited near half the biexciton energy. In this case, the pump beam approaches the upper anomaly of figure 1 and creates biexcitons by two-photon absorption. The resulting dispersion is shown in figure 2. We observe a strong variation of the dispersion around $\frac{1}{2}E_{\mathrm{bi}}$. Outside the biexciton resonance, biexcitons are only virtually created, but the variation of the dispersion is still quite strong.

By hyper-Raman scattering (Grun *et al.* 1983) and non-degenerated four-wave mixing (Hönerlage *et al.* 1983), we have been able to study the energy and intensity dependence of these anomalies for different polarizations of the light sources. We have concluded that the nonlinear refractive index is, at the maximum, of the order of 4×10^{-5} cm^2 kW^{-1}. It is therefore comparable to the results found in GaAs.

After this discussion on the physical origin of these nonlinearities, let us now consider the optical bistability, which has been predicted for CuCl by Koch *et al.* (1981) and Hanamura (1981) and observed experimentally by our group (Levy *et al.* 1983) and by Peyghambarian *et al.* (1983).

III. OPTICAL BISTABILITY

We have worked with a single exciting beam, using the auto-renormalization shown in figure 2 as the source of nonlinearity. When the laser frequency is tuned through the bi-exciton resonance, exciting polaritons are subject to this anomaly. Therefore, one might hope that this process could lead to optical bistability if enough feedback is provided. When

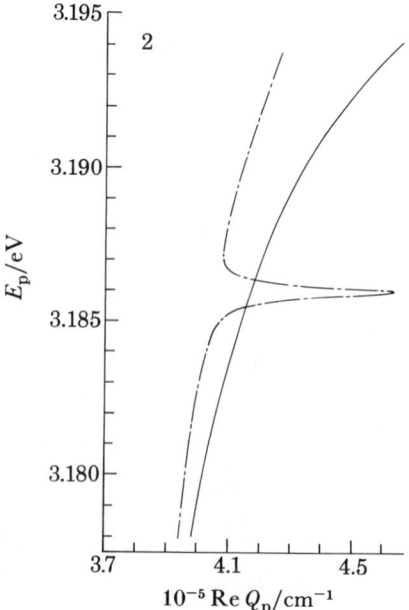

FIGURE 2. Dispersion $E_p(Q_p)$ in CuCl in a one-oscillator model when a single beam ($E_t = E_p$) is considered: full line, without renormalization; chained line, our model with $n_p = 10^{15}$ cm^{-3}.

biexcitons are created only virtually in the process, the switching time of the device should not depend on the lifetimes of the elementary excitations and therefore may be very short.

A CuCl sample of 10–30 μm thickness is placed between mirrors that provide an optical feedback. Since the sample is in direct contact with the mirrors, this is also the length of the Fabry–Perot etalon (F.P.), which is excited perpendicular to its surface. The mirrors are thin glass plates coated with platinum films. Their reflectivity is 90%. Platinum has been chosen since most other metals react chemically with CuCl. When using natural or directly coated surfaces, no sufficient feedback was obtained. Although the mirrors are highly reflective, only a maximum intensity variation of 40% is obtained. This is due to the absorption of the crystal near the 1S exciton line and to a lack of parallelism of the crystal surfaces. The F.P. and the sample are cooled down to low temperatures.

Figure 3 shows our experimental arrangement. A XeCl laser (Lambda Physik) pumps a dye laser containing a diluted solution of αNND in ethanol. The emission is amplified in a second dye cell also filled with αNND and pumped by the same excimer laser. Great care is taken to keep the intensity of the super-radiant emission small compared to the laser emission. The shape of the pulses is kept well-defined in time with a width of about 3 ns (fwhm). The laser emission has a spectral width of 0.05 meV. After passing through a diaphragm, neutral density filters and a glass polarizer, the beam is split into two parts by a glass plate.

One part of the beam is focused onto the F.P. containing the crystal, in a spot of 100 μm diameter. Power intensity can be varied up to 50 MW cm^{-2}. The transmission of the laser pulses through the F.P is detected by a fast photocell and is analysed in time by an oscilloscope operating at gigahertz frequencies. The overall time resolution is less than 500 ps. The transmission spectrum could equally well be analysed by a spectrograph and an O.M.A. system. The outer part of the beam passes, after an optical delay of ɩ.9 ns, through neutral density filters and is then detected by the same photocell. This beam is used as a reference for the shape

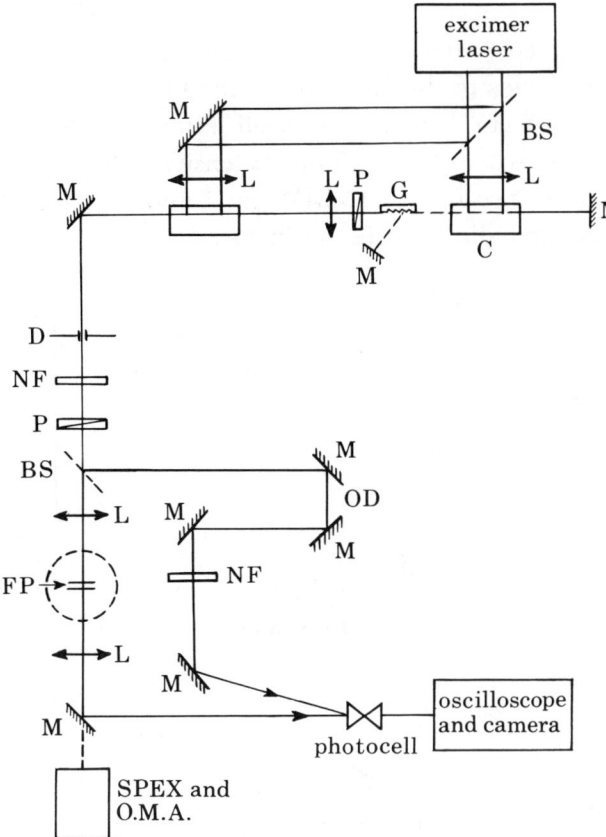

FIGURE 3. Experimental arrangement: M, mirror; BS, beam splitter; L, lens; C, dye cell; P, polarizer; D, diaphragm; NF, neutral density filter; F.P., Fabry–Perot; G, grating; OD, optical delay; O.M.A., optical multichannel analyser; SPEX, spectrograph.

of the pulse. The signal, observed on the oscilloscope is then photographed with a camera, so that single shots may be analysed.

Figure 4 shows the transmitted and reference laser pulses analysed by the oscilloscope when the F.P. is excited with a photon energy close to the biexciton energy; δ is the time delay between transmitted and reference pulses. At this energy and intensity of excitation, the transmitted pulse is clearly deformed compared to the reference pulse. The intensity of the transmitted pulse is now analysed in detail.

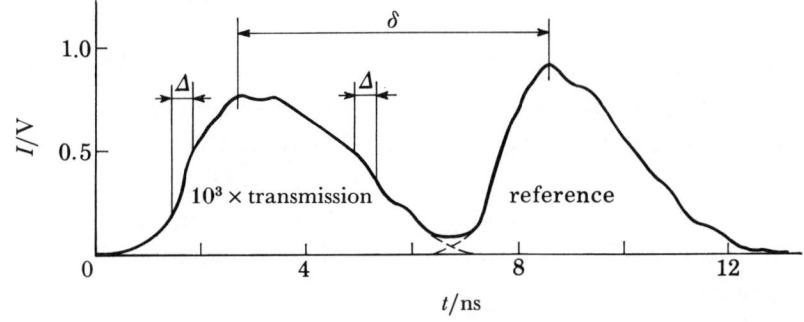

FIGURE 4. Picture of the transmitted and incident dye laser pulses (the transmitted pulse is attenuated by a factor of 1000 by the F.P). Δ, time resolution (ca. 500 ps); δ, optical delay between the two pulses.

[43]

Figure 5 shows the output against input intensity curves for different maximum intensities of excitations. Open circles correspond to increasing intensities and crosses to decreasing intensities. Figure 5a corresponds to the figure when the bistability is clearly observed with a 'switch-on' intensity of 15 MW cm⁻² and a 'switch-off' intensity of 5 MW cm⁻². The switching point stays at a fixed intensity when the maximum intensity is decreased (figure 5b). In figure 5d, no bistability is observed since $I_{\max} < 15$ MW cm⁻² and, in figure 5e, we had no sample inside the cavity. Both corresponding curves are linear.

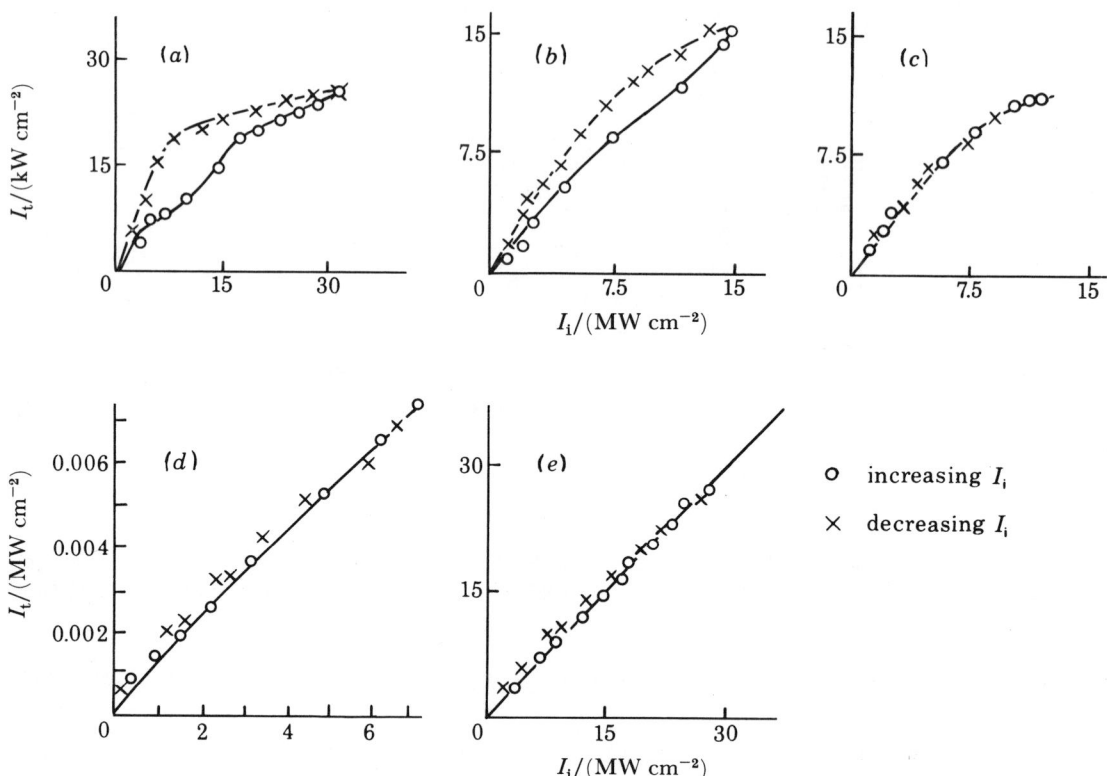

FIGURE 5. Output (I_t) against input (I_i) intensity curves for different maximum intensities (a)–(d), (e) no sample inside the cavity.

Let us now investigate the dependence of the hysteresis loop on the photon energy of the exciting beam. In figure 6, we have plotted the areas of the bistability loop as a function of the photon energy of the exciting beam. This quantity is proportional to the energy stocked inside the F.P. cavity. Here we obtain a structure with two maxima. The first maximum is situated at an energy corresponding to a maximum of transmission of the F.P. at 3.1855 eV. The second maximum is situated near $\frac{1}{2}E_{\mathrm{bi}} = 3.186$ eV. This doublet structure depends on the position of the maximum of transmission of the F.P. relative to $\frac{1}{2}E_{\mathrm{bi}}$. In this case they are well separated, but we can also obtain excitation spectra with one maximum. In any case, we have shown that the switching behaviour is different depending on whether the crystal is excited at the biexciton resonance or not.

Figure 7a shows a hysteresis cycle that is typically observed when we excite near the first maximum. The 'switch-on' and 'switch-off' times are not resolved by the apparatus and they

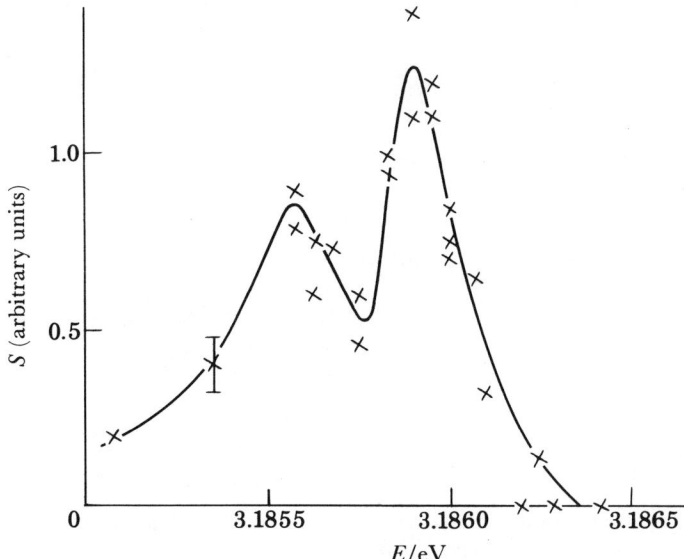

FIGURE 6. Areas of the bistability loop as a function of the photon energy of the exciting beam.

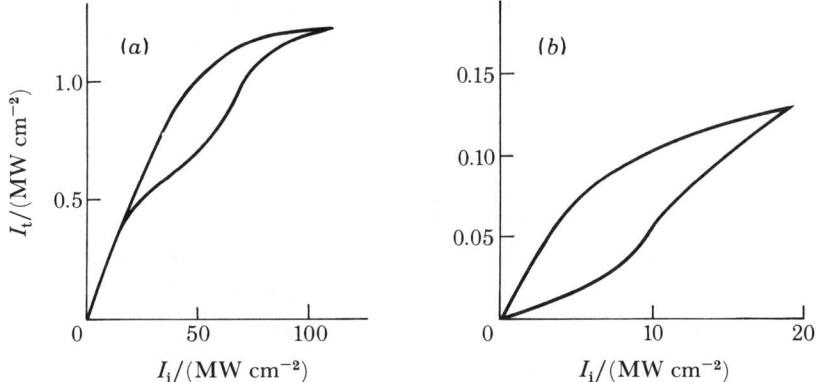

FIGURE 7. (a) A typical loop when exciting near the first maximum of figure 6, i.e. 3.1855 eV. (b) Hysteresis loop when exciting near the second maximum of figure 6, i.e. 3.186 eV.

are both below 500 ps. This can be concluded from the fact that increasing and decreasing pulses follow the same straight line near the origin (Hönerlage *et al.* 1983).

If, however, the energy is near $\frac{1}{2}E_{bi}$, the hysteresis cycle (figure 7 b) has a completely different shape than the one shown in figure 7 a. As our pulses are not long enough, we estimate the commutation time from the 'on' to the 'off' state to be of the order of 3 ns.

These results lead us to suppose that, in the first case, the commutation time is small because biexcitons are created only virtually. In the second, biexcitons are really created. The index change may then be due to the number of excitons and biexcitons created. This would imply that their radiative lifetime affects the switch-off time of the device.

We have used our theoretical model discussed in connection with figure 2 to calculate the transmission characteristic of a 10 μm thick sample in a F.P with a coefficient of reflection of 95 %.

Near $\frac{1}{2}E_{bi}$ we obtain an important hysteresis loop, as shown in figure 8 a. If we tune the energy

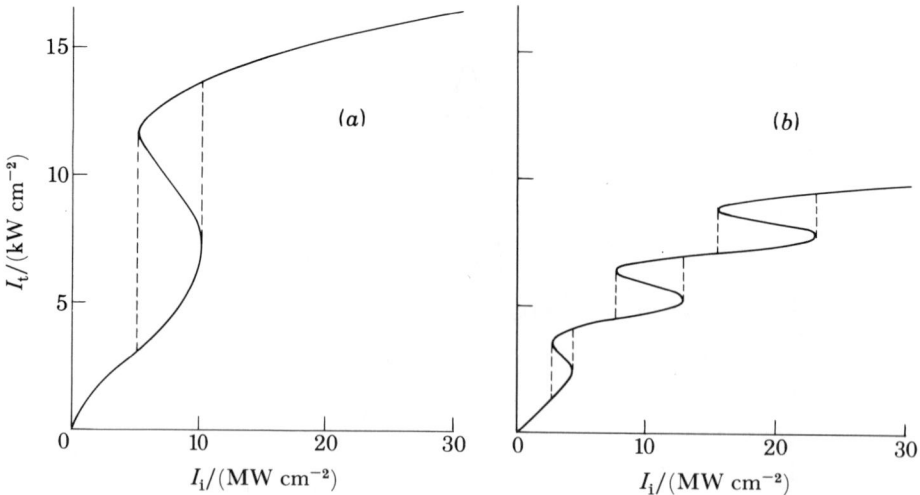

FIGURE 8 (a) Calculated hysteresis cycle for a photon energy of 3.185 eV. (b) Calculated hysteresis cycles for a photon energy of 3.1855 eV.

of the exciting beam closer to $\frac{1}{2}E_{bi}$, more F.P. fringes are then covered. A typical result is shown in figure 8b. Here many hysteresis loops overlap and only the envelope may be observed experimentally. Using this model, we can qualitatively explain the excitation spectrum given in figure 6. A more detailed study is under way.

CONCLUSION

In conclusion, we have given experimental evidence for optical bistability due to biexcitons in CuCl at both 1.9 K and 60 K. We have shown that the phenomenon is strongly resonant at half the biexciton energy, which allows us to state that the nonlinearity is due to biexciton two-photon absorption. Working with virtual states, the switching times are shorter than the experimental time resolution, which is 500 ps. However, the required input intensity is rather high (*ca.* 15 MW cm^{-2}). At the resonance, switching times are longer.

The authors are grateful to Mr J. Y. Bigot and Mr F. Tomasini for many stimulating discussions. This work was supported by a contract with the Ministère des P.T.T. of France, Direction Générale des Télécommunications, Direction des Affaires Industrielles et Internationales. It has been made in the framework of an operation launched by the Commission of the European Communities under the experimental phase of the European Community Stimulation Action (1983–1985).

REFERENCES

Bigot, J. Y. & Hönerlage, B. 1984 *Physica Status Solidi* B **121**, 649–659.
Gibbs, H. M., McCall, S. L. & Venkatesan, T. N. C. 1976 *Phys. Rev. Lett.* **36**, 1135–38.
Gibbs, H. M., McCall S. L., Venkatesan, T. N. C., Goss, A. C., Passner, A. & Wiegman, W. 1979 *Appl. Phys. Lett.* **35**, 451–453.
Grun, J. B., Hönerlage, B. & Levy, R. 1982 In *Excitons* (ed. E. J. Rashba & M. D. Sturge), pp. 459–504. Amsterdam. North-Holland.
Grun, J. B., Hönerlage, B. & Levy, R. 1983 *Solid St. Commun.* **46**, 51–54.
Hanamura, E. 1981 *Solid St. Commun.* **36**, 939–942.
Hönerlage, B. & Bigot, J. Y. 1984 *Physica Status Solidi* B **123**, 201.

Hönerlage, B., Bigot, J. Y. & Levy, R. 1984 In *Optical Bistability*, vol. 2 (ed. C. M. Bowden, H. M. Gibbs & S. L. MacCall) New York: Plenum Press (In the press.)
Hönerlage, B., Bigot, J. Y., Levy, R., Tomasini, F. & Grun, J. B. 1983 *Solid St. Commun.* **48**, 803–806.
Hönerlage, B., Bivas, A. & Phach, V. D. 1978 *Phys. Rev. Lett.* **41**, 49–52.
Hopfield, J. J. 1969 *Phys. Rev.* **182**, 945–952.
Koch, S. W. & Haug, H. 1981 *Phys. Rev. Lett.* **46**, 450–452.
Levy, R., Bigot, J. Y., Hönerlage, B., Tomasini, F. & Grun, J. B. 1983 *Solid St. Commun.* **48**, 705–708.
März, R. Schmitt-Rink, S. & Haug, H. 1980 *Z. Phys.* **B40**, 9–14.
May, V., Henneberger, K. & Henneberger, F. 1979 *Physica Status Solidi* B **94**, 611–620.
Miller, D. A. B., Smith, S. D. & Johnston, A. 1979 *Appl. Phys. Lett.* **35**, 658–660.
Peyghambarian, N., Gibbs, H. M., Rashford, M. C. & Weinberger, D. A. 1983 *Phys. Rev. Lett.* **51**, 1692–1695.
Phach, V. D., Bivas, A., Hönerlage, B. & Grun, J. B. 1978 *Physica Status Solidi* B **86**, 159–168.

16-2

Phil. Trans. R. Soc. Lond. A **313**, 239–244 (1984)

Printed in Great Britain

Multiple quantum well optical nonlinearities: bistability from increasing absorption and the self electro-optic device

By D. A. B. Miller

AT&T Bell Laboratories, Holmdel, New Jersey 07733, *U.S.A.*

This paper briefly reviews the nonlinear optical and electro-optical properties and applications of GaAs–AlGaAs multiple quantum wells, all of which are compatible with laser diodes or semiconductor electronics or both. They show large nonlinear absorption and refraction (associated with their strong room-temperature exciton resonances) applicable to all-optical devices. They also show large electroabsorptive effects, some of which are unique to the quantum wells at any temperature, which are so strong that optical modulators can be made with micrometre dimensions. A new class of optical bistability (due to increasing absorption) is also reviewed; combination of this with the electroabsorptive effects enables a new mirrorless optical switch called a Self Electro-optic Effect Device (SEED), which sets a new standard for optical switching energies, with energy densities reduced by a factor of six by comparison to even the best resonant cavity devices.

1. Introduction

One problem central to all nonlinear optical processing devices and systems is that of finding suitable nonlinear materials. An ideal material would have a very large and at least reasonably fast nonlinearity, would be easy to use in practical device structures and would be compatible with readily available light sources. Semiconductors at wavelengths near the band edge are attractive because they offer large and moderately fast nonlinearities in relatively convenient materials, although most of the nonlinearities require low temperature operation.

Multiple quantum well (m.q.w.) structures consisting of alternate thin (*ca.* 10 nm) layers of two different semiconductors (for example GaAs and AlGaAs) have recently shown remarkable nonlinear optical effects at room temperature at wavelengths, power levels and timescales compatible with diode lasers (Miller *et al.* 1982, 1983*a, b*; Chemla *et al.* 1984; Gibbs *et al.* 1982; Tarng *et al.* 1984). Some of these effects arise from the existence of exceptionally strong room-temperature exciton resonances. These exciton resonances are particularly easy to saturate, giving large nonlinear absorption and nonlinear refraction that make the m.q.w. interesting for all-optical devices. Additionally, it has recently been discovered that m.q.w.s show large electroabsorptive effects at room temperatures with fields both parallel and perpendicular to the layers (Chemla *et al.* 1983; Wood *et al.* 1984, Miller *et al.* 1984*a*). These effects are so large that high speed optical modulators can be made with only micrometres of material (Wood *et al.* 1984). The small energy drive requirements of these modulators makes them also compatible with low power electronic devices.

Recently, a class of mirrorless bistability has been identified (Miller 1984; Miller *et al.* 1984*b*) which relies only on an optical absorption that increases as a material is excited. This has very recently been applied to make a novel hybrid optically bistable device that uses the m.q.w. perpendicular field electro-absorptive effect; this device uses the same piece of m.q.w. simultaneously as both modulator and detector and has consequently been called a Self

Electro-optic Effect Device (SEED) (Miller *et al.* 1984*c*). One remarkable feature of the SEED is the extremely low total switching energy per unit area (*ca.* 20 fJ μm^{-2}). This is achieved without a cavity and represents a sixfold improvement even over the best cavity devices at comparable wavelengths.

In the rest of this paper, the key results of the work on m.q.w. nonlinearities, optical bistability due to increasing absorption and the SEED will be briefly summarized.

2. Nonlinear absorption and refraction

GaAs–AlGaAs is thought to show exciton resonances at room temperature (Ishibashi *et al.* 1982; Miller *et al.* 1982) because the confinement of the exciton in the layer increases its binding energy and phonon broadening is not correspondingly increased (Miller *et al.* 1982; Chemla *et al.* 1984). Excitonic absorption saturation arises from physical processes fundamentally different from those of interband absorption saturation. In the m.q.w., nonlinear absorption and refraction due to excitonic effects can be observed at excitation levels much lower than those required for interband effects (Miller *et al.* 1982; Chemla *et al.* 1984) and the effects can be clearly modelled as being due to changes in exciton broadening, oscillator strength and energy. Detailed measurements were made of nonlinear absorption and degenerate four-wave mixing (d.f.w.m.); from these nonlinear absorption cross section and change in refractive index per excited carrier shown in figure 1 can be deduced. The nonlinearity is so large that, in an

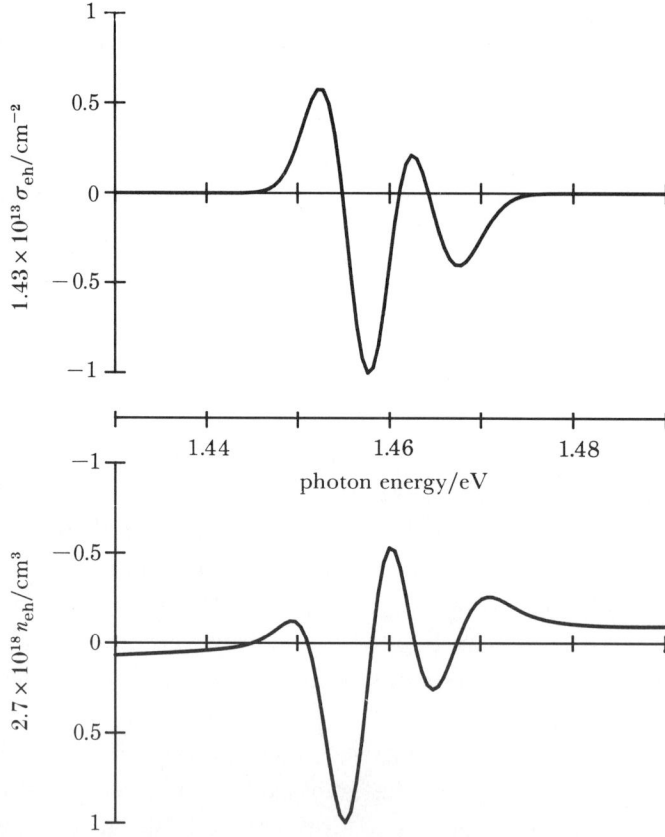

FIGURE 1. Absorption saturation cross section, σ_{eh} and change in refractive index per excited carrier pair n_{eh} (in the vicinity of the exciton resonances), as deduced from nonlinear absorption and degenerate four-wave mixing spectra (Chemla *et al.* 1984).

important practical demonstration, it was possible to achieve d.f.w.m. in a 1.26 μm thick sample by using only a continuous wave laser diode. These nonlinear effects have recovery times of up to 30 ns, which may be shortened by diffusion of carriers or otherwise (Miller *et al.* 1983 *a, b*; Chemla *et al.* 1984). They are attractive for d.f.w.m., saturable absorption and optically bistable devices (Gibbs *et al.* 1982; Tarng *et al.* 1984) although it has proved difficult to exploit them to their full potential for optical bistability, probably because of high linear absorption in the band tail region. The physical mechanism of the excitonic nonlinear absorption is thought to be screening of potential excitons by free carriers; the free carriers may be created directly or by the phonon ionization of excitons (Miller *et al.* 1982; Chemla *et al.* 1984).

3. ELECTROABSORPTION

Figure 2 shows the effect of electric fields applied (*a*) parallel and (*b*) perpendicular to the m.q.w. layers (Chemla *et al.* 1983; Wood *et al.* 1984; Miller *et al.* 1984*c*). For parallel fields the dominant effect appears to be Stark broadening, i.e. the exciton is field-ionized by the applied field, reducing its lifetime and hence broadening the optical transition (Miller *et al.* 1984*c*). This broadening is very significant with electric fields of *ca.* 10^4 V cm^{-1}, which corresponds to a potential drop of greater than one binding energy (*ca.* 9 meV) across the exciton diameter (*ca.* 12 nm); this is a severe perturbation but is relatively easily applied with high purity material. For perpendicular fields, the dominant effect is a shift of the spectrum to lower energies (Wood *et al.* 1984; Miller *et al.* 1984*c*); the small apparent broadening may be due to field inhomogeneity in the sample. This shift is readily seen with fields over 3×10^4 V cm^{-1} and is thought to arise (Miller *et al.* 1984*c*) from a shift in the confined energy levels of the carrier sub-bands in the wells (Bastard *et al.* 1982; Miller *et al.* 1984*c*) with a small

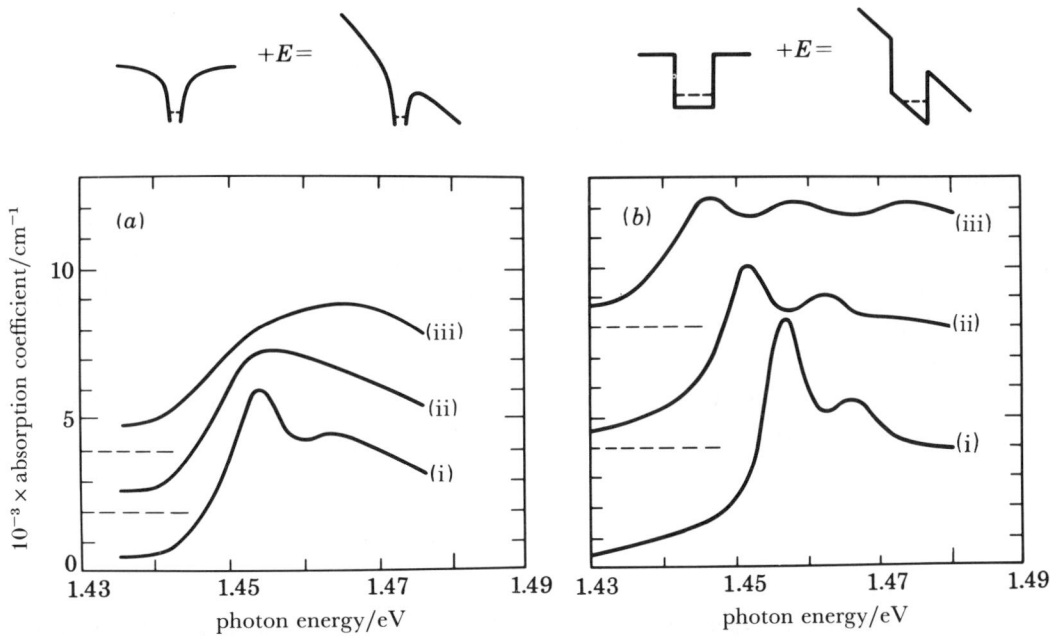

FIGURE 2. Absorption spectra with applied field, *E*. (*a*) Field parallel to the layers: (i) 0 V cm^{-1}; (ii) 1.6×10^4 V cm^{-1}; (iii) 4.8×10^4 V cm^{-1}. Sketch above: schematic diagram of Coulomb potential of exciton. (*b*) Field perpendicular to the layers: (i) *ca.* $1 \times 10^+$ V cm^{-1}; (ii) *ca.* 4.7×10^4 V cm^{-1}; (iii) *ca.* 7.3×10^4 V cm^{-1}. Sketch above: schematic diagram of quantum well potential (Miller *et al.* 1984*c*).

correction for the change in exciton binding energy (Miller *et al.* 1984*c*). (The exciton does not field ionize because it is held together by the walls of the quantum well.) This effect is distinct from and much larger than the Franz–Keldysh effect normally seen in bulk semiconductors because it arises from perturbations of large (*ca.* 10 nm) envelope functions with small associated energies (*ca.* 5–30 meV) and is unique to the m.q.w. at any temperature.

Both parallel and perpendicular field effects produce large changes in transmission (about 50 % in samples only micrometres thick). They are not limited by recombination times; in practice they are limited so far by the time taken to apply the electric field (for example, by *RC* time constants) and the fundamental limit is likely to be an uncertainty time up to 1 ps. A novel high speed modulator (up to 2 ns) has already been demonstrated (Wood *et al.* 1984).

4. OPTICAL BISTABILITY DUE TO INCREASING ABSORPTION

It has recently been pointed out (Miller *et al.* 1984*a*; Miller 1984) that a common principle underlies several previously independent theoretical discussions (Kaplan 1982; Hajto & Janossy 1983; Hopf *et al.* 1984) and experimental demonstrations (Hajto & Janossy 1983; Bohnert *et al.* 1983) of mirrorless optical bistability in diverse, specific, physical systems. The principle relies on an increase of absorption as the material becomes more excited. Increased incident optical power gives increased absorbed power and hence increased excitation of the material; this gives increased absorption, which gives further increased excitation and hence further increased absorption. Under the right conditions (Miller *et al.* 1984*a*; Miller 1984), this positive feedback leads to switching and bistability. Good agreement is found between theory and experiment in a simple thermal system (Miller *et al.* 1984*a*). The theory is particularly simple and general, is readily interpreted graphically (Hajto & Janossy 1983, Miller *et al.* 1984*a*), and simple conclusions can be drawn about limiting switching powers and the width of the bistable region (Miller 1984).

This class of bistability shows several exceptional features (Miller *et al.* 1984*a*; Miller 1984): (i) it requires no mirrors or other optical feedback and hence does not require coherent light; (ii) the switching is to a *higher* absorption (and hence *lower* transmission) state; (iii) it requires a material whose absorption increases as the real state of excitation of the material is increased. Although the 'state of excitation' may be parameterized in many different ways (for example, temperature rise (Miller *et al.* 1984*a*, Hajto & Janossy 1983), kinetic energy (Kaplan 1982), population inversion (Hopf *et al.* 1984), band gap renormalization (Bohnert *et al.* 1983) and, as discussed later, voltage change (Miller *et al.* 1984*b*)), the nonlinearity must be 'dynamic', and no susceptibility rigorously expandable in a power series only of the optical electric field (with constant coefficients) can give this bistability.

This bistability is important in its own right, but also happens to be the class exploited in the hybrid device now discussed.

5. THE OPTICALLY BISTABLE SEED

The samples used to measure the perpendicular field electroabsorption (Miller *et al.* 1984*c*) and to demonstrate fast modulators (Wood *et al.* 1984) used a p-i-n diode structure in which the m.q.w. material was contained in the intrinsic region. All other parts of the structure were transparent. Such a structure also serves as an efficient photodetector (Miller *et al.* 1984*c*), with the responsivity depending on the absorption of the m.q.w., which in turn depends on the voltage of the m.q.w. If the photocurrent alters the voltage, then modulator and detector feed back on one another, giving a Self Electro-optic Effect Device (SEED) (Miller *et al.* 1984*b*).

Connecting the p-i-n diode through a large series resistor to a constant reverse bias supply can give an optically bistable SEED. The wavelength of the light source shining through the SEED is chosen near the zero-field heavy-hole exciton-peak position. With no incident optical power, the diode is reverse biased, the exciton peak is moved to lower energy and the absorption is comparatively low. With increasing power, a photocurrent is generated that drops a voltage across the resistor, reducing the voltage across the diode and thereby shifting the exciton absorption back towards the operating wavelength and increasing the absorption and photocurrent. Optical bistability is then seen through the mechanism described in §4. Figure 3 shows the observed bistability together with a theroetical model, based on measured transmission and responsivity, with no fitted parameters.

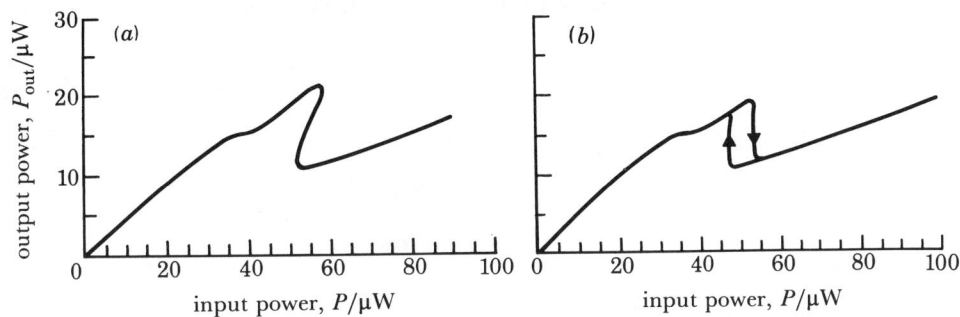

FIGURE 3. THEORY (*a*) and experiment (*b*) for SEED optical bistability (Miller *et al.* 1984*b*) and with 1 MΩ resistor and 20 V bias supply at 851.7 nm.

The resulting optical device can be operated over a very wide range of parameters. With a 600 μm diameter device, switching powers of 670 nW (at 1.5 ms switching time) to 3.7 mW (at 400 ns switching time) have been observed between 850 and 860 nm. Switching times are *RC* time-constant limited, and are expected to be proportionally smaller for smaller area devices. However, the most remarkable feature of the device is the switching energy per unit area (4 fJ μm^{-2} for incident optical and up to 14 fJ μm^{-2} for electrical). Even the total switching energy per unit area is lower than any previously reported optical switching energy per unit area at a comparable wavelength (*ca.* 120 fJ μm^{-2}) by a factor of over six, despite the fact that the SEED uses no cavity to reduce switching energy. The reason for this fundamentally lower energy stems from the very small volume and moderate field requirements of the m.q.w. perpendicular field electroabsorptive effect; the switching energy reduces mainly to the stored electrostatic energy in the reverse-biased diode.

6. CONCLUSIONS

In summary, the m.q.w. shows itself to be a remarkable material for optical devices, both for all-optical nonlinearities and electro-optical effects. In electroabsorption it opens up new opportunities not accessible with any other material at any temperature. All these phenomena are seen at room temperature at power levels, wavelengths, voltages and timescales compatible with laser diodes and other semiconductor optoelectronic and electronic devices. A new class of optical bistability has also been identified and applied to demonstrate the SEED by using the quantum wells to give a device that sets new standards in optical switching energy and represents a significant advance towards practical optical switching and processing.

REFERENCES

Bastard, G., Mendez, E. E., Chang, L. L. & Esaki, L. 1982 *Phys. Rev.* B 28, 3241–3245.

Bohnert, K., Kalt, H. & Klingshirn, C. 1983 *Appl. Phys. Lett.* 43, 1088–1090.

Chemla, D. S., Damen, T. C., Miller, D. A. B., Gossard, A. C. & Wiegmann, W. 1983 *Appl. Phys. Lett.* 42, 864–866.

Chemla, D. S., Miller, D. A. B., Smith, P. W., Gossard, A. C. & Wiegmann, W. 1984 *IEEE J. Quantum Electron.* 20, 265–275.

Gibbs, H. M., Tarng, S. S., Jewell, J. L., Weinberger, D. A., Tai, K., Gossard, A. C., McCall, S. L., Passner, A. & Wiegmann, W. 1982 *Appl. Phys. Lett.* 41, 221–222.

Hajto, T. & Janossy, I. 1983 *Phil. Mag.* B47, 347–366.

Hopf, F. A., Bowden, C. M. & Louisell, W. 1984 *Phys. Rev.* A 29, 2591–2596.

Ishibashi, T., Tarucha, S. & Okamoto, H. 1982 *Inst Phys. Conf. Ser.* 63, 587–588.

Kaplan, A. E. 1981 *Phys. Rev. Lett.* 48, 138–141.

Miller, D. A. B. 1984 (Submitted to *J. opt. Soc. Am.* B.)

Miller D. A. B., Chemla, D. S., Eilenberger, D. J., Smith, P. W., Gossard, A. C. & Tsang, W. T. 1982 *Appl. Phys. Lett.* 41, 679–681.

Miller, D. A. B., Chemla, D. S., Eilenberger, D. J., Smith, P. W., Gossard, A. C. & Wiegmann, W. 1983*a* *Appl. Phys. Lett.* 42, 925–927.

Miller, D. A. B., Chemla, D. S., Smith, P. W., Gossard, A. C. & Wiegmann, W. 1983*b* *Optics Lett.* 8, 477–479.

Miller, D. A. B., Gossard, A. C. & Wiegmann, W. 1984*a* *Optics Lett.* 9, 162–164.

Miller, D. A. B., Chemla, D. S., Damen, T. C., Gossard, A. C., Wiegmann, W., Wood, T. H. & Burrus, C. A. 1984*b* *Appl. Phys. Lett.* 45, 13–15.

Miller, D. A. B., Chemla, D. S., Damen, T. C., Gossard, A. C., Wiegmann, W., Wood, T. H. & Burrus, C. A. 1984*c* (In preparation.)

Tarng, S. S., Gibbs, H. M., Jewell, J. L., Peyghambarian, N., Gossard, A. C., Venkatesan, T. & Wiegmann, W. 1984 *Appl. Phys. Lett.* 44, 360–361.

Wood, T. H., Burrus, C. A., Miller, D. A. B., Chemla, D. S., Damen, T. C., Gossard, A. C. & Weigmann, W. 1984 *Appl. Phys. Lett.* 44, 16–18.

Phil. Trans. R. Soc. Lond. A **313**, 245–248 (1984)

Printed in Great Britain

Semiconductor nonlinear etalons

By H. M. Gibbs[1], J. L. Jewell[1]†, N. Peyghambarian[1], M. C. Rushford[1], K. Tai[1],
S. S. Tarng[1]‡, D. A. Weinberger[1], A. C. Gossard[2], W. Wiegmann[2]
and T. Venkatesan[3]

[1] *Optical Sciences Center, University of Arizona, Tucson, Arizona 85721, U.S.A.*
[2] *AT&T Bell Laboratories, Murray Hill, New Jersey 07974, U.S.A.*
[3] *Bell Communications Research, Murray Hill, New Jersey 07974, U.S.A.*

Previous observations of optical bistability in nonlinear etalons of ZnS, CuCl and GaAs are summarized. Emphasis is placed upon recent results in GaAs: similar room-temperature optical bistability at powers under 10 mW in bulk and multiple quantum-well etalons; room-temperature bistability achieved with a diode laser as a light source; NOR and other logic operations; optical fibre signal regeneration and continuous wave operation.

Semiconductors are attractive for use as nonlinear media in practical, intrinsic optical bistable devices. Laser light of photon energy slightly below the band edge interacts with these materials via sharp absorption resonances (such as free and bound excitons and biexcitons), which permit fast, room-temperature, low power optical bistability. Furthermore, the absorption coefficients α are of the order of 10 cm^{-1}, which allows $\alpha L \approx 1$ to be achieved in short interaction lengths and thus makes possible the realization of devices 1 μm thick. Here we review our recent progress in semiconductor optical bistability. Results of optical bistability in ZnS, CuCl, and GaAs achieved before the Topical Conference on Optical Bistability (Bowden *et al.* 1984) are summarized briefly. Subsequent results in GaAs are described in greater detail.

Zinc sulphide

Apparently, the first semiconductor to exhibit optical bistability was ZnS (Karpushko & Sinitsyn 1978, 1982). Karpushko and Sinitsyn attribute the nonlinear refraction to a two-photon photorefractive effect, which is large in evaporated thin films and much smaller or non-existent in bulk crystals (Karpushko & Sinitsyn 1978, 1982). Their ZnS etalon was simply a narrow-band interference filter with an intermediate layer of ZnS about λ/n_0 thick (λ is the vacuum wavelength and n_0 is the refractive index of ZnS at λ). The short length and simple fabrication process make ZnS etalons attractive for parallel signal processing. Unfortunately attempts (Weinberger *et al.* 1982; Rushford *et al.* 1983) to reproduce the fast switching times (*ca.* 10 μs) and low intensities (*ca.* 300 W cm^{-1}) claimed (Karpushko & Sinitsyn, personal communication) have achieved only thermal bistability with *ca.* 1 ms switching times and *ca.* 10 kW cm^{-2} intensities. So far we have tried only commercial interference filters, but a study of bistability as a function of growth parameters is under way.

---

† Present address: AT & T Bell Laboratories, Holmdel, New Jersey 07733, U.S.A.
‡ Present address: Martin Marietta Laboratory, Baltimore, Maryland 21227, U.S.A.

Copper chloride

Optical bistability achieved with the biexciton resonance (Koch & Haug 1981; Hanamura 1981) in CuCl is of particular interest because of its potential (Hanamura 1981) for picosecond switch-off times. Recently, two groups have reported bistability in CuCl in which the switching times were shorter than the *ca.* 500 ps resolution of the detection systems (Peyghambarian *et al.* 1983; Levy *et al.* 1983, and this symposium; see also Bowden *et al.* 1984). The switch-on intensity of *ca.* 10 MW cm^{-2} for 10 to 30 μm long etalons with *ca.* 90% reflectivity is considerably higher than first predicted (Koch & Haug 1981). However, it is consistent with a recent numerical calculation (Sarid *et al.* 1983), which includes two effects previously neglected: intensity broadening of the biexciton resonance and background absorption from the wing of the free exciton resonance. Our research is now focused on a streak camera study of the dependence of the switching times upon laser–biexciton detuning to see if switch-off times under 10 ps occur. The high-intensity and low-temperature requirements appear to make CuCl biexcitonic bistability unattractive for signal processing. However, it is noteworthy that if a device could be fabricated with a transverse dimension equal to one wavelength in the material (λ/n_0) it would need only several femtojoules of switching energy if the switching time is in fact only 1 ps. This corresponds to only a few thousand photons, a number approaching the statistical limit (i.e. the minimum number of photons needed to avoid occasional failure of switching due to statistical fluctuations).

Gallum arsenide: room-temperature bistability

Optical bistability in a 4 μm GaAs etalon was reported from 5 to 120 K; it was suggested that a superlattice might permit room-temperature operation because of the increased binding energy of the free exciton (Gibbs *et al.* 1979). Room-temperature bistability was achieved in a multiple quantum well (m.q.w.) etalon consisting of 33.6 nm GaAs layers alternated with 40.1 nm AlGaAs layers for *ca.* 2 μm of GaAs (Gibbs *et al.* 1982). Such a wide well increases the free exciton binding energy to only 6 meV, not that much larger than the 4.2 meV bulk value (Miller *et al.* 1981). Consequently, the bulk sample was tried at room temperature, and bistability was observed; presumably, this was primarily due to better focusing with the dewar removed (Gibbs *et al.* 1982). The bistability was better (lower switch-on power, wider loops) in the m.q.w. etalon.

Bulk bistability compared with m.q.w. bistability

Subsequently, we have compared bulk and m.q.w. devices (well thicknesses 5.3, 15.2, 29.9 and 33.6 nm) including bistability at powers as low as *ca.* 5 mW. This limiting power is so consistent throughout our samples that we believe it provides an accurate measure for present materials and devices. Since the intensity is 3 kW cm^{-2} or less, the optical nonlinearity must be excitonic, not band-to-band. We are studying the apparent contradiction that the bulk and m.q.w. bistabilities look so similar (and the low intensity suggests an exciton nonlinearity), but the m.q.w. exciton features are more prounced because of the increased binding energies. One possible explanation is that unsaturable background absorption from the band tail forces a large detuning below the exciton resonance in both cases; since the maximum change in refractive index, assuming complete saturation, is proportional to the product of the peak and the width of a two-level transition, the smaller, broader bulk exciton resonance may be as effective as the larger, narrower m.q.w. exciton feature. An understanding of this similarity might result in lower power operation closer to the m.q.w. exciton resonance. Both m.q.w. and bulk crystals have the unusual property of having exciton binding energies E_x less than kT at room temperature, but larger than or comparable to the exciton linewidth Γ_x determined by optical phonon broadening. The condition $E_x > \Gamma_x$ is needed to see the exciton resonance, but the smaller the E_x the smaller the saturation intensity (smaller E_x implies a larger exciton Bohr radius and thus fewer excitations are required to produce one exciton, or carrier, per exciton volume to screen the exciton). GaAs apparently has an unusually small Γ_x permitting low-power room-temperature exciton features, especially pronounced in a m.q.w. crystal. (For a discussion of the nonlinear refraction of a GaAs–AlGaAs m.q.w. crystal see D. A. B. Miller (this symposium).

BISTABILITY BY USING A DIODE LASER

Room-temperature bistability of a m.q.w. etalon with a diode laser as the only light source emphasizes that these GaAs devices are coming closer to practical operating powers. A m.q.w. crystal consisting of 300 periods of 5.3 nm GaAs and 5.6 nm $Al_{0.3}Ga_{0.7}As$ was grown by molecular beam epitaxy (Gossard 1979) and sandwiched between two mirrors of reflectivity 0.9 and 0.98 to form the Fabry-Perot etalon. A power of 6 mW from a Hitachi HLP 1400 diode laser at 830 nm switches on the etalon at room temperature (Tarng et al. 1984).

NOR GATE OPERATING ON A PICOSECOND TIMESCALE

A power of 10 mW per pixel and 10^6 pixels per square centimetre would imply 10 kW cm^{-2} and require much better cooling than that used in electronic systems. It is hoped that further reductions in power will occur when the present limitations are understood. However, such an etalon could be used for picosecond decision-making at a gigahertz repetition rate without thermal problems (Jewell et al. 1984 and this symposium). The idea is to use two control pulses followed by a probe pulse. The etalon peak wavelength can be shifted rapidly so that the logic output could occur in a picosecond. For example, if the etalon is initially tuned to the probe wavelength, a NOR gate operation results if each of the control pulses is able to detune the etalon from the probe wavelength (Jewell et al. 1984 and this symposium). The carriers must then recombine, which may take a nanosecond or more, and may take longer if thermal or other restrictions require such, but during this dead time there is no light on the etalon to generate heat.

DATA REGENERATION THROUGH FIBRES

Another precursor of actual applications is the demonstration of data regeneration through fibres. Recently the regeneration of a pseudorandom sequence of optical pulses was demonstrated with a GaAs etalon and three fibres 1 km long (Venkatesan et al. 1984). A train of rectangular bias pulses was sent through one fibre while the data sequence of picosecond pulses (synchronized with the bias pulses) travelled down a second fibre and both were focused onto the etalon. The transmitted signal was sent through the third fibre and monitored by a photodetector. The following accomplishments were made in the experiment: (i) completely remote operation and readout of a bistable device; (ii) all-optical regeneration of optical data; (iii) conversion of data from one wavelength to another; (iv) control of the pulse width of the output by varying the phase delay of the picosecond switching pulses relative to the bias pulses.

QUASI-CONTINUOUS WAVE BISTABILITY

One impediment to continuous wave applications of GaAs etalons is the regenerative pulsations arising from a competition between thermal and excitonic index changes (Jewell *et al.* 1981). Quasi-continuous wave operation has been obtained by using a diamond heat sink (polished and coated for $R \approx 0.9$) as one mirror of the GaAs etalon. The device stayed on for the entire *ca.* 100 ms duration of the input for most traces, but some showed brief (*ca.* 1 µs) randomly located periods in the lower state. The cause of the undesired switching has not been identified, but it may have been laser mode hopping.

In summary, GaAs continues to appear very promising for optical signal processing because bistability can be seen at room temperature with low powers (*ca.* 10 mW) and fast switching times (switch-on time under 200 ps and probably *ca.* 1 ps; switch-off time under 10 ns, depending upon carrier lifetime). Neither the holding power nor the switch-off time has been minimized, so further improvements are expected. The search for and construction of even better nonlinear materials and devices should continue.

The portion of this research made at the University of Arizona was supported by the U.S. Air Force Office of Scientific Research, the U.S. Army Research Office and the National Science Foundation.

REFERENCES

Bowden C. M., Gibbs, H. M. & McCall, S. L. (eds) 1984 *Optical Bistability*, vol. 2. New York: Plenum Press.
Gibbs, H. M., Tarng, S. S., Jewell, J. L., Weinberger, D. A., Tai, K., Gossard, A. C., McCall, S. L., Passner, A. & Wiegmann, W. 1982 *Appl. Phys. Lett.* **41**, 221–222.
Gibbs, H. M., Venkatesan, T. N. C., McCall, S. L., Passner, A., Gossard, A. C. & Wiegmann, W. 1979 *Appl. Phys. Lett.* **34**, 511–513.
Gossard, A. C. 1979 *Thin solid films* **57**, 3–13.
Hanamura, E. 1981 *Solid St. Commun.* **38**, 939–942.
Jewell, J. L., Gibbs, H. M., Tarng, S. S., Gossard, A. C. & Wiegmann, W. 1981 *Appl. Phys. Lett.* **40**, 291–293.
Jewell, J. L., Rushford, M. C. & Gibbs, H. M. 1984 *Appl. Phys. Lett.* **44**, 172–174.
Karpushko, F. V. & Sinitsyn, G. V. 1978 *J. appl. Spectrosc. (U.S.S.R.)* **29**, 1323–1326
Karpushko, F. V. & Sinitsyn, G. V. 1982 *Appl. Phys.* B **28**, 137.
Koch, S. W. & Haug, H. 1981 *Phys. Rev. Lett.* **46**, 450–452.
Levy, R., Bigot, J. Y., Hönerlage, B., Tomasini, F. & Grun, J. B. 1983 *Solid St. Commun.* **48**, 705–708.
Miller, R. C., Kleinman, D. A., Tsang, W. T. & Gossard, A. C. 1981 *Phys. Rev.* B **24**, 1134–1136.
Peyghambarian, N., Gibbs, H. M., Rushford, M. C. & Weinberger, D. A. 1983 *Phys. Rev. Lett.* **51**, 1692–1695.
Sarid, D., Peyghambarian, N. & Gibbs, H. M. 1983 *Phys. Rev.* B. **28**, 1184–1186.
Tarng, S. S., Gibbs, H. M., Jewell, J. L., Peyghambarian, N., Gossard, A. C. & Wiegmann, W. 1984 *Appl. Phys. Lett.* **44**, 360–362.
Venkatesan, T. N. C., Lemaire, P. J., Wilkens, B., Soto, L., Gossard, A. C., Wiegmann, W., Jewell, J. L. Gibbs, H. M. & Tarng, S. S. 1984 *Postdeadline paper PD-10 at OSA 1983. (In the press.)*
Weinberger, D. A., Gibbs, H. M., Li, C. F. & Rushford, M. C. 1982 *J. opt. Soc. Am.* **72**, 1769.
Rushford, M. C., Weinberger, D. A., Gibbs, H. M., Li, C. F. & Peyghambarian, N. 1983 *Paper ThB5, Topical Meeting on Optical Bistability.* Washington: Optical Society of America.

Phil. Trans. R. Soc. Lond. A **313**, 249–256 (1984)

Printed in Great Britain

InSb devices: transphasors with high gain, bistable switches and sequential logic gates

By A. C. Walker, F. A. P. Tooley, M. E. Prise, J. G. H. Mathew,
A. K. Kar, M. R. Taghizadeh and S. D. Smith, F.R.S.

Department of Physics, Heriot-Watt University, Riccarton, Currie, Edinburgh EH14 4AS, U.K.

InSb etalons operated at 77 K and illuminated by CO lasers (5.5 μm) exhibit continuous wave (c.w.) optical bistability. A wide range of experiments have been performed to further the basic characterization of these devices and to demonstrate their various potential applications. The latter include signal amplification, modulation and, with external switching, the construction of logic gates. Two devices on a single etalon have now been coupled to form a simple all-optical circuit.

New results have also been obtained with InSb at room temperature with pulsed CO_2 lasers (10.6 μm).

1. Introduction

The purpose of this paper is to review the recent experimental results in the optical bistability (o.b.) project at Heriot–Watt University obtained with InSb as the nonlinear medium. Liquid-nitrogen cooled InSb and CO lasers have proved to be an extremely useful combination in both the study of optical bistability and the demonstration of practical bistable devices. In addition, InSb with CO_2 laser illumination has permitted the demonstration of bistable switching at room temperature.

2. Basic measurements

From an experimental point of view we are motivated in making measurements of the nonlinearities of the basic material by the desire to predict in advance the input–output characteristic of any bistable-type device we choose to fabricate. It has been clear for some time that the simplest plane-wave theory is inadequate, as illustrated in the comparison with experiment shown in figure 1 *a, b*. Before resorting to the complexity of a full 2D theory (taking into account the laser focal-spot irradiance profile, diffraction, refraction and carrier diffusion (see, for example, Firth *et al.*, this symposium)) it is worth considering how far simple improvements to the uniform-illumination plane-wave theory can take us towards realistic modelling of experimental characteristics.

Firstly, the dependence of Δn, the radiation-induced change in refractive index, upon irradiance, I, should be generalized beyond more than the simple $n_2 I$ linear dependence. Density-dependent recombination rates and saturation of the contribution of each excess carrier-pair to Δn are both mechanisms by which this dependence will become more complex. $\Delta n(I)$ could either be calculated from a full microscopic theory (see, for example, Wherrett, this symposium) or, alternatively, deduced from an experimental o.b. characteristic, provided that *all* other experimental parameters are known (including the temperature and absorption effects discussed below).

[59]

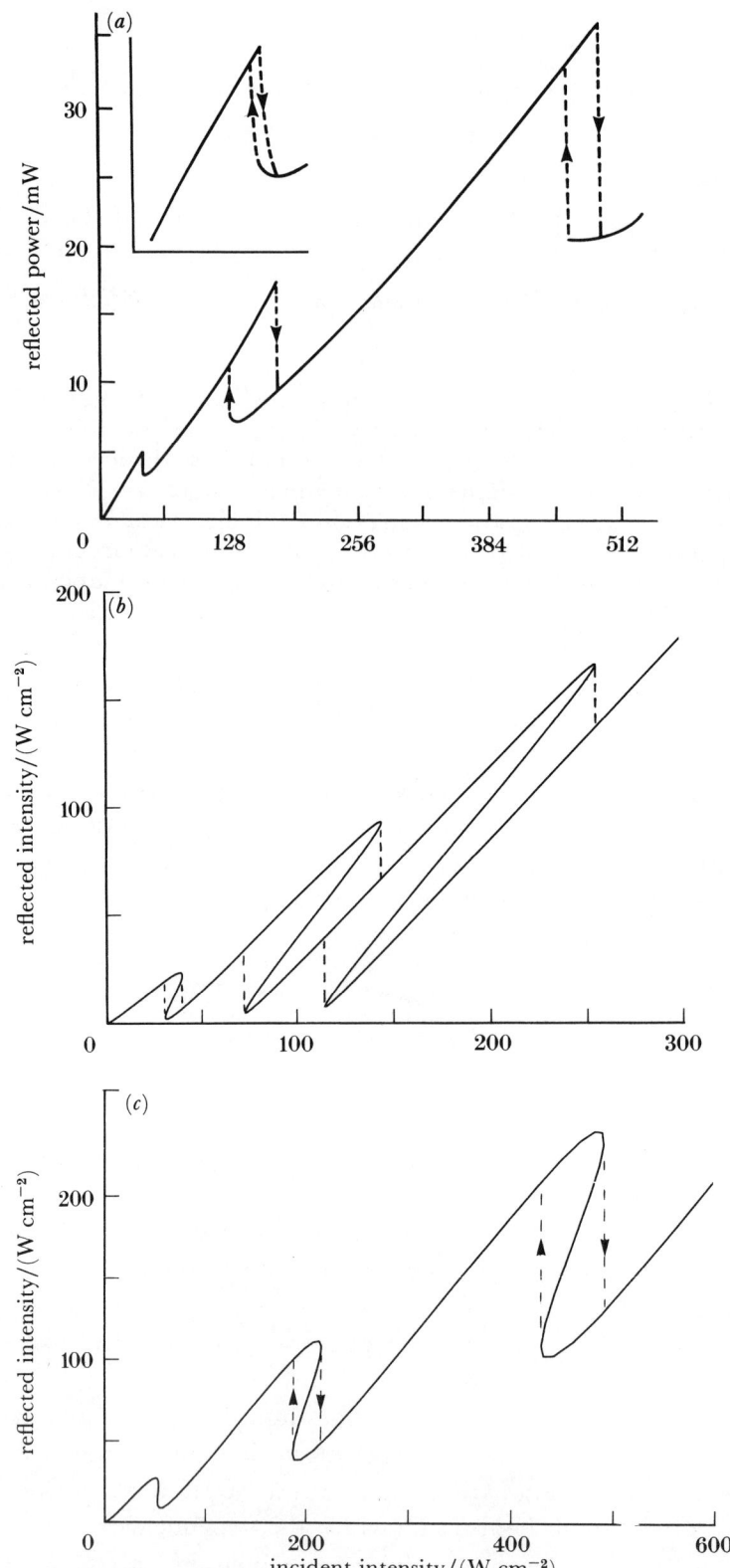

FIGURE 1. Input–output characteristics (reflection) for an InSb etalon: (*a*) experimental result (temperature 77 K, cavity length 210 μm, natural reflectivity 36%, laser line 1818.8 cm^{-1}); (*b*) calculated from simple plane-wave theory; (*c*) calculated from the improved plane-wave theory described in the text, with $\sigma = 1.9 \times 10^{-18}$ cm^3. The inset in (*a*) shows first-order bistability magnified.

Secondly, the irradiance dependence of the absorption coefficient, α, should be included. This may be a saturation phenomenon, associated with the microscopic origins of the refractive nonlinearity itself, or an induced absorption effect, for example due to rapidly increasing free-carrier absorption.

Finally, thermal effects should be taken into account. In addition to the linear refractive index's being temperature dependent, the position of the band edge also moves with temperature. Thus for a band-gap resonantly enhanced nonlinearity both $\Delta n(I)$ and α can change significantly for a temperature change of only a few kelvins.

We are currently working towards including these three factors in a modified plane-wave model of InSb optically bistable devices. The following considers each in turn.

A compromise assumption for determining $\Delta n(I)$ is that saturation of the contribution of each excess carrier-pair can be neglected and hence

$$\Delta n(I) = \sigma \Delta N(I). \tag{1}$$

That is, Δn is directly proportional to the excess carrier-pairs, ΔN, with a constant of proportionality σ (Miller et al. 1981); $\Delta N(I)$ then remains to be determined. This requires a knowledge of carrier recombination rates. A review of published experimental data for excess-carrier lifetimes in n-InSb at 77–100 K shows considerable spread of values. However, there is a clear trend showing shortening lifetimes, T_1, at higher carrier densities (over 5×10^{15} cm^{-3}) and an empirical relation can be deduced of the form

$$T_1^{-1} = r_1 + r_2(N_0 + \Delta N), \tag{2}$$

where N_0 is the dark carrier density and ΔN the excess carrier density. This relation implies a monomolecular, e.g. trap, recombination at low carrier densities, evolving to a bimolecular, e.g. radiative, recombination process at higher densities. A fit to the data can be obtained by using $r_1 = 1.5 \times 10^6$ s^{-1} and $r_2 = 1.5 \times 10^{-10}$ cm^3 s^{-1}, both rates being accurate to about $\pm 50\%$.

Using (2) for T_1, the equilibrium excess carrier density can be calculated for any internal irradiance, I (in watts per square centimetre), from $\Delta N = \alpha_0 T_1 I/h\nu$. Thus $\Delta N(I)$, and hence the refractive index change, can be obtained from

$$\Delta N(I) = \frac{[(r_1 + r_2 N_0)^2 + 4r_2 \alpha_0 I/h\nu]^{\frac{1}{2}} - (r_1 + r_2 N_0)}{2r_2}, \tag{3}$$

where α_0 is the carrier generating absorption coefficient (in reciprocal centimetres) and $h\nu$ the photon energy (in joules).

It is assumed in the above that α_0 is not significantly saturated at these irradiance levels. This is consistent with experimental measurements of absorption as a function of irradiance performed on other samples from the batch currently being used to fabricate bistable devices. These results, shown in figure 2, demonstrate an increase in transmission losses with increasing irradiance, and appear to be consistent with a simple model based on additional absorption being induced by the generated free-carrier pairs. Assuming equal electron and hole concentrations, free-carrier absorption is dominated by the direct intra-valence-band hole transition. The hole absorption cross section is $\sigma_p = 2.5 \times 10^{-15}$ cm^2 at 77 K in InSb (Kurnick & Powell 1959). The total absorption coefficient is then given by

$$\alpha = \alpha_0 + \Delta N(I)\,\sigma_p. \tag{4}$$

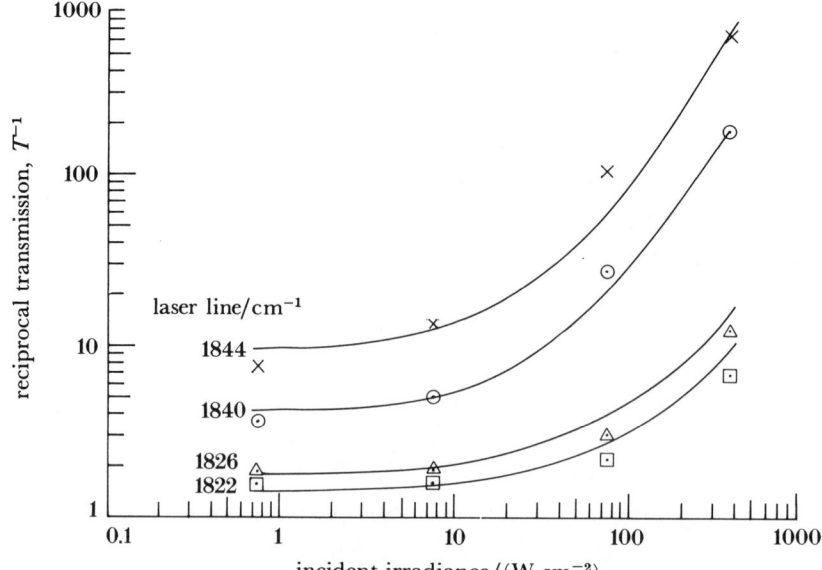

FIGURE 2. Transmission of a 0.5 mm InSb sample at 77 K as a function of incident irradiance. Points are experimental results, curves are calculated (free carrier absorption and temperature).

To simplify the integration across the sample thickness, the relation $\Delta N(I) = \alpha_0 \, T_{\mathrm{eff}} \, I_0/h\nu$ was used, where T_{eff} is an effective carrier lifetime for the average ΔN values deduced from the full equation (3). Thus the transmission, T, through a sample of thickness L can be obtained:

$$T = \frac{\exp\left(-\alpha_0 L\right)}{1 + \{1 - \exp\left(-\alpha_0 L\right)\} I_0 \, T_{\mathrm{eff}} \, \sigma_{\mathrm{p}}/h\nu}, \tag{5}$$

where I_0 is the incident irradiance.

By using this relation, plus a small correction allowing for (measured) heating effects, the curves in figure 2, showing an approximate fit with the experimental results, were calculated. It is important to note that this calculation assumes that every photon absorbed, as a result of the band tail, creates an electron–hole pair. This is despite the photon energy deficit relative to the band gap. The mechanism for this band-tail absorption process in the region commonly used in these bistability experiments (typically $\alpha_0 \approx 5$ to $40\ \mathrm{cm}^{-1}$) is not understood. Furthermore, in contrast to this, absorption *saturation* is the dominant absorption nonlinearity near 77 K for photon energies *above* the band gap (Nurmikko 1976) whereas below 5 K this is true both below and above the band gap (Miller *et al.* 1978; Lavallard *et al.* 1976). Although induced absorption of the type observed here has been reported (Miller *et al.* 1978) it has not been previously measured this close to the band edge.

The remaining factor to be considered is the effect of sample heating caused by the incident laser beam. This is discussed more fully by Tooley *et al.* (this symposium) in a paper on incoherent–coherent switching. Detailed modelling of thermal effects is complicated by the need for full knowledge of the heat diffusion and conduction properties of the sample–mount assembly and is not considered further here.

It has been found, by using (3) to calculate the excess carrier density, and hence both Δn from (1) and the total absorption from (2), that the input–output characteristic calculated from plane-wave theory gives a much closer fit to the experimental result. This is shown in figure

[62]

$1c$, where σ has been taken to be 1.9×10^{-18} cm³ (equivalent to $n_2 = 0.3$ cm² kW⁻¹ at low ΔN). Further improvements in modelling such characteristics will probably require a fuller 2D calculation.

3. TRANSPHASOR ACTION

By adjusting the initial detuning from resonance of the InSb etalon the transmission characteristic can be made to have a steeply sloped single-valued region, giving high differential gain. This permits the construction of a transphasor amplifier. By using a separate 3 μW, 5.5 μm wavelength chopped beam as a probe, pulse amplification of up to 1.3×10^4 has been observed from a single device (Tooley *et al.* 1983). It should be noted that there is no requirement for the signal to be coherent with the bias beam in this application, as was demonstrated by using orthogonally polarized beams with equal success.

4. RESPONSE SPEEDS

The signal frequency in the high-gain transphasor experiment was limited to *ca.* 1 kHz by mechanical chopping rates. It is important that the upper bandwidth limit of the InSb transphasor be determined. Direct observations of the switching times obtained by very slowly sweeping the input power give an upper limit of *ca.* 2 μs for both switch-on and switch-off. (Much faster switch-on times should be possible with more intense, fast-rising pulses, as implied by successful switching with 35 ps Nd:YAG laser pulses (Seaton *et al.* 1983).) To investigate the higher frequency response of a transphasor-type device, a 1.3 μm laser diode, capable of being modulated at over 10 MHz, has been used as the signal source. The experimental arrangement is shown in figure 3, together with the reflection characteristic employed. No

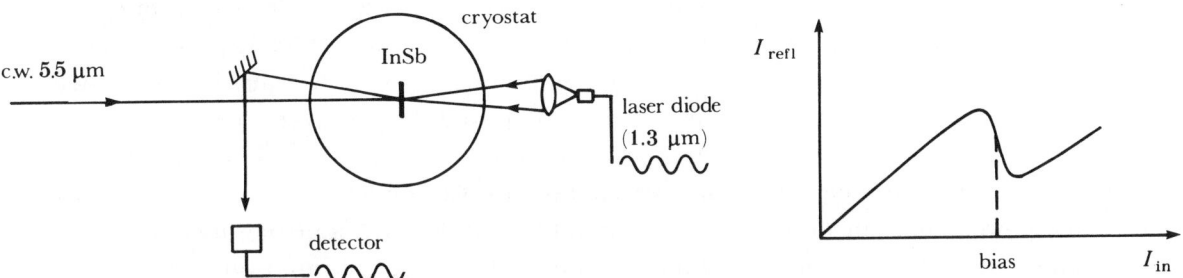

FIGURE 3. Experimental layout and operating condition for the demonstration of high-frequency cross-wavelength modulation (1.3 and 5.5 μm). A gain of *ca.* 1 is observed between 0.2 and 1.0 MHz.

attempt was made to achieve gain in this instance. With the device biased to the centre of the negative slope region, the signal on the 1.3 μm beam incident on the rear face of the sample induced linearly proportional modulation on the reflected 5.5 μm beam. From initial experiments only it appears that no significant roll-off in response occurs in the frequency band so far studied, up to *ca.* 1 MHz.

5. CROSS-WAVELENGTH MODULATION AND SWITCHING

In addition to the 1.3 μm laser diode a wide range of other sources have now been used to switch or modulate the 5 μm InSb o.b.-switch or transphasor. These include visible wavelengths, which are of particular interest to potential image-processing applications. For example, with the laser diode in figure 3 replaced by a 10 mW He–Ne laser, both switching and modulation of the 5.5 μm reflected beam by coherent visible radiation have been demonstrated. Alternatively, by using a photographic flash unit, incoherent–coherent switching with white light has been studied (Tooley *et al.* this symposium). Table 1 summarizes these and other experiments.

TABLE 1. CROSS-WAVELENGTH MODULATION AND SWITCHING OF AN InSb o.b. DEVICE PUMPED AT 5.5 μm

wavelength, etc. (source)	operation mode	gain	frequency
5.5 μm, 3 μW (CO laser)	switching	1.3×10^4	*ca.* kilohertz
1.3 μm, 3 mW (laser diode)	switching	*ca.* 1	single pulse
	modulation	*ca.* 1	*ca.* megahertz
1.06 and 0.53 μm, 5 nJ (Nd:YAG laser)	switching	—	single pulse (35 ps)
0.633 μm, 10 mW (He–Ne laser)	switching ⎱ modulation ⎰	*ca.* 1	3 kHz
white light (camera flash)	switching	—	single pulse

6. DEMONSTRATIONS OF SEQUENTIALLY COUPLED LOGIC ELEMENTS

An experiment in which the transmission change through one bistable InSb device was used to switch a second has been reported (Smith & Tooley 1984). Recently we have coupled two devices together working in their reflection mode and, more significantly, with both gates operating adjacent to each other on a single InSb etalon. The experimental geometry is shown in figure 4, together with the results obtained. The two gates were addressed from opposite sides of the sample and were separated by about 0.5 mm, i.e. about 2.5 focal-spot diameters. The reflection from gate 1 was directed at gate 2. By biasing gate 2 to just below its switch point the following sequence was demonstrated, as reproduced in figure 4. Firstly, as the input power to gate 1 is increased the reflected power becomes sufficient to switch on gate 2 (low reflection state). A further increase, however, causes gate 1 to switch on. The consequent drop in reflected power simultaneously causes gate 2 to turn off. Finally, a further increase in input power can eventually turn gate 2 back on once more. The range of input powers over which gate 2 remains on is determined by its initial bias condition. This device represents an XNOR gate and is also close to a flip–flop configuration, the latter simply requiring the reflection from gate 2 to be directed back at gate 1. Finally, by using transmission feedback, or by adding a third switch, an oscillator could be constructed.

Altogether, the feasibility of a wide range of logic devices has been demonstrated. For example, with a single active element only, both AND and OR gates can be made by using the transmission mode, while NAND and NOR gates are obtained by operating in the reflection mode. The device described above demonstrated a two-element XNOR gate, while the first element alone acted as an XOR gate. Finally, of course, a bistable characteristic provides a memory element. We have now reached the point where we can start to build simple all-optical circuits.

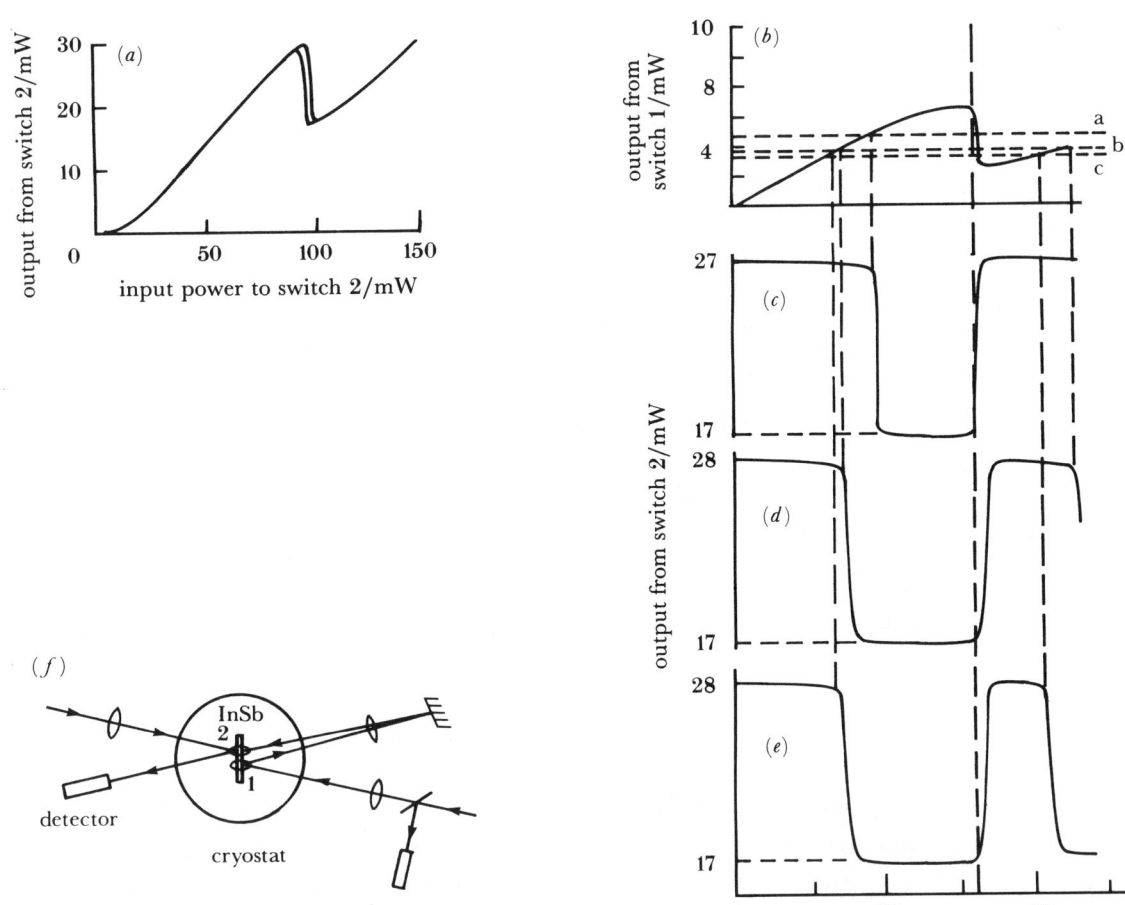

FIGURE 4. Two coupled o.b. switches on a single InSb element. (*f*) Experimental layout. (*a, b*) Input–output characteristics of the two switches; (*c–e*) input–output characteristics of the combination, showing the result of increasing approach of the initial bias of switch 2 to its switch point.

7. ROOM-TEMPERATURE InSb RESULTS

The results of initial optical bistability studies on room-temperature InSb have already been published (Kar *et al.* 1983). In these experiments free carriers are excited across the 0.18 eV band gap with 10.6 μm (CO_2 laser) radiation by a two-photon process. Recent results have been obtained with longer, *ca.* 3 μs, duration pulses to minimize the dynamic effects of the rise and fall of incident power. In addition, small-area pinholes mounted directly onto the samples have been used to define an area of uniform illumination at a specific point on the device. This should permit more direct comparison of experimental results with plane-wave theory. Controlled variation of the initial detuning is achieved by angular adjustment of the sample. The pinhole ensures that all measurements are made on the same part of the sample, avoiding any uncertainties caused by material or surface-finish non-uniformity. Figure 5 includes an example of the transmission characteristic, which shows clear hysteresis. Care must be taken in claiming true bistability before completion of a full analysis of *all* the dynamic effects. Figure 5 also plots the switching irradiance as a function of initial detuning and shows, at least qualitatively, the expected trend.

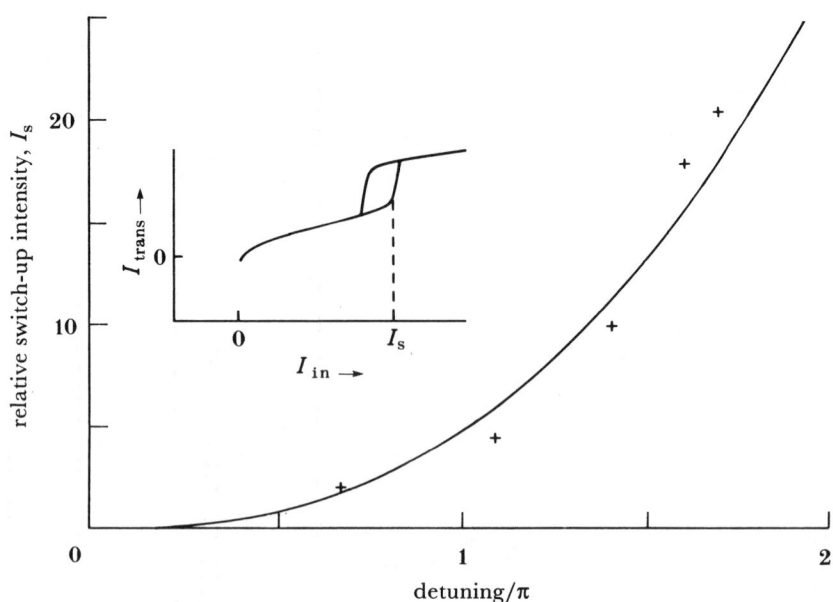

FIGURE 5. Switch-up irradiance as a function of initial detuning for a room-temperature InSb device operating at 10.6 μm wavelength. Inset shows a typical input–output characteristic (transmission).

8. CONCLUSIONS

This experimental programme can be summarized under three main headings: the physics of the nonlinearity and associated parameters, e.g. the form of $\Delta n(I, T)$ and $\alpha(I, T)$; the development and characterization of single devices, e.g. switches, transphasors and modulators, and the coupling of devices to develop photonic logic. Future work is directed at taking these InSb devices further, including the investigation of possible waveguide geometries.

It now seems possible to iterate a significant number of all-optical circuit elements and to demonstrate a simple processor.

The authors acknowledge the assistance of Dr H. A. Mackenzie and Mr J. Reid in these experiments, the contribution of Mr N. Ross to sample preparation and the general support of Dr B. S. Wherrett. The 1.3 μm laser diodes were generously provided by S.T.C. Ltd., Harlow.

REFERENCES

Kar, A. K., Mathew, J. G. H., Smith, S. D., Davis, B. & Prettl, W. 1983 *Appl. Phys. Lett.* **42**, 334.
Kurnick, S. W. & Powell, J. M. 1959 *Phys. Rev.* **116**, 597.
Lavallard, P., Bichard, R. & Benoit á la Guillaume, C. 1976 *Phys. Rev.* B **16**, 2804.
Miller, D. A. B., Mozolowski, M. H., Miller, A. & Smith, S. D. 1978 *Optics Commun.* **27**, 133.
Miller, D. A. B., Seaton, C. T., Prise, M. E. & Smith, S. D. 1981 *Phys. Rev. Lett.* **47**, 197.
Nurmikko, A. V. 1976 *Optics Commun.* **16**, 365.
Seaton, C. T., Smith, S. D., Tooley, F. A. P., Prise, M. E. & Taghizadeh, M. R. 1983 *Appl. Phys. Lett.* **42**, 131.
Smith, S. D. & Tooley, F. A. P. 1984 In *Proc. Topical Meeting on Optical Bistability, Rochester, U.S.A.* (In the press.)
Tooley, F. A. P., Smith, S. D. & Seaton, C. T. 1983 *Appl. Phys. Lett.* **43**, 807.

Phil Trans. R. Soc. Lond. A **313**, 257–264 (1984)

Printed in Great Britain

Bistability experimentally observed at three milliwatts in indium arsenide and theoretically predicted for a new class of nonlinear dielectrics

By E. Garmire, C. D. Poole and J. A. Goldstone

*Centre for Laser Studies, University of Southern California, Los Angeles,
California 90089–1112, U.S.A.*

The results of an experimental investigation of nonlinearities and bistability at the band gap in InAs are presented. Bistability was observed in reflection at 3 mW with a HF laser by using the large nonlinearity at the band-gap resonance in InAs. The measured nonlinear refractive index is shown quantitatively to agree with the dynamic Burstein–Moss shift as the mechanism, as long as both the full Fermi–Dirac statistics and the effect of light holes are included in the calculation. In a separate study a theoretical analysis has investigated general conditions for intrinsic bistability without cavities. It is shown that bistability can result from a nonlinear constitutive relation. Specific calculations are presented for bistability due to nonlinear polarization.

This paper is a discussion of recent developments in optical bistability at the University of Southern California, which includes experimental investigations as well as theoretical studies of optical bistability. The results of an experimental investigation of nonlinearities and bistability at the band gap in InAs are presented. Bistability was observed in reflection at 3 mW with a HF laser by using the large nonlinearity at the band gap resonance in InAs. The measured nonlinear refractive index is shown quantitatively to agree with the dynamic Burstein–Moss shift as the mechanism, as long as both the full Fermi–Dirac statistics and the effect of light holes are included in the calculation. In a separate study a theoretical analysis has investigated general conditions for intrinsic bistability without cavities. It is shown that bistability can result from a nonlinear constitutive relation. Specific calculations are presented for bistability due to nonlinear polarization. The experimental discovery of optical bistability in InAs will be discussed first, and the theoretical discussions will be presented in the later parts of the paper.

Bistability in indium arsenide

Indium arsenide is a III–V semiconductor with a band gap at 3 μm, between the band gap of InSb at 5 μm and GaAs at 1 μm, both of which have previously been shown to exhibit optical bistability near their band gap (Miller *et al.* 1979; Gibbs *et al.* 1979). The reason for investigating InAs is the coincidence of its band gap with the lines of the HF laser, near 3 μm. The results presented here used the 3.096 μm line of the HF laser, which matches the band gap of 2.6×10^{-16} n-type InAs at 77 K. Optical bistability was observed in reflection from a 170 μm thick etalon of InAs, which was polished plane and parallel, and whose back surface was turned into a mirror with silver (Poole & Garmire 1984a). The observation of bistability in reflection is easier than the observation of bistability in transmission, primarily because of the simplicity of applying 100% reflection to the back face, compared to enhancing the natural 31% Fresnel reflection with multilayer coatings. An additional advantage of studying optical bistability in reflection is the ease with which the sample is heat-sunk. The devices were mounted on a stress-relieving InAs wafer and thence on a copper block. No thermal heating effects from the incident HF laser radiation were observed when this configuration was used. Figure 1 shows experimental results for optical bistability in InAs. In figure 1a, switch-on

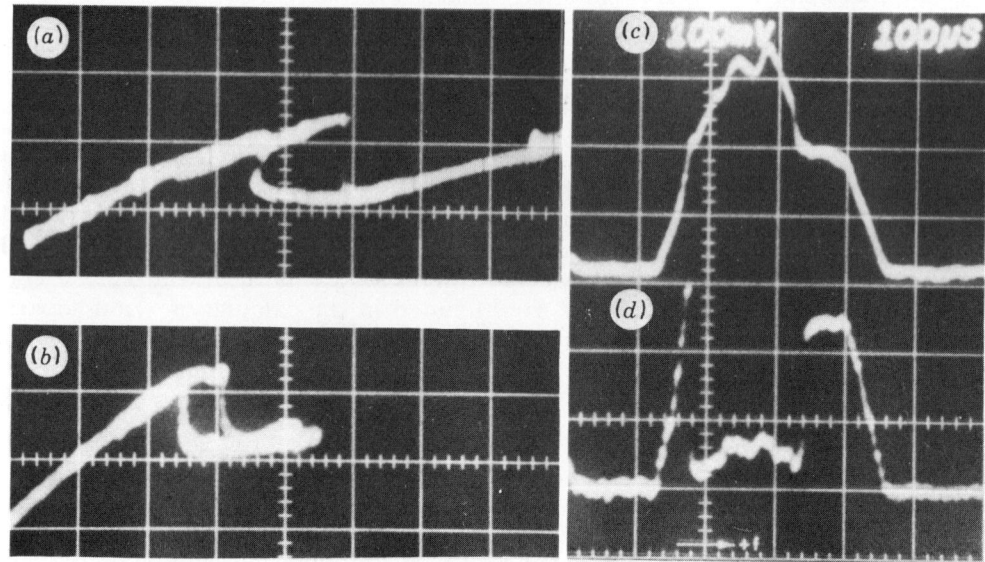

FIGURE 1. Experimental results on whole-beam bistability in InAs. (a), (b) Reflected power against incident power for two different detunings; horizontal scale is 1.5 mW per division, vertical scale is 1 mW per division. (c), (d) Input and reflected signals, respectively, as a function of time.

occurred at 7 mW and is indicated by a drop in the reflected light level. Switch-off occurred at 5 mW and is manifested by a sudden increase in reflectivity of the device. The resulting hysteresis is characteristic of optical bistability. The detuning could be varied by lateral displacement of the etalon, since it was slightly wedge-shaped. The minimum power level at which bistability was observed was 4 mW and is shown in figure 1b. The intensity level corresponding to this power level is 75 W cm^{-2}, at the centre of a Gaussian focal spot of width 72 μm. The hysteresis curves were largest when the output power was monitored at the centre of the beam only. This was achieved by imaging the near field of the etalon onto a 25 μm pinhole, placed before the detector. Whole-beam switching is shown here.

The variation in input intensity was obtained by natural pulsations in the HF laser output due to instabilities in its plasma discharge. This led to pulses several hundred microseconds long, as shown in figure 1c. The reflected signal as a function of time is shown in figure 1d, and the rapid switching of the bistable device to a low-reflecting state is evident. Switching was faster than the time constant of the detector and is expected to be of the order of the lifetime of carriers in InAs, approximately 300 ns.

Nonlinear refractive index in indium arsenide

The nonlinear index can be inferred from the threshold for bistability and is 3×10^{-5} cm^2 W^{-1}. The mechanism that is believed to be responsible for this nonlinear index is the same as that for InSb (Miller *et al.* 1981) and HgCdTe (Hill *et al.* 1982). That is, the dynamic Burstein–Moss shift, or the change in the absorption edge of the band gap due to a filling of the lower levels of the conduction band (Miller *et al.* 1981; Moss 1980). Through the Kramers–Krönig relation, this change in absorption leads to a change in refractive index. Since this change is intensity-dependent, it leads to a nonlinear refractive index.

An accurate experimental determination of the nonlinear refractive index was made by measuring transmission as a function of intensity for an InAs etalon with Fresnel reflection from both surfaces. Fitting these curves to the numerical analysis by using plane wave theory for the transmission of a nonlinear Fabry–Perot etalon with a single value of nonlinear reflective index and linear absorption, for a variety of input phase conditions, allowed a determination of the nonlinear refractive index and the linear absorption. Data for nonlinear transmission curves, their numerical fits and values for n_2 and α inferred from these data are shown in figure 2 for a number of temperatures, which correspond to varying differences of the incident photon

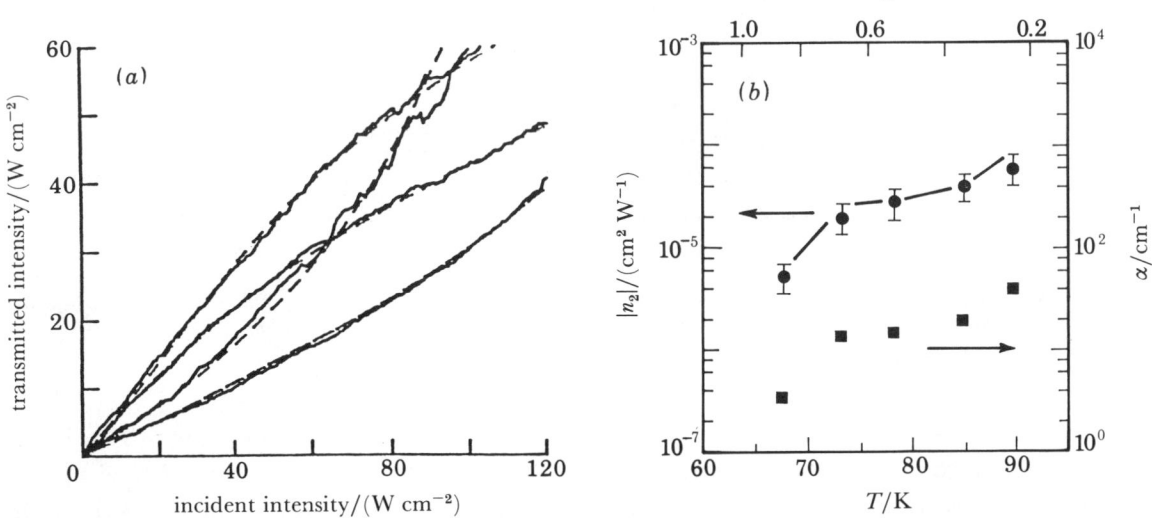

Figure 2. Measurements of the nonlinear index in InAs at 3.096 μm. (*a*) Nonlinear transmission through an etalon for several detunings and the plane-wave theoretical fit. (*b*) Summary of results at several temperatures for the nonlinear index n_2 and the linear absorption α. Solid curve is obtained by using (1) with no free parameters. The top scale gives the position of the photon energy relative to the band gap energy ($\epsilon = (E_G - \hbar\omega)/KT$].

energy from the band gap. It is interesting to note that the figure of merit, which is given by n_2/α, is essentially independent of distance from the band gap. This fact will be used later in predicting expected bistability thresholds for other geometries.

To confirm the mechanism for the large nonlinear index, a calculation was made of the nonlinear index at the band gap of direct-band semiconductors. This analysis follows that of Miller *et al.* (1981), but includes the complete Fermi–Dirac statistics required since the semiconductor is degenerate, and also includes the effects of the light-hole band. The absorption edge is described by the Kane analysis (Kane 1957) and leads to an expression for the figure of merit given by (Poole & Garmire 1984b)

$$\frac{n_2}{\alpha} = \frac{-2\pi\tau e^2}{3n_0\,kTE_G^3}\left(\frac{\mu_{hh}}{m_e}\right)^{\frac{3}{2}}P^2K,\tag{1}$$

where τ is the lifetime of the carriers, P is the momentum element (Kane 1957), μ_{hh} is the mobility of the heavy holes, m_e is the effective mass of the electrons and e is their charge, E_G is the energy of the band gap, n_0 is the linear refractive index, kT is the temperature in energy units, and K is a dimensionless parameter that depends on the carrier concentration and the distance from the band edge, and which has a value close to one. K is shown in figure 3 as

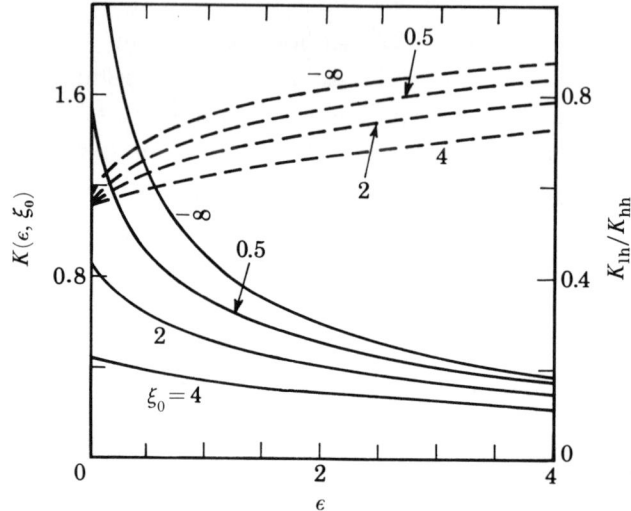

FIGURE 3. $K(\epsilon, \xi_0)$ (solid curves, left scale) and ratio of light-hole contribution to heavy-hole contribution, K_{lh}/K_{hh} (broken curves, right scale) against distance from band gap in units of $\epsilon = (E_G - \hbar\omega)/kT$ for several values of the Fermi energy ξ_0 (in units of kT).

a function of distance from the band gap for various values of the Fermi energy. The Fermi energy is related to the carrier concentration in the semiconductor. The larger the Fermi energy, the more impurities in the material. For the impurity levels in the samples used in the experiments reported here, the figure of merit was essentially independent of band gap. This can be seen from figure 2, since in each case $n_2/\alpha = (1.7 \pm 0.3) \times 10^{-6}$ cm^3 W^{-1}. Also included in figure 3 is the ratio of the light hole to the heavy hole contribution, which may reach a value of almost 90%, demonstrating the importance of this term.

To compare experimentally measured nonlinear refractive indices with the theoretical prediction of (1), it is necessary to know the carrier lifetime τ. This was explicitly determined in the sample reported here by monitoring the photoconductivity as a function of time after excitation by a pulse from a Nd:YAG laser; this value was 330 ns.

The excellent agreement between experimental results and theoretical modelling indicates that the dynamic Burstein–Moss shift is indeed the mechanism for the large nonlinear index in InAs at the band gap.

Possibilities for optical computation

From the theory for the nonlinear index, the figure of merit can be used to predict room-temperature operation based on an estimate of the carrier lifetime at room temperature and on the change of band gap with temperature. There is considerable uncertainty in the literature as to the temperature dependence of the lifetime of carriers in InAs. The most optimistic values indicate that liquid phase epitaxial material has a carrier lifetime at room temperature equal to 300 ns (Dalal 1974). This is comparable to the value measured in bulk material at 77 K. The band gap at room temperature corresponds to a wavelength of 3.7 μm. Since the figure of merit is proportional to τ/E_G^3, the smaller room-temperature band gap and comparable carrier lifetime predict a room-temperature threshold for bistability in l.p.e. InAs even lower than at 77 K in bulk InAs.

Observation of room-temperature bistability requires a laser whose wavelength matches the InAs band gap. The DF laser has lines at 3.7 μm, appropriate to room-temperature operation. This laser is too expensive to operate to provide practical room temperature bistability, however. Another approach would be the HeNe laser at 3.39 μm, which matches the bandgap at 180 K, and operation would be practical using a thermoelectric cooler. Alternatively, the 3.6 μm line in the xenon laser may provide for near-room temperature operation.

To estimate the effectiveness of InAs for optical computations, it is necessary to calculate the minimum power required for switching in an etalon. In the theory for optimization of a nonlinear etalon (Miller 1981), it was found that the critical intensity in the high finesse limit is given by

$$I_c = \lambda \alpha L / (n_2/\alpha), \tag{2}$$

where it is assumed that the loss per pass equals the loss upon reflection; so the thinner the etalon the better. The experiments presented here were for an etalon of 170 μm, optimized according to the calculations of D. A. B. Miller. Under these conditions, the threshold for bistability was 75 W cm^{-2}. An etalon 1.7 μm thick gives a predicted threshold of 0.75 W cm^{-2}. Assuming that the beam can be focused to within a wavelength of light, the minimum predicted threshold power is 100 nW, which gives a capability of 10^6 parallel bits with a 100 mW laser!

In conclusion, the observation of bistability in InAs and comparison with theory suggests the possibility of an extremely impressive room-temperature computational capability.

Bistability due to nonlinear constitutive relations

In addition to experimental studies in InAs, the result of a theoretical study on a new type of optical bistability is outlined. In particular, we wish to point out that optical bistability is possible whenever a medium can be represented by a suitable nonlinear constitutive relation. The constitutive relation is a functional relation between the electric field and some scalar or vector parameters of the system, or both. In the usual formulation of the constitutive relations, material parameters are expressed as functions of the electric field (typically in power series expansions). This formulation necessarily precludes the description of this type of bistability. However, if the electric field is expressed in terms of the material parameters, bistability may be predicted for suitable nonlinear relations. The technique of inverting the expression to obtain the second branch is well known from the theory of bistability in resonant cavities, since the single-valued expression of the input as a function of the output leads to a multivalued expression for the output as a function of the input.

In our analysis we consider polarization as the material parameter and show that if the electric field is expressed in terms of the polarization, bistability may result. This does not require any explicit feedback, merely the ability to define a constitutive relation. Other examples of this general relation have been presented at this symposium. They include thermal bistability that may result when absorption is the material parameter (M. Dagenais, D. A. B. Miller, C. Klingshirn et al.) and bistability that may occur as a result of a change of state within the material, as, for example, in liquid crystals (Y. R. Shen).

BISTABILITY DUE TO NONLINEAR POLARIZATION

To illustrate the general concept we consider here the polarization that results from nonlinear oscillations in molecular or atomic systems. The usual phenomenological description of the polarization as a power series in the electric field obviates the possibility of describing bistability since the electric field is treated as the independent parameter. Thus the second branch cannot be found. A nonlinear model for the general constitutive relation is required before bistability can be postulated. For the purposes of demonstration we use the classical Duffing oscillator as the model of electron oscillation to derive a bistable polarization.

The Duffing oscillator model may be described by the following differential equation for the microscopic material response:

$$m\ddot{x} + \gamma\dot{x} = -Kx - \beta x^2 + eE/m. \tag{3}$$

This is a differential form of the constitutive relation, since the polarization relates to the atomic (or molecular) vibration through $P = Nex$ (for low molecular densities), and the above equation relates x to the electric field. Under these conditions bistability on the microscopic level implies bistability on the macroscopic level. The bistability of the Duffing oscillator is well known and can be calculated by inverting the relation between the amplitude of the oscillation and the force driving the oscillation. For polarization, the driving force is the electric field. We thus look for an expression for the electric field in terms of the amplitude of the oscillation x. Following Duffing, a solution exists at the fundamental frequency given by

$$\{m(\omega_0^2 - \omega^2) - i\gamma\omega + \tfrac{3}{4}\beta|x|^2\} x = eE. \tag{4}$$

There is an additional relation for the third harmonic, which is here ignored. The cubic form of (4) yields three solutions for the polarization over a range of impressed fields for appropriate parameter choices. Of these three, two are locally stable and one is unstable. Hence a choice of P implies a unique E but the inverse is not true, i.e. $E = E(P)$, but $P \neq P(E)$ exclusively.

From (4), the constitutive relation for the polarization amplitude $P = Ne|x|$ and the magnitude of the field squared can be written as

$$\left\{(\omega_0^2 - \omega^2 + bP^2)^2 + \frac{\gamma^2\omega^2}{m^2}\right\} P^2 = \frac{e^4 N^2}{m^2}|E|^2 \equiv S, \tag{5}$$

where $b = 3\beta/4N^2e^2m$.

The condition for the existence of bistability is the requirement that $\partial S/\partial P \to 0$, which becomes

$$(\omega^2 - \omega_0^2)/b > 0 \quad \text{and} \quad C \equiv 3\gamma^2\omega^2/m^2(\omega_0^2 - \omega^2)^2 < 1. \tag{6}$$

In this case the critical polarizations that lead to zero slope are

$$bP_c^2 = \tfrac{1}{3}(\omega^2 - \omega_0^2)\{2 \pm (1-C)^{\frac{1}{2}}\}. \tag{7}$$

The minimum switch-on power occurs when the damping is very small and $C \ll 1$. In this case the critical power for switching to the higher branch is

$$bS_{\min} = 12/(\omega^2 - \omega_0^2).$$

However, since the maximum value for $\omega^2 - \omega_0^2 = m/\sqrt{3}\,\gamma\omega$, the minimum critical power is $S_{\min} = 12\sqrt{3}\,\gamma\omega/mb$.

Expression (5) for the magnitude of the flux of the electric field S as a function of the polarization is single-valued and shown in figure 4. This result can be seen to lead to a bistable

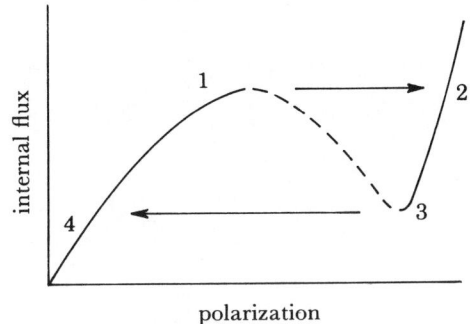

FIGURE 4. Nonlinear constitutive relation that results from the Duffing oscillator. Field squared, inside the medium, plotted as a function of the internal polarization. The field is single-valued in the polarization, but the polarization is not single-valued in the field.

polarization as a function of field, since for a given value of the field there are several values of the polarization. Bistability (or multistability) in the polarization manifests itself in a large number of observable macroscopic effects. Among these are phase bistability, which may be converted to an intensity bistability, and a bistability in both the reflected intensity and refracted angle. Further details about these macroscopic manifestations are shown in the other contribution by Garmire & Goldstone (this symposium).

The realization that expressing the electric field as a function of the macroscopic polarization may produce bistability encourages us to look in more detail at nonlinear constitutive relations from first principles. This has been done effectively in absorptive cases and for changes of state that are electric-field driven. In this paper we have shown the possibility of bistability in the polarization by using the Duffing oscillator model. Quantum mechanical formulations should also demonstrate bistability under appropriate conditions.

This work was supported by the National Science Foundation.

REFERENCES

Dalal, V. L., Hicinbothem, W. A. Jr, Kressel, H. 1974 *Appl. Phys. Lett.* **24**, 184–5.
Gibbs, H. M., McCall, S. L., Venkatesan, T. N. C., Gossard, A. C., Passner, A., Wiegmann, W. 1979 *Appl. Phys. Lett.* **35**, 451–453.
Hill, J. R., Parry, G., Miller, A. 1982 *Optics Commun.* **43**, 151–153.
Kane, E. O. 1957 *J. phys. Chem.* **1**, 249–254.

[73]

Miller, D. A. B. 1981 *J. quant. Electron.* **17**, 306–311.
Miller, D. A. B., Seaton, C. T., Prise, M. E., Smith, S. D. 1981 *Phys. Rev. Lett.* **4**, 197–199.
Miller, D. A. B., Smith, S. D., Johnston, A. 1979 *Appl. Phys. Lett.* **35**, 658–661.
Moss, T. S. 1980 *Physica Status Solidi* B **101**, 1980–1982.
Poole, C. D., Garmire, E. 1984*a* *Appl. Phys. Lett.* **44**, 364–366.
Poole, C. D., Garmire, E. 1984*b* *Optics Lett.* (Submitted.)

Phil. Trans. R. Soc. Lond. A **313**, 265–268 (1984)

Printed in Great Britain

Giant nonlinearities and low power optical bistability in cadmium sulphide platelets

By M. Dagenais

GTE Laboratories Incorporated, 40 *Sylvan Road, Waltham, Massachusetts* 02254, *U.S.A.*

Both whole-beam and transverse continuous wave optical bistability are observed at milliwatt power levels near the band gap in thin uncoated cadmium sulphide platelets. The whole-beam bistability is free from cavities and requires an absorption coefficient that increases with intensity.

Introduction

This paper reports the observation of both whole-beam and transverse optical bistability in thin uncoated cadmium sulphide (CdS) platelets at milliwatt power levels when a continuous wave laser is tuned near the band gap. The observed whole-beam bistability is thermally induced and can be relatively fast due to the high thermal conductivity of CdS at low temperatures. It is free from cavities, does not require external feedback and exhibits a very large contrast ratio (over 10) between the 'on' and 'off' states. In samples of small surface area, whole-beam bistability is observed for temperatures varying between 5 and 50 K. The transverse bistability measurements were made with larger samples immersed in superfluid helium. A large induced absorption of thermal origin is detected and is believed to be responsible for our transverse bistability results. Very large transverse effects are observed in the far field. As the input intensity is varied, we observe large hysteresis loops in the centre part of the beam. When the whole beam is monitored, no switching is observed. At high intensities, very regular self-pulsations with a 10 µs period appear. The pulsations are attributed to thermal effects. To better understand the origin of the large observed nonlinear behaviour, the intensity and the time dependence of the total beam absorption was studied. Sharp induced absorption lines appearing in less than 10 ns are detected near the band gap. The optical saturation of the I_2 bound exciton is also observed.

Whole-beam bistability

For the whole-beam bistability measurements, a good optical quality CdS sample, 10 µm thick with a surface area of less than 0.01 cm², is held electrostatically to a glass slide and then mounted in a cryogenic dewar. The laser is detuned a few wavenumbers below the I_2 bound exciton resonance, an exciton bound to a neutral donor. We observe large steady-state inverted hysteresis loops in the whole-beam transmission with contrast ratios between the on and off states greater than 10. Only a few milliwats of input power are required to observe this bistability. As the detuning below the bound exciton resonance is increased, larger hysteresis loops are observed and higher input intensities are required to observe optical bistability. Bistability is seen for detunings as large as 20–30 cm⁻¹ below the resonance and at input intensities up to about 20 mW. For very small detunings below the resonance, and for detunings above the resonance, we do not observe any hysteresis (except at high intensities). Whole-beam bistability is observed for sample temperatures (with the laser off) as low as 5 K and as high as 50 K. For samples of larger surface areas, we are not always able to observe whole-beam bistability. These results can be explained in the following way. When the sample is illuminated with light whose frequency is less than the bound exciton resonance, the sample heats up locally as the intensity increases. This causes the resonance to shift to lower energies (closer to the laser frequency). As it shifts toward lower energies, more heat is generated in the sample owing to the increased absorption and the exciton resonance frequency continues to shift until it slightly overshoots the frequency of the laser. The transmission suddenly switches from a high state to a low state due to this rapid shift in the excitonic frequency. The laser light is then on the high-energy side of the resonance and is substantially absorbed. This point is a stable point since, if the resonance was to continue to move away from the laser, the absorption would decrease, the sample would cool, and the resonance would move back toward the laser (negative feedback). When the intensity is reduced, switching back to a high state is obtained at a much reduced intensity because heat is stored in the sample.

We have recently developed the theory of this effect (Dagenais & Sharfin 1984) and have obtained good qualitative agreement (see figure 1). Transient measurements with a mechanical

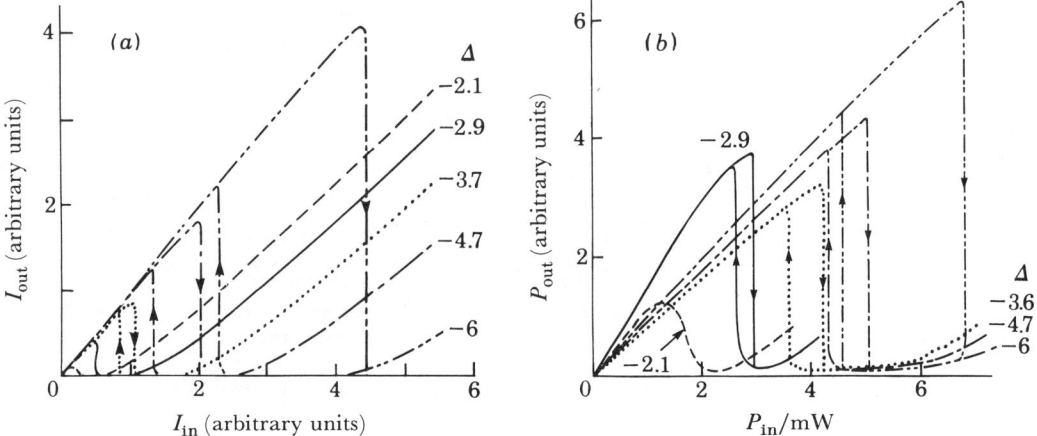

FIGURE 1. Whole-beam bistability. (a) Theoretical predictions; (b) experimental observations; $T = 7$ K.

chopper indicate switch-on and off times faster than 20 μs, our actual time resolution; switch-off times (high to low transmission) can be much faster. Switch-on times are expected to be slower and depend on the rate at which the heat is taken away from the sample. It can be improved by immersing the sample in liquid helium and by optimizing the size of the sample. This type of optical bistability is particularly suited for XOR, AND and inverted logic gate operations. It combines ease of sample preparation, low power operation and high reproducibility.

TRANSVERSE BISTABILITY

For the transverse bistability measurements, a 20 μm thick CdS sample is immersed in superfluid helium. Optical bistability can be seen for detunings below and above the bound exciton resonance. Here, we present our results for detunings of about 2 cm⁻¹ below the resonance (Dagenais & Winful 1984 a). At very low intensities, the far-field profile of the output beam from the sample is observed to be Gaussian. As the intensity is increased, the beam diameter increases and a circular fringe pattern develops. By increasing the input intensity further, the number of rings increases at the same time as the fringe diameter decreases continuously. This continues with increasing intensity until a sudden discontinuity in the position of the rings in the far field occurs. No simultaneous change of the total transmitted power through the sample is observed (see figure 2). From this we deduce that the laser beam

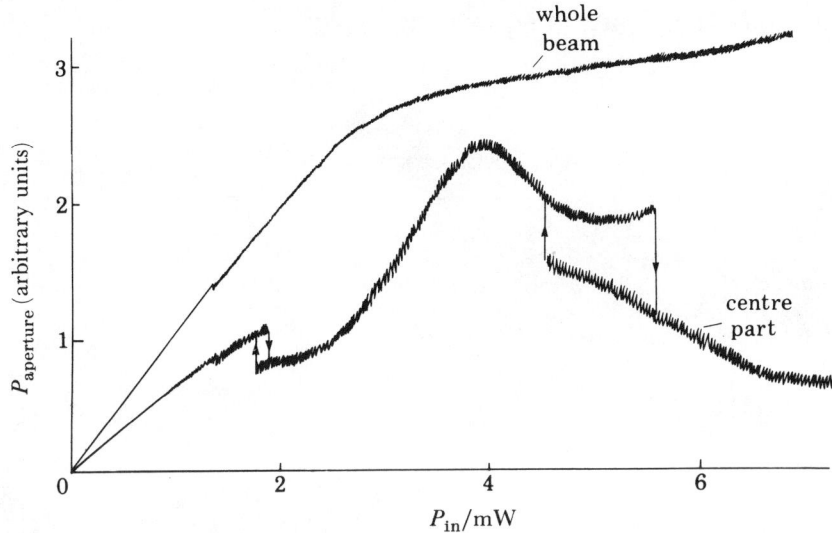

FIGURE 2. Observation of transverse optical bistability; $\nu = 20\,529.5$ cm^{-1}, $T = 2$ K.

profile within the sample must change discontinuously without a concomitant change of the total transmitted power. Such an effect has recently been predicted by Firth & Wright (1982) in connection with optical bistability in a low-finesse cavity. By monitoring the intensity of the central spot only, using an aperture and a detector in the far field, it is found that the transmitted intensity through the aperture increases monotonically with the input intensity to a point where it suddenly drops. As the input intensity is reduced, the transmitted light decreases and remains below the previous transmission until it switches back to its initial value at a lower input intensity than when it switched previously. Large hysteresis loops are observed and more than one hysteresis loop can be recorded as the intensity is varied (see figure 2). Optically induced refractive index changes as large as 0.15 are deduced from measurements of the transverse ring profile. As the incident intensity is increased, the transmitted intensity can exhibit highly regular pulsations (Dagenais & Winful 1984 b) with a period of the order of 10 μs. If the whole transmitted beam is collected with a large aperture lens, the pulsation is rather washed out. The pulsations are believed to be due to an interplay between thermal effects and the intrinsic nonlinear effects. Detailed absorption measurements were made and the intensity dependence of the absorption was studied. The saturation of the bound exciton is observed (Dagenais 1983). Sharp induced absorption lines are also seen near both the bound and the free exciton. The transmission through the crystal can be changed by a factor of three when the laser intensity is increased from a low value to a high one (*ca.* 10 kW cm^{-2}). Gated absorption measurements reveal that an important fraction of the induced absorption can appear in 10 ns or less. This demonstrates that this is a highly nonlinear system with a potentially fast response time.

Conclusion

In conclusion, large nonlinear effects at low incident power are seen near the band gap of CdS. The observed whole-beam bistability does not require a cavity or external feedback and combines ease of sample preparation, low power operation and high reproducibility. Clear evidence for spatial bistability without whole-beam bistability is also reported.

The author acknowledges W. F. Sharfin and H. G. Winful for informative, stimulating discussions. The author is indebted to Dr D. C. Reynolds of Wright Patterson Air Force Base for his generosity in providing very high quality cadmium sulphide platelets.

References

Dagenais, M. 1983 *Appl. Phys. Lett.* **43**, 742–744.
Dagenais, M. & Sharfin, W. F. 1984 *Appl. Phys. Lett.* (In the press.)
Dagenais, M. & Winful, H. G. 1984*a* *Appl. Phys. Lett.* **44**, 574–576.
Dagenais, M. & Winful, H. G. 1984*b* In *Optical Bistability*, vol. 2 (ed. C. M. Bowden, H. M. Gibbs & S. L. McCall), pp. 267–272. New York: Plenum Press.
Firth, W. J. & Wright, E. 1982 *Optics Commun.* **40**, 233–238.

Phil. Trans. R. Soc. Lond. A **313**, 269–275 (1984)

Printed in Great Britain

Nonlinear and optically bistable effects in cadmium sulphide

By C. Klingshirn, K. Bohnert, K. Kempf and H. Kalt

Physikalisches Institut der Universität, Robert-Mayer-Straße 2-4,
D-6000 Frankfurt am Main, F.R.G.

Among the numerous optical nonlinearities known for CdS, we select here those connected with the transition from a low-density gas of excitons to an electron–hole plasma. The properties of the plasma are reviewed and it is demonstrated that the variations of the indices of absorption and refraction, which occur in this phase transition, allow for the realization of optical bistability.

1. Introduction

The invention of lasers gave a strong impetus to the investigation of nonlinear optical phenomena in semiconductors and insulators. The results of the first years have, for example, been reviewed by Bloembergen (1965). The common availability of spectrally narrow tunable dye lasers allowed in the last decade for detailed investigations in the spectral region of electronic resonances close to the absorption edge. In this region, the optical properties of semiconductors are dominated by exciton states. Excitons are bound electron–hole pair states analogous to the positronium or hydrogen atoms. For more details see, for example, Klingshirn & Haug (1981) and the references therein.

The reasons for the optical nonlinearities are manifold, but they can roughly be classified in three categories (for CdS see, for example, Kempf *et al.* (1981) and Kreissl *et al.* (1982)):

(i) excitation-induced modifications of resonances that exist already at low intensities, like a broadening of exciton resonances;

(ii) the appearance of new resonances with increasing excitation like those connected with two-photon transitions to the excitonic molecule;

(iii) the phase transition of a low-density gas of excitons to an electron–hole plasma.

In this paper we shall concentrate on the last point. We have selected CdS as a typical example for a direct-gap semiconductor. In §2 we summarize the properties of the electron–hole plasma, in §3 we present the variations of the optical properties connected with its formation and their application in the formation of laser-induced gratings, in §4 we present optical bistability and in §5 we give an outlook to further research activities and possible technical applications.

2. Properties of the electron–hole plasma

If electron–hole pairs are excited in a semiconductor at low density and temperature, they will form excitons. If the generation rate is increased, the density of excitons may become so high that their mean separation is comparable to their Bohr radius. Then excitons are no longer individual quasi-particles, but form a new collective phase, the electron–hole plasma. This transition from the insulating exciton gas to the metallic plasma-phase is also referred to as Mott transition.

This phase transition has consequences for the band structure. It has been found (see, for example, the literature cited by Klingshirn & Haug (1981)) that the width of the forbidden gap E_g is a monotonical decreasing function of the electron–hole pair density in the plasma n_p. The chemical potential of the plasma μ (the energetic distance between the quasi-Fermi levels of electrons and holes) depends strongly on n_p and the plasma temperature T_p. An analysis of the functions $E_g(n_p)$ and $\mu(n_p, T_p)$ shows that under thermodynamic quasi-equilibrium conditions there is a first-order phase transition below a critical temperature T_c from a low density exciton gas to an electron–hole plasma liquid. This phase transition is similar to the gas–liquid transition of real gases and has been analysed in great detail for the indirect-gap semiconductors Ge and Si. See, for example, the reviews written by Rice (1977) and by Hensel et al. (1977). In these materials the evolution of the liquid-like phase in the form of plasma droplets is favoured by the long lifetime of electrons and holes.

In direct-gap materials, and especially in CdS, it has been found that a plasma is formed, which, however, does not reach its liquid-like state because at the short lifetime of carriers of only about 100 ps (see Bohnert et al. (1981) and Kempf & Klingshirn (1984); Kempf (1984)). These authors find, in contrast to what is expected for a liquid-like plasma state, that n_p increases at constant T_p with increasing excitation intensity I_{exc} and at constant I_{exc} with increasing T_p. Because of the high optical gain $g(g \gtrsim 10^4 \text{ cm}^{-1})$ connected with a degenerate electron–hole plasma in direct-gap semiconductors, only small volumes of some hundreds of cubic micrometres can be filled with plasma. Some authors speculated about a fast expansion of the plasma at the border of the excited volume over distances d of 100 μm and with expansion velocities v_d beyond 10^7 cm s^{-1}. For CdS it has been shown by Kempf & Klingshirn (1984) that the values of v_d and d are considerably smaller. A consistent set of plasma parameters for lattice temperatures around 5 K, which also fulfils the necessary condition

$$n_p = I_{exc} \tau_p / d\hbar\omega_{exc},$$

where τ_p is the lifetime of the carrier pairs in the plasma, and $\hbar\omega_{exc}$ is the photon energy of the exiting laser with intensity $I_{exc}(\hbar\omega_{exc} > \mu)$, is the following:

$$0.5 \text{ MW cm}^{-2} \lesssim I_{exc} \lesssim 5 \text{ MW cm}^{-2};$$
$$10^{18} \text{ cm}^{-3} \lesssim n_p \lesssim 3 \times 10^{18} \text{ cm}^{-3};$$
$$10 \text{ K} \lesssim T_p \lesssim 35 \text{ K};$$
$$\tau_p = (150 \pm 50) \text{ ps};$$
$$v_D \approx 2 \times 10^6 \text{ cm s}^{-1} \quad \text{and} \quad d = (3 \pm 2) \text{ μm}.$$

This set of data has been deduced by Kempf & Klingshirn (1984) from spatially resolved transmission and reflection experiments. Deviations from this set of parameters will lead to internal contradictions. It should be noted that the observed values of n_p in CdS are slightly smaller than the calculated equilibrium density of the plasma, n_0, of $5 \times 10^{18} \text{ cm}^{-3}$ (Rösler & Zimmermann 1977). In CdSe, the value of n_0 is about one order of magnitude smaller. Here it is possible to reach values $n_p > n_0$ before crystal damage occurs. Preliminary results of spatially resolved gain spectroscopy indicate values of d of several micrometres (Majumder et al. 1984). Experiments with a spatially confined plasma help to avoid problems connected with plasma expansion, as has been demonstrated for $Ga_{1-x}Al_xAs$ by Capizzi et al. (1983).

3. Variations of the optical properties in the plasma

The variations of the absorption and refraction spectra around the absorption edge owing to the transition from a low-density exciton gas to an electron–hole plasma have been investigated in detail for CdS by Kreissl et al. (1982) for fixed $\hbar\omega_{exc}$.

For purposes of optical bistability, it is necessary to know the variations at the photon energy of the exciting laser itself. The corresponding experiments and their results are presented in this section. The measurement of the absorption coefficient at the spectral position of the incident laser is rather trivial. The variation of the refractive index is determined by a weak probe beam, which is sent on the sample in the centre of the excitation spot. In contrast to the experiments of Kreissl, samples with a thickness between 0.5 and 2.5 μm are used, which are homogeneously excited within this range . The transmitted probe beam is nicely modulated by the Fabry–Perot interference structure produced by the platelet type samples. From the position of the interference maxima and minima and their shift with increasing excitation it is possible to deduce $n(\omega)$ and $\Delta n(\omega_{exc}, I_{exc})$. Obviously this technique works only in the spectral region where the sample is transparent. Figure 1a shows the variations of the absorption coefficient $\Delta\alpha(\hbar\omega_{exc})$ for $I_{exc} = 2$ MW cm^{-2}. Since the unexcited sample is transparent below 2.55 eV, $\Delta\alpha(\hbar\omega_{exc}) = \alpha(\hbar\omega_{exc})$. Figure 1b gives the refractive index $n(\omega)$ of the unexcited sample, while figure 1c shows its variation $\Delta n(\omega_{exc})$ for $I_{exc} = 2$ MW cm^{-2}. One finds that Δn is negative due to the disappearance of the exciton resonance; however, with the onset of a resonance-like structure if $\hbar\omega_{exc}$ approaches the onset of absorption.

If two coherent monochromatic laser beams of equal frequency and intensity are falling on the sample under a small angle (see insert of figure 2), they produce an interference pattern. In the regions of constructive interference, the intensity is very high, producing the variation $\Delta\alpha$ or Δn shown in figure 2, while the intensity is small in between. Thus a laser-induced grating is formed. Figure 2 gives the efficiency of this grating for the first order as a function of $\hbar\omega_{exc}$. By comparing figure 1a and c and figure 2 it is obvious that an absorptive grating is induced for $\hbar\omega_{exc}$ between 2.545 eV and 2.540 eV, while a phase grating is formed below; however, with lower efficiency. More details about this experiment and about the influence of the other nonlinearities, mentioned in §1, on laser induced gratings will be published by Kalt et al. (1984).

4. Optical bistability

The excitation-induced renormalizations of the refraction and absorption coefficients shown in figure 1 give rise to two different types of optical bistability. Bohnert et al. (1983) found first experimental indications of a dispersive optical bistability in Fabry–Perot-type CdS platelets at photon energies slightly below 2.54 eV and a new type of intrinsic absorptive bistability above. Dispersive optical bistabilities in semiconductors have been predicted by a number of authors (for example, Koch et al. 1981; Hanamura 1981) and are treated in several contributions to this meeting. We therefore concentrate here on the absorptive bistability. First hints have been found by Rossmann et al. (1983) in terms of an optical memory effect, but without a clear interpretation. Detailed experimental investigations, with rather poor temporal resolution, led Bohnert et al. (1983) to the correct qualitative interpretation, which will be given here together with new experiments. First quantitative calculations have been made by H. Haug and coworkers (see, for example, Schmidt et al. (1984) and Haug, this symposium). Later, Henneberger & Rossmann (1984) also adapted this interpretation.

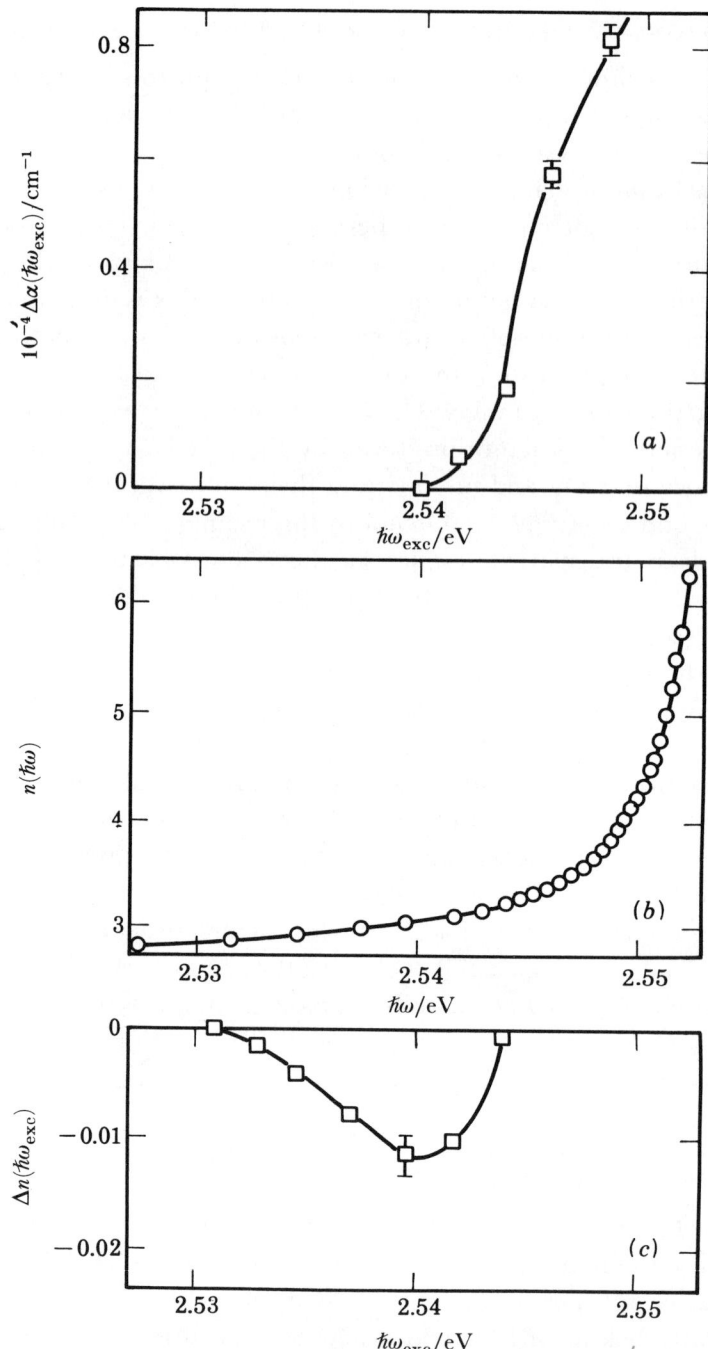

FIGURE 1. The variation of the absorption coefficient $\Delta\alpha$ of CdS as a function of photoenergy of the exciting laser $\hbar\omega_{\text{exc}}$ for constant excitation intensity $I_{\text{exc}} = 2$ MW cm^{-2} (a); the refractive index $n(\hbar\omega)$ of an unexcited sample (b); its variation $\Delta n(\hbar\omega_{\text{exc}})$ for $I_{\text{exc}} = 2$ MW cm^{-2} (c). $(\boldsymbol{E}\perp\boldsymbol{c}, \boldsymbol{k}\perp\boldsymbol{c}, T_{\text{L}} \approx 5$ K.)

The basic physical mechanism is now described. An incident laser pulse $I_0(t)$ incident on the sample is situated in the spectral region where the crystal is transparent for low excitation but becomes absorbing at high excitation. According to figure 1 this is, for example, the region between 2.55 eV and 2.54 eV. When the pulse rises, the sample is transparent. Owing to weak absorption in the tail of the absorption edge or owing to two-photon absorption, carriers are

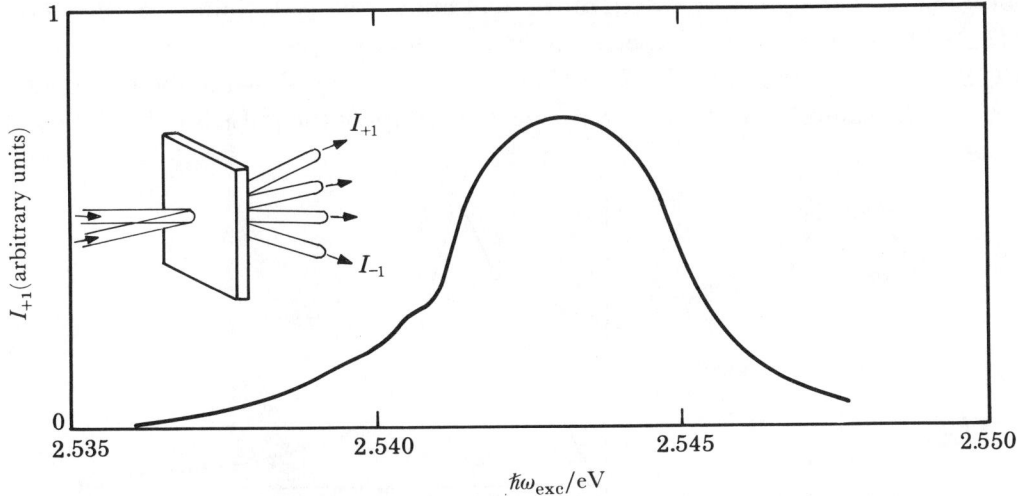

FIGURE 2. The spectral dependence of the efficiency of a laser-induced grating in CdS ($T = 5$ K, $E \perp c$, $k \perp c$, $I_{exc} = 1$ MW cm^{-2}).

created, but with low efficiency. If their density becomes sufficiently high, an electron–hole plasma is formed and strong one-photon absorption sets in because of the reduction of the band gap. The sample now becomes opaque and the transmissivity is switched down to low values. Since the one-photon absorption allows pumping of the plasma to be much more effective than the two-photon absorption, the sample remains in the state of low transmission down to excitation intensities, which are considerably below those necessary to produce the transition from high to low transmissivity. Thus the sample has two stable states, one of high and one of low transmissivity, i.e. optical bistability is achieved. This absorptive type of bistability needs no external feedback, for example in the form of a cavity. The 'feedback' occurs intrinsically in the electronic system.

Figure 3 shows the temporal shape of an incident and a transmitted pulse, $I_0(t)$ and $I_t(t)$, respectively, of an excimer-laser pumped dye-laser. The pulses are registered by an Imacon

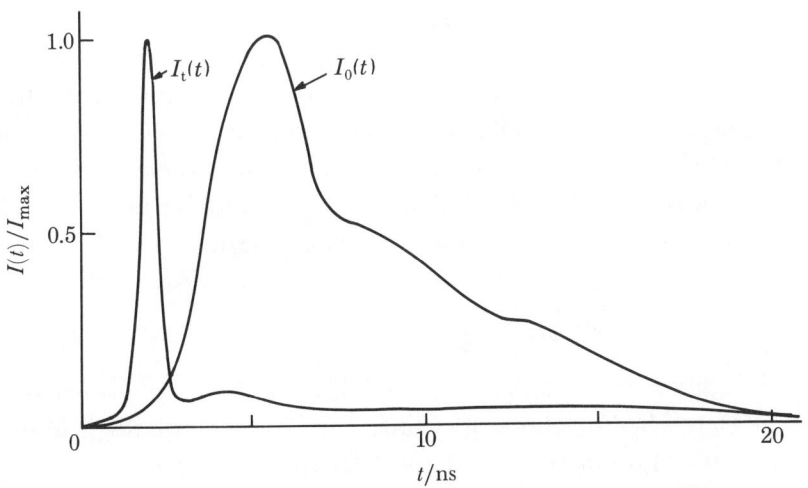

FIGURE 3. The temporal shape of incident and transmitted pulses $I_0(t)$ and $I_t(t)$. Thickness of the CdS sample is 2 μm; pulse-height normalized; $T = 5$ K, $\hbar\omega_{exc} = 2.548$ eV, $E \perp c$, $k \perp c$, $I_{0max} = 5$ MW cm^{-2}.

500 streak camera. The apparent switch-off time of about 200 ps is limited by the detection system. The real switching time is expected to be considerably lower.

From $I_0(t)$ and $I_t(t)$ a time-free plot $I_t = f(I_0)$ can be deduced, showing the hysteresis loop. Figure 4 gives two examples. In figure 4a, the peak intensity of the pulse is only slightly above

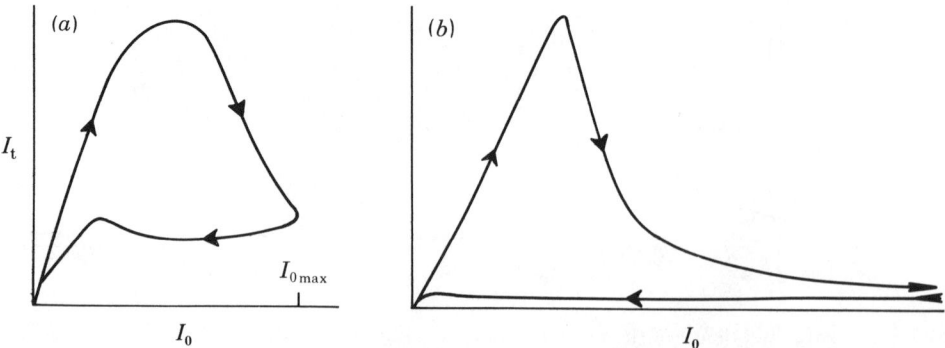

FIGURE 4. Hysteresis loops due to intrinsic optical bistability for (a) $I_{max} = 0.25$ MW cm^{-2} and (b) 10 MW cm^{-2}. Intensities are given on all axes in arbitrary linear units (CdS, $T = 5$ K, $\hbar\omega_{exc} = 2.548$ eV, $E \perp c$, $k \perp c$).

the value for switching 'off'. In figure 4b the peak intensity of the pulse is more than an order of magnitude larger. It should be noted that the hysteresis loop is revolved clockwise in both cases, this means in the opposite direction to that for dispersive bistability or for an absorptive bistability, which is due to bleaching of an absorption band. Both cases need the feedback of a cavity.

5. OUTLOOK

Although the principal physical mechanism of this type of intrinsic absorptive optical bistability is rather clear now, a lot of details have to be investigated in the future, among others the actual switching times, the conditions for down switching, the holding power for the state of low transmission, the upper temperature limit at which the device works or the influence of the sample thickness compared to the diffusion length of the carriers in the plasma.

It should be pointed out that this optical device is presumably of great technical interest. The bistable operation can be used for fast optical storage. If the width of the hysteresis loop is made small, it can be used for the realization of the logic connection NOT, while the dispersive bistability realizes AND and OR. As has been mentioned already, there are indications for the latter phenomenon at longer wavelength. So one device could be used for different logic connections depending on the wavelength. If only one wavelength is available, the use of mixed crystals or graded-gap materials of the type $CdS_{1-x}Se_x$ or $Zn_{1-x}Cd_xS$ will allow for various functions depending on x.

This work is a project of the Sonderforschungsbereich 65, Festkörperspektroskopie, financed by special funds of the Deutsche Forschungsgemeinschaft. The high quality samples have been grown in the Kristall- und Materiallabor of the University of Karlsruhe.

REFERENCES

Bloembergen, N. 1965 *Nonlinear optics*. New York: W. A. Benjamin Inc.

Bohnert, K., Anselment, M., Kobbe, G., Klingshirn, C., Haug, H., Koch, S. W., Schmitt-Rink, S. & Abraham, F. F. 1981 *Z. Phys.* B**42**, 1.

Bohnert, K., Kalt, H. & Klingshirn, C. 1983 *Appl. Phys. Lett.* **43**, 1088.

Capizzi, M., Frova, A., Martelli, F., Modesti, S., Quagliano, L. G., Staehli, J.-L., Guzzi, M., Logan, R. A. & Chiaretti, G. 1983 *Physica* **117**B/**118**B, 333.

Hanamura, E. 1981 *Solid St. Commun.* **38**, 939.

Henneberger, F. & Rossmann, H. 1984 *Physica Status Solidi* B **121**, 685.

Hensel, J. C., Philips, T. G. & Thomas, G. A. 1977 *Solid St. Physics* **32**, 88.

Kalt, H., Lyssenko, V. G., Renner, R. & Klingshirn, C. 1984 *Solid. St. Commun.* (In the press.)

Kempf, K. 1984 Ph.D. thesis, University of Frankfurt.

Kempf, K. & Klingshirn, C. 1984 *Solid St. Commun.* **49**, 23.

Kempf, K., Schmieder, G., Kurtze, G. & Klingshirn, C. 1981 *Physica Status Solidi* B**107**, 297.

Klingshirn, C. & Haug, H. 1981 *Physics Rep.* **70**, 315.

Koch, S. W. & Haug, H. 1981 *Phys. Rev. Lett.* **46**, 450.

Kreissl, A., Bohnert, K., Lyssenko, V. G. & Klingshirn, C. 1982 *Physica Status Solidi* B**114**, 537.

Majumder, F., Swoboda, E., Kempf, K. & Klingshirn, C. 1984 (Submitted).

Rice, T. M. 1977 *Solid St. Phys.* **32**, 1.

Rösler, M. & Zimmermann, R. 1977 *Physica Status Solidi* B**83**, 85.

Rossmann, H., Henneberger, F. & Voigt, J. 1983 *Physica Status Solidi* B**115**, 63.

Schmidt, H. E., Haug, H. & Koch, S. W. 1984 *Appl. Phys. Lett.* **44**, 787.

Phil. Trans. R. Soc. Lond. A **313**, 277–284 (1984)

Printed in Great Britain

Fast and sensitive nonlinear processes: bistability in CdHgTe

By A. Miller and G. Parry†

Royal Signals and Radar Establishment, St Andrews Road, Great Malvern WR14 3PS, U.K.

We discuss the origin and characteristics of the very large intensity-dependent changes in the refractive index of the semiconductor cadmium mercury telluride at 10.6 μm. With continuous wave (c.w.) radiation and sample temperatures of 77 K nonlinear Fabry–Perot effects can be observed at powers as low as 1 mW. The Auger dominated recombination process and the thermal contributions to the refractive index make observations of optical bistability difficult with c.w. radiation. However, the use of a pulsed CO_2 laser and two-photon absorption of the radiation does lead to the observation of bistability at room temperature. We compare these results with the reported work on InSb.

1. Introduction

The alloy semiconductor $Cd_x Hg_{1-x} Te$ offers a variation of band gap with composition from 0 to 1.5 eV (Dornhaus & Nimtz 1976). This wide range thus offers energy gaps resonant with a number of important laser frequencies and in particular it is one of the few semiconductors suitable for the study of band-gap resonant optical nonlinearities in the CO_2 laser output band at around 10 μm (Hill *et al.* 1982). CdHgTe crystallizes in the zincblende structure so that small-gap alloys of this material have energy bands closely related to that of InSb, a semiconductor that shows extremely large band-filling optical nonlinearities at 5 μm (D. A. B. Miller *et al.* 1978, 1981). This makes CdHgTe an obvious choice as a material in which the band-gap resonant nonlinearities discovered in InSb and subsequently used to demonstrate various optically bistable and related devices (Walker *et al.*, this symposium) can be studied with the flexibility of varying the band-gap energy as a function of both alloy composition and temperature. In this paper we discuss the origin of large optical nonlinearities in CdHgTe at 10 μm, report nonlinear Fabry–Perot response under different experimental conditions and make some comparisons with InSb.

With the use of a low-power c.w. CO_2 laser, a $Cd_{0.23}Hg_{0.77}Te$ etalon at 77 K showed nonlinear features at input powers as low as 1 mW. With pulsed lasers, optical switching and bistability could be observed in the transmission of a room-temperature etalon. As predicted by theory, the resonant nonlinearity was found to be larger than for InSb for this smaller band gap semiconductor, but we also found that the carrier recombination mechanism and thermal properties of CdHgTe give an intensity-dependent etalon response quite distinct from that of InSb.

2. Optical nonlinearities in CdHgTe

In semiconductors, numerous mechanisms may result in optical nonlinearities (A. Miller *et al.* 1981; Jain 1982), and a number of different nonlinear effects have been observed in CdHgTe under a variety of different conditions. Initial interest centred on the stimulated spin-flip Raman

---

† Present address: Electronic and Electrical Engineering Department, University College London, Torrington Place, London WC1E 7JE, U.K.

process which can be described in terms of a third-order nonlinear susceptibility, $\chi^{(3)}$. Kruse *et al.* (1979) achieved tunable laser action by using a Q-switched CO_2 laser to pump a band-gap resonant sample of $Cd_{0.23}Hg_{0.77}Te$ at 4 K in a high magnetic field. Under similar conditions, but making use of two different laser frequencies, ω_1 and ω_2, resonant four-wave mixing of type $\omega_3 = 2\omega_1 - \omega_2$ could be observed when $\omega_2 - \omega_1$ equalled the energy separation of spin-up and spin-down states of a given Landau level, and a value of $\chi^{(3)} = 10^{-4}$ e.s.u.† was deduced for this spin-resonant mixing process (Kruse 1981).

More recently, interest has turned to phase conjugation by degenerate four-wave mixing (d.f.w.m.). Initially, high-power pulsed lasers were employed. Khan *et al.* (1980) observed 9% phase conjugate reflectivities of 10.6 µm radiation at 12, 77 and 295 K with n-type $Cd_xHg_{1-x}Te$ with $x = 0.216$ and 0.232. Although some of these conditions were band-gap resonant, interpretation was made in terms of a non-resonant nonlinearity arising from conduction-band non-parabolicity and values of $\chi^{(3)}$ in the range 5×10^{-8} to 4×10^{-7} e.s.u. were deduced. Independently, Jain & Steel (1980) achieved similar results at room temperature under band-gap resonant conditions. They concluded that this was consistent with a phase grating produced by the generated free-carrier plasma refraction with a value of $\chi^{(3)} \approx 5.4 \times 10^{-6}$ e.s.u. Limiting of the reflected d.f.w.m. pulses at 10% (Khan *et al.* 1980; Jain & Steel 1980) has been studied recently by Yuen & Becla (1983) in p-type CdHgTe under band-gap resonant conditions and interpreted in terms of the effect that bleaching the inter-band absorption has on the refractive index change. Also, non-degenerate four-wave mixing of type $\omega_3 = 2\omega_1 - \omega_2$ with two CO_2 lasers by Yuen (1982) in $Cd_{0.265}Hg_{0.735}Te$ at 2 °K (band gap 0.183 eV) showed a variation of $\chi^{(3)}$ with $\Delta\omega = \omega_1 - \omega_2$. For small $\Delta\omega$, $\chi^{(3)}$ was considered to be due to band non-parabolicity for free electrons generated by two-photon absorption, whereas for $\Delta\omega > 10$ cm^{-1}, the origin of the nonlinearity was thought to be due to bound electrons with $\chi^{(3)} \approx 3 \times 10^{-8}$ e.s.u.

Much larger free carrier plasma nonlinearities, $\chi^{(3)} \approx 5 \times 10^{-2}$ e.s.u., were discovered by Jain *et al.* (1981, 1982) in d.f.w.m. experiments by lowering the temperature of $Cd_xHg_{1-x}Te$ ($x = 0.21$–0.23) to tune the band gap to the output of a c.w. CO_2 laser. A lower temperature also has the effect of increasing the Auger-dominated lifetime of the free carrier plasma, thus enhancing the magnitude of the nonlinearity. Similar d.f.w.m. experiments by Khan *et al.* (1981) gave $\chi^{(3)} \approx 3 \times 10^{-2}$ e.s.u.

Even larger optical nonlinearities were observed by Hill *et al.* (1982) in band-gap resonant self-defocusing studies of $Cd_{0.21}Hg_{0.79}Te$ at 175 K. Refractive index changes were found to vary according to $\Delta n = -7 \times 10^{-3}I^{\frac{1}{3}}$ (I in watts per square centimetre) and changes in beam profiles in the far-field were observable at 1 W cm^{-2} corresponding to an effective $\chi^{(3)} \approx 6$ e.s.u. at this intensity. This extremely large nonlinearity arises from both a free carrier plasma contribution and the effect of band filling, and the power dependence is explained by the role of Auger recombination limiting the accumulation of generated carriers as the intensity is increased (§3).

Preliminary evidence of nonlinear Fabry–Perot transmission in CdHgTe has been reported by Jain & Steel (1982) and Khan *et al.* (1983).

† 1 e.s.u. = 1 cm^3 erg^{-1} $\equiv 1.4 \times 10^{-8}$ m^2 V^{-2}.

3. Band-gap resonant nonlinear refraction in CdHgTe

For photon energies less than the band gap energy of a semiconductor, a change in refractive index, Δn, may be expected to result when an excess density of electron–hole pairs, ΔN, is created. In CdHgTe, a large refractive index change can arise from the free carrier plasma. This can be adequately described by the Drude expression derived from standard dispersion theory (Jain & Klein 1979):

$$\sigma_p = -e^2/2\epsilon_0 n_0 m^* \omega^2, \tag{1}$$

where σ_p gives the refractive index change per electron–hole pair per unit volume, e is the electronic charge, ϵ_0 the free space dielectric constant, n_0 the linear refractive index, m^* the conduction band effective mass and ω the photon frequency. For a semiconductor with small band gap this is a particularly large effect because of the low effective mass and small photon energy. An additional important contribution to the nonlinear refraction for photon energies close to the band gap is filling of the conduction band states resulting in blocking of interband transitions. This effect has been analysed in some detail in InSb (D. A. B. Miller *et al.* 1981) and found to be consistent with the expression derived by using the Kramers–Kronig integration to relate the changes in refraction with changes in absorption:

$$\sigma_s = \sigma_p \left\{ \frac{2}{3\sqrt{\pi}} \frac{m^*}{m} \left(\frac{mP^2}{\hbar} \right) \frac{1}{kT} J\left(\frac{\hbar\omega - E_g}{kT} \right) \right\}, \tag{2}$$

where P is the momentum matrix element and

$$J(a) = \int_0^\infty \frac{\mathrm{d}x \, x^{\frac{1}{2}} e^{-x}}{(x-a)}.$$

Sub-band excitation can cause the creation of excess carriers either via band-tail states or by two-photon absorption. For the 77 K, $Cd_{0.23}Hg_{0.77}Te$ samples employed in the experiments described in §4, the photon energy is within $3.5 \, kT$ of the band gap energy for 10.6 µm radiation and should result in $\sigma_p \approx -1.1 \times 10^{-18} \text{ cm}^3$ and $\sigma_s \approx -1.2 \times 10^{-18} \text{ cm}^3$. The same composition at room temperature ($E_g = 0.2$ eV) is suitable for two photon excitation and should give $\sigma_p \approx -7.4 \times 10^{-19} \text{ cm}^3$ and $\sigma_s \approx -3.7 \times 10^{-19} \text{ cm}^3$ at 10.6 µm.

To a first approximation, these free carrier nonlinearities are proportional to the density of excess carriers so that

$$\Delta n = (\sigma_p + \sigma_s) \, \Delta N. \tag{3}$$

The density of carriers in a semiconductor under optical excitation is governed by the dynamic balance of generation, recombination and diffusion. Under the conditions employed in our experiments, diffusion can be neglected. Recombination in CdHgTe at both 77 and 295 K is dominated by Auger processes, which cause a strong concentration-dependence for the carrier lifetime. This therefore has a significant effect on the intensity dependence of the nonlinearity and in turn the form of the nonlinear Fabry–Perot characteristics. For our 77 K samples ($N_D - N_A = 1 \times 10^{15} \text{ cm}^{-3}$), the small-modulation Auger lifetime is about 2.5 µs and to a good approximation we should expect the refractive index change to be governed by

$$\Delta n = \gamma I^{\frac{1}{3}}, \tag{4}$$

where $\gamma = -5 \times 10^{-3} \, (\text{cm}^2 \, \text{W}^{-1})^{\frac{1}{3}}$. This intensity dependence was confirmed in self-defocusing measurements in which the $I^{\frac{1}{3}}$ dependence is seen in the power dependence of far-field beam widths (figure 1) (Hill *et al.* 1982). It has the effect of making optical bistability more difficult to achieve because spacing of the cavity orders with intensity becomes larger and the slope of the transmission curve with cavity intensity is shallower in all orders than for a nonlinearity in which the refractive index change is proportional to intensity (Miller & Parry 1984).

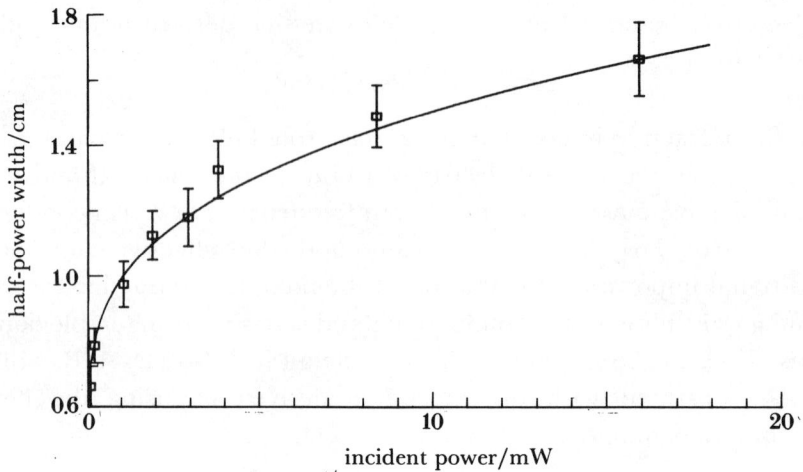

FIGURE 1. Power dependence of far-field beam widths for 10.6 μm radiation after transmission through 330 μm thick $Cd_{0.21}Hg_{0.79}Te$ at 175 K placed just beyond a beam waist of 200 μm (f.w.h.m.). The solid line shows a one-third power dependence for comparison (Hill *et al.* 1982).

4. LOW-POWER NONLINEAR FABRY–PEROT

We have found that a 200 μm thick 77 K etalon of $Cd_{0.23}Hg_{0.77}Te$ with polished faces gives a nonlinear response at a c.w. CO_2 laser power level less than 1 mW (75 μm f.w.h.m. spot diameter), in good agreement with the expected band-gap resonant nonlinearities (§3) (Parry *et al.* 1984). We have also studied a sample with a high-reflectivity metal coating applied to the rear surface to enhance the etalon finesse (Miller *et al.* 1984). This gave minimum and maximum reflectivities of 6 and 62 % respectively in the 10.6 μm region. A 20:1 mark–space ratio chopper was placed before the sample to reduce thermal contributions. Figure 2 shows the total reflected power plotted against the input power at a wavelength of 10.48 μm for a given initial cavity tuning adjusted by making use of the small variation of optical thickness across the sample. The negative slope at 30–40 mW predicts an optical gain of 2.2. This nonlinear characteristic occurs at incident powers consistent with that predicted by a theoretical analysis of band-gap resonant nonlinearities including Auger recombination. An estimate of the nonlinear coefficient can be deduced from the incident intensities at which the maximum and minimum reflected powers occur, giving $\gamma = -3 \times 10^{-3}$ $(cm^2\ W^{-1})^{\frac{1}{3}}$, in good agreement with the calculated value of -5×10^{-3} (§3).

There are two principal difficulties in attaining optical bistability in CdHgTe with a c.w. laser. Auger recombination severely limits the size of the nonlinearity above about 10 W cm^{-2}, which causes higher-order resonances to occur at relatively high powers. The low thermal conductivity of CdHgTe means that these higher orders are difficult to reach. The second difficulty relates to the negative sign for the temperature coefficient of the refractive index. This is opposite to most other semiconductors and has the result that as a cavity approaches resonance due to the negative electronic nonlinearity, the rapidly increasing power absorbed in the etalon pushes the cavity further into resonance. Thus even a small thermal contribution will always force the cavity into resonance. This is avoided in the case illustrated in figure 2 because the chopper allows the heat to dissipate between successive pulses. However, altering the initial tuning to give steeper nonlinear characteristics resulted in the etalon's always being swept into resonance (low reflectivity) by the thermal contribution. Nevertheless it should be possible to demonstrate differential gain, modulation transfer and optical logic functions at 10 μm with the characteristic shown in figure 2 if synchronously chopped beams are employed.

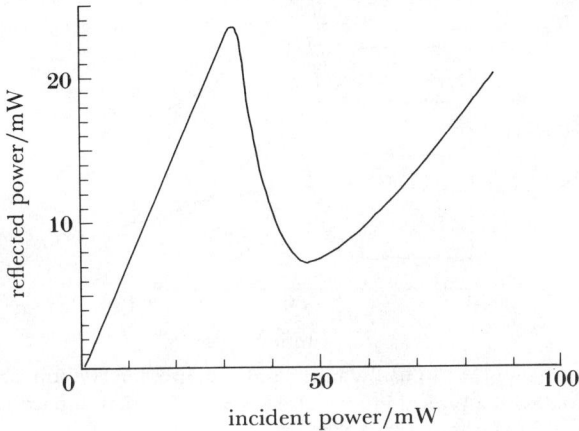

FIGURE 2. Input–output characteristic in reflection of 10.49 μm radiation with the use of a 208 μm thick $Cd_{0.23}Hg_{0.77}Te$ etalon at 77 K with a high reflectivity coating on the rear face (Miller *et al.* 1984).

5. TWO-PHOTON INDUCED OPTICAL SWITCHING AND BISTABILITY

The excitation of excess carriers by two-photon absorption has the attraction of causing refractive index changes at wavelengths where the linear absorption of the material is low. Optical switching by this mechanism has been reported recently in the transmission of a room-temperature InSb etalon at a few hundred kilowatts per square centimetre (Kar *et al.* 1983). The increase in band gap energy with temperature for CdHgTe makes the alloy composition used for the studies described in §4 ideal for room-temperature two-photon excitation. At room temperature $Cd_{0.23}Hg_{0.77}Te$ is intrinsic with concentration 1.5×10^{16} cm^{-3} and has a low-excitation Auger lifetime of 90 ns. An uncoated sample 200 μm thick was studied in transmission by using a short-cavity TEA CO_2 laser (Mathew *et al.* 1984). An estimated two-photon absorption coefficient of 5 cm MW^{-1} and the calculated nonlinear refraction (§3) predicts that 2.3×10^{16} cm^{-3} excess carriers (requiring a cavity intensity of *ca.* 100 kW cm^{-2}) should cause a single-pass phase change of π. At this density, the carrier lifetime will be approximately 20 ns, i.e. close to the laser pulse width.

Figure 3 shows the input pulse temporal profile (*a*) and transmitted profiles (*b–f*) for increasing peak incident intensities. Only the central region of the beam profiles were monitored with fast small-area detectors. Nonlinear Fabry–Perot effects are observed in these profiles at 100 kW cm^{-2} peak intensities as a dynamic modulation of the transmission as the intensity varies within the pulse. The maximum carrier concentration (and thus the maximum phase change) will occur after the peak of the input pulse owing to the effect of carrier lifetime. Thereafter the cavity will retrace its transmission state during the tail of the laser pulse on a timescale

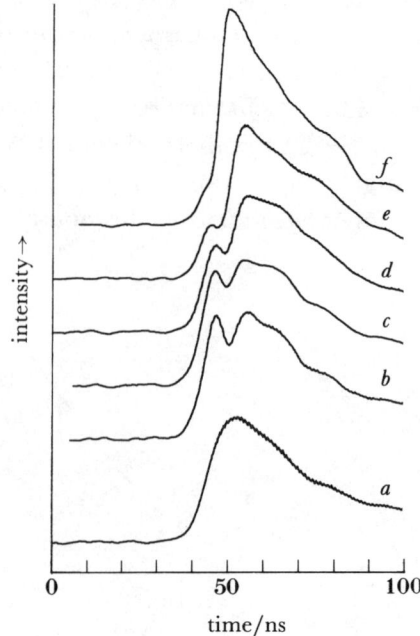

FIGURE 3. Transmitted 10.6 μm temporal (30 ns f.w.h.m.) pulse shapes for 208 μm thick, 300 K, $Cd_{0.23}Hg_{0.77}Te$ with polished faces compared with a typical input pulse (a). Peak incident intensities: (b) 116, (c) 130, (d) 165, (e) 323, and (f) 448 kW cm⁻² (Mathew *et al.* 1984).

determined principally by the carrier recombination rate. At higher peak intensities, the first-order feature occurs earlier in the pulse and another feature appears at the peak of the pulse. At above 400 kW cm⁻², this second-order feature switches the transmission abruptly from low to high transmission in a time close to the resolution of the detection equipment. This switching is a rather complex dynamical effect due to the comparable pulse width and carrier lifetime.

Optical bistability was demonstrated by using longer duration pulses to give quasi-c.w. conditions (Craig *et al.* 1984). An uncoated sample at room temperature of the same composition and 363 μm in thickness was studied by using a hybrid TEA CO_2 laser producing 1.5 μs (f.w.h.m.) pulses. Figure 4a shows incident and transmitted temporal pulse profiles at 500 kW cm⁻² peak intensity. In this case, a 100 μm diameter pinhole placed in contact with the exit face of the sample reduced the effect of etalon inhomogeneities. The plot of output against input signals (figure 4b) extracted from the temporal pulse shapes shows hysteresis due to optical bistability.

6. CONCLUSIONS

Low-temperature CdHgTe shows an extremely large optical nonlinearity at 10.6 μm for samples with band-gap energies close to the photon energy. This is significant for all-optical device applications in the 10 μm region because of the importance and widespread use of CO_2 lasers. As predicted, the nonlinearity is larger than that measured in InSb at 5 μm. However, CdHgTe has several disadvantages for optically bistable devices. It is appropriate here to compare these two materials.

1. CdHgTe is an alloy that is difficult to grow in a homogenous crystal. A Fabry–Perot is

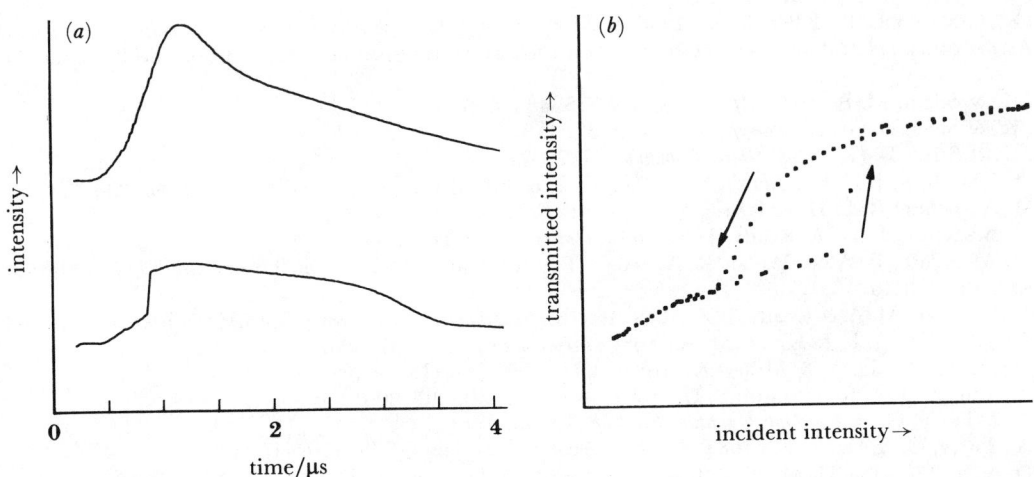

FIGURE 4. (a) Incident and transmitted temporal (1.5 µs f.w.h.m.) pulse shapes for 363 µm thick, 300 K, $Cd_{0.23}Hg_{0.77}Te$ with polished faces at 500 kW cm^{-2}. (b) Corresponding input–output characteristics showing optical bistability (Craig et al. 1984).

very sensitive to variations in refractive index with the result that etalon fringes can easily be washed out for moderate optical beam sizes unless the crystal is of very high quality.

2. The recombination mechanism at low temperature in CdHgTe is dominated by the non-radiative Auger mechanism, which is very dependent on carrier density, whereas low-temperature recombination in InSb is governed by a Shockley-Read process independent of carrier density at 77 K. Recombination in InSb at 77 K should however, become Auger-dominated once approximately 5×10^{15} cm^{-3} electron–hole pairs are excited.

3. The principal reason why c.w. optical bistability is very difficult to achieve in CdHgTe relates to the band-gap dependence with temperature. The negative refractive index change for both electronic and thermal contributions gives a positive feedback, which tends to sweep the cavity into resonance at a fixed input power. This is opposite to most other semiconductors in which the thermal refractive index coefficients are positive, and indeed the thermal effects will tend to stabilize the output on a steep nonlinear Fabry–Perot characteristic in these materials.

4. Pulsed inputs can overcome thermal problems and CdHgTe etalons behave in a similar way to InSb in two-photon induced optical bistability at room temperature. Carrier recombination in both materials is governed by Auger processes and the lifetimes are determined principally by the band-gap energies. A nonlinear refraction dependence according to $I^{\frac{2}{3}}$ results. Here CdHgTe has the advantage of the flexibility it offers in band-gap variation with composition to optimize the two-photon absorption coefficient, refractive index change per excited carrier and carrier lifetime.

We thank Mullard Limited for providing CdHgTe samples.

REFERENCES

Craig, D., Mathew, J. G. H., Kar, A. K. & Miller, A. 1984 (In preparation.)
Dornhaus, R. & Nimtz, G. 1976 In *Springer tracts in modern physics*, vol. 78, pp. 1–119, Berlin: Springer-Verlag.
Hill, J. R., Parry, G. & Miller, A. 1982 *Optics Commun.* **43**, 151–156.

Jain, R. K. 1982 *Opt. Engng* **21**, 199–218.

Jain, R. K., Giuliano, C. R., Klein, M. B., Lind, R. C. & Steel, D. G. 1981 In *Proceedings of the International Conference on Excited States and Multiresonant Nonlinear Optical Processes in Solids, Aussois* (Les Éditions de Physique, Orsay), pp. 4–5.

Jain, R. K. & Klein, M. B. 1979 *Appl. Phys. Lett.* **35**, 454–456.

Jain, R. K. & Steel, D. G. 1980 *Appl. Phys. Lett.* **37**, 1–3.

Jain, R. K. & Steel, D. G. 1982 *Optics Commun.* **43**, 72–77.

Kar, A. K., Mathew, J. G. H., Smith, S. D., Davis, B. & Prettl, W. 1983 *Appl. Phys. Lett.* **42**, 334–336.

Khan, M. A., Bennet, R. L. H. & Kruse, P. W. 1981 *Optics Lett.* **6**, 560–562.

Khan, M. A., Kruse, P. W. & Ready, J. F. 1980 *Optics Lett.* **5**, 261–263.

Khan, M. A., Kruse, P. W. & Wood, R. A. 1983 Presented at Conference on Lasers and Electro-Optics, May 1983, paper ThH6.

Kruse, P. W., Khan, M. A. & Ready, J. F. 1981 In *Wavefront distortions in power optics* (*S.P.I.E.* no. 293), pp. 183–189.

Kruse, P. W., Ready, J. F. & Khan, M. A. 1979 *Infrared Phys.* **19**, 497–506.

Mathew, J. G. H., Craig, A. & Miller, A. 1984 *Appl. Phys. Lett.* (In the press.)

Miller, A., Miller, D. A. B. & Smith, S. D. 1981 *Adv. Phys.* **30**, 697–800.

Miller, A. & Parry, G. 1984 *Opt. Quantum Electron.* **16**, 339–348.

Miller, A., Parry, G. & Daley, R. 1984 *IEEE Jl Quantum Electron.* **QE20**, 710–715.

Miller, D. A. B., Mozolowski, M. H., Miller, A. & Smith, S. D. 1978 *Optics Commun.* **27**, 133–136.

Miller, D. A. B., Seaton, C. T., Prise, M. E. & Smith, S. D. 1981 *Phys. Rev. Lett.* **47**, 197–200.

Parry, G., Miller, A. & Daley, R. 1984 In *Optical Bistability II* (ed. C. Bowden, H. Gibbs & S. McCall), pp. 289–296. New York: Plenum.

Yuen, S. Y. 1982 *Appl. Phys. Lett.* **41**, 590–592.

Yuen, S. Y. & Becla, P. 1983 *Optics Lett.* **8**, 356–358.

Phil. Trans. R. Soc. Lond. A **313**, 285–290 (1984)

Printed in Great Britain

Dynamic effects in optical bistability

P. Mandel† AND T. Erneux‡

† *Université Libre de Bruxelles, Service de Chimie-Physique II, Campus de la Plaine, C.P. 231, Bruxelles 1050, Belgium*

‡ *Department of Engineering Sciences and Applied Mathematics, Northwestern University, Evanston, Illinois 60201, U.S.A.*

In the first part we give a brief review of some theoretical problems that still hinder the practical use of bistable optical devices in logic circuits. In the second part we study the influence of a time-dependent control parameter on the solution of differential equations. We prove that even for small sweeping rates large delays can be expected, resulting in the dynamical stabilization of unstable solutions. Examples are given for dynamically induced optical bistability in a laser and dynamical stabilization in a laser with saturable absorber.

INTRODUCTION

Optical bistability (o.b.) has been for many years a stable attractor for theoretical and experimental physics (Bowden *et al.* (eds) 1981, 1984; Abraham & Smith 1982; Lugiato 1984). It has now reached a development such that one begins to think seriously of using bistable optical devices ('biodes') to construct all-optical logic circuits and hence all-optical computers. The topics covered by this Discussion Meeting illustrate in many respects the state of the art in this domain. In the first part of this paper, we review some of the theoretical problems that still hinder the practical realization of an all-optical computer (for each topic we have selected a few references; no attempt has been made to provide an exhaustive list of references). In the second part we study more carefully one problem, related to the time-dependence of the input field.

REVIEW

Bistability is a property that is often found in Nature as soon as the phenomena involve a nonlinear mechanism. Electronic circuits, chemical reactions, hydrodynamics, biological systems and quantum optics display quite a number of situations in which bistability has been predicted or observed, or both. In all-optical arrangements, o.b. can be achieved by using either active or passive systems. The passive system most widely studied (Miller *et al.* 1979; Gibbs *et al.* 1979) is a Fabry–Perot resonator pumped by a coherent beam; inside the cavity a semiconductor provides the necessary nonlinearity. O.b. can also be realized in active systems in which the inversion of a population is created by some incoherent mechanism. In fact the earliest proposal to create a biode is due to Lasher (Lasher 1964), who suggested coupling two semiconductor lasers in a single cavity to make a laser with saturable absorber. In such a system one part of the device is acting as a normal laser whereas the other part provides the nonlinear mechanism. This laser with a saturable absorber displays o.b. and passive Q-switching (Harder *et al.* 1982).

O.b. refers to a situation where two stable states coexist, implying a hysteresis effect. Each of these two states can be steady or periodic. Although chaotic attractors have also been found and are very important for fundamental research, they are of little interest for the topics of this Discussion Meeting. One way to achieve o.b. is to have a resonator whose length is a function of the input intensity. This can be achieved, in passive systems, by allowing an intensity dependence of either the physical length or the optical length of the resonator. In the first case, one uses the pressure of the input radiation to modify the position of a suspended mirror (Dorsel *et al.* 1983), the cavity being empty. In the second case the mirrors are fixed but the cavity contains a medium whose complex susceptibility is significantly intensity-dependent (i.e. a nonlinear medium). When the input field, the material medium and the cavity have nearly equal frequencies, the dominant mechanism will be nonlinear absorption (i.e. absorptive o.b. (see Bonifacio & Lugiato 1976; Szöke *et al.* 1969; Weyer *et al.* 1981)). When there are large detunings we can still observe o.b. (in this case dispersive o.b.) involving nonlinear refraction coupled to linear absorption (Gibbs *et al.* 1976; Marburger & Felber 1978). The physics of absorptive and dispersive o.b. are quite different and therefore lead to different experimental constraints and optimization schemes. Nevertheless these schemes show a number of problems that need to be resolved if we wish to build an all-optical computer. Some of these problems have been carefully analysed by Fork (Fork 1982). Let us consider a thin sample of nonlinear material with a large section. A number of beams at constant intensity are aimed at the sample and at each entry point a probe beam or pulse is added to obtain a logic operation by means of o.b. in the bulk of the sample (Smith 1984). The obvious main constraints to be imposed are that each pixel (in this case the volume used by the beam within the sample) should be stable before and after the switching and should be independent of its neighbours. The following factors, to be discussed at this meeting, affect the stablity of and the cross-talk between pixels.

(i) *Transverse effects*. A real input beam has a finite diameter and a transverse intensity profile. This affects both steady-state and dynamic responses of a biode. At steady state the width of the hysteresis domain is generally reduced (Drummond 1981) and can even vanish, owing to transverse effects. Moreover the domain for stable steady states is usually reduced by transverse effects (Moloney *et al.* 1982). More important is the dynamics of the switching process, which is deeply affected because not all parts of the beam will switch simultaneously (Rosanov & Semenov 1981). The resulting increased gradient of intensity may enhance the self-focusing or defocusing in the beam. Transverse effects will be discussed by Lugiato & Narducci at this meeting (see also Firth & Wright 1982).

(ii) *Diffusion*. Another limitation to dense packing of pixels is diffusion cross-talk. It is characterized by the recombination time of the free carriers, to which the nonlinear refraction index is directly proportional. If N is the density of carriers, the recombination time is proportional to N^a, depending on the dominant mechanism through which the carriers recombine, e.g. the direct trapping of carriers yields $a = 1$, the Auger process yields $a = 3$ (direct three-body interaction) and radiative recombination gives $a = 2$. The avoidance of diffusion cross-talk requires a good understanding of the microscopic mechanism on which o.b. is based. This question is discussed in details by many authors in this meeting.

(iii) *Diffraction*. Another source of cross-talk is diffraction. It is intimately connected with transverse effects and competes with diffusion against dense packing of pixels. It is already better understood than diffusion because it does not involve a refined understanding of the light–matter interaction and can therefore be treated at a macroscopic level. In a nonlinear medium it will induce self-focusing or defocusing, both effects being potentially disastrous. The relative weight of diffraction and diffusion is analysed by Firth and his colleagues at this meeting.

(iv) *Noise*. A pixel is created by the presence of a holding beam with an intensity located in the bistable domain. This domain may be fairly small, or we may wish to have a holding intensity near the switch-up point, or both. In both cases the stability of the biode will be affected by unavoidable noise in the holding beam. Here too a knowledge of the noise tolerance is required to realize a reliable biode (Schmidt *et al.* 1983; Willis 1983; DelleDonne *et al.* 1981). Noise problems are considered by Arecchi at this meeting.

(v) *Dynamics*. There are relatively few results on the dynamical properties of biodes because their discussion requires a solution of time-dependent equations. Most results are therefore numerical or experimental. Among the dynamical properties let us mention:

the dependence of switch-up and switch-down times upon the cavity and atomic decay rates (Mandel & Erneux 1982; Erneux & Mandel 1983; Moloney & Gibbs 1982; Hopf & Meystre 1979; Bonifacio & Meystre 1978; Bischofberger & Shen 1979);
critical slowing down (Bonifacio & Meystre 1979; Garmire *et al.* 1979);
overshoot switching (Goldstone *et al.* 1981);
self-pulsing (Bonifacio & Lugiato 1978; Lugiato & Milani 1983).

Although some of these topics are reviewed at this meeting for specific materials, a reasonable theory has still to be worked out.

BIFURCATIONS WITH TIME-DEPENDENT PARAMETERS

In most problems that we face in quantum optics, there occur bifurcation points at which two states coalesce. The three most common critical points are:
(i) steady bifurcations, where a stationary state emerges from another steady state;
(ii) Hopf bifurcatons, where a time-periodic state emerges from a steady state;
(iii) limit points, where a solution ceases to exist.

For technical reasons, it is often necessary to investigate these bifurcation points experimentally by sweeping a suitable parameter (the bifurcation parameter) across the transition domain. It is usually argued that if the sweeping rate is small enough, the dynamical effects associated with the time-dependence of the bifurcation parameter will be negligible and the system will somehow 'adiabatically follow' the states described with a constant bifurcation parameter. We shall see that this assumption may be quite wrong.

Let us first take a pedagogical example. Consider the equation

$$z_t = zA - z^2, \tag{1}$$

which corresponds to the cubic approximation of the standard laser equations when the atomic variables have been adiabatically eliminated. Here z is the field intensity and A is the pump parameter plus one. When A is constant, (1) has two steady states: $z = 0$ and $z = A$. The trivial solution $z = 0$ is stable for negative values of A whereas $z = A$ is stable for positive A. Hence $A = 0$ is a steady bifurcation point.

When A is time-dependent, the situation is quite different. First we note that even if $A_t \neq 0$, the trivial state $z = 0$ remains an exact solution of (1). Second, if the initial condition verifies the inequality $z(0) = z_i \ll 1$, we can analyse the stability of $z = 0$ by linearizing (1) around the trivial solution

$$z_t = Az. \tag{2}$$

The solution of (2) is

$$z(t) = z_i \exp \int_0^t A(s)\,\mathrm{d}s. \tag{3}$$

It is clear that (3) will diverge (i.e. z will become unstable) when

$$\int_0^{t^*} A(s)\,\mathrm{d}s = 0, \tag{4}$$

which is the dynamical bifurcation equation. If we define $\check{t}$ as the time at at which the steady bifurcation is reached, i.e. $A(\check{t}) = 0$, then we can decompose (4) into

$$\int_0^{\check{t}} A(s)\,\mathrm{d}s = -\int_{\check{t}}^{t^*} A(s)\,\mathrm{d}s, \tag{5}$$

which expresses the balance between the stability accumulated from 0 to $\check{t}$ (where $A(s)$ is negative) and the instability accumulated from $\check{t}$ and t^* (where $A(s)$ is positive). An obvious inequality is $t^* > \check{t}$, implying that $A(t^*)$ is necessarily delayed compared with the steady value $A(\check{t})$. For a linear dependence,

$$A(t) = A(0) + bt, \quad A(0) < 0, \quad b > 0, \tag{6}$$

it is elementary to solve (4) and to find

$$A(t^*) = -A(0). \tag{7}$$

In other terms, the distance between the dynamical and the steady bifurcations $(A(t^*) - A(\check{t}))$ is equal to the distance between the steady bifurcation and the initial value $(A(\check{t}) - A(0))$ and independent of the sweeping rate, b. This counterintuitive result is a long way from the assumed adiabatic following of the steady states.

Of course, (7) was derived under the assumption that we may neglect the nonlinear term in (1). We can remove this limitation by studying the exact solution of (1), which is

$$z(t) = \frac{\exp \int_0^t A(s)\,\mathrm{d}s}{z_i^{-1} + \int_0^t \left\{ \exp \int_0^{t'} A(s)\,\mathrm{d}s \right\} \mathrm{d}t'}. \tag{8}$$

For constant A this reduces to

$$z(t) = A\{(A/z_i)\,\mathrm{e}^{-At} + 1\}^{-1}.$$

The solution (8) gives qualitatively the same results as (7) as long as $0 < b \ll 1$.
We now give two examples where delayed bifurcaton leads to new effects.
(i) *Transient. o.b.* We first consider the semi-classical laser equations

$$\begin{aligned}
E_t &= -E + Av, \\
v_t &= d(-v + EF), \\
F_t &= d_{\parallel}(-F + 1 - Ev),
\end{aligned} \tag{9}$$

for which a similar analysis has been performed (Mandel & Erneux 1984). In figure 1 we indicate the result of a numerical integration of (9) with $d = d_{\parallel} = 10$, $A(t) = -0.5 + 10^{-2}t$. As

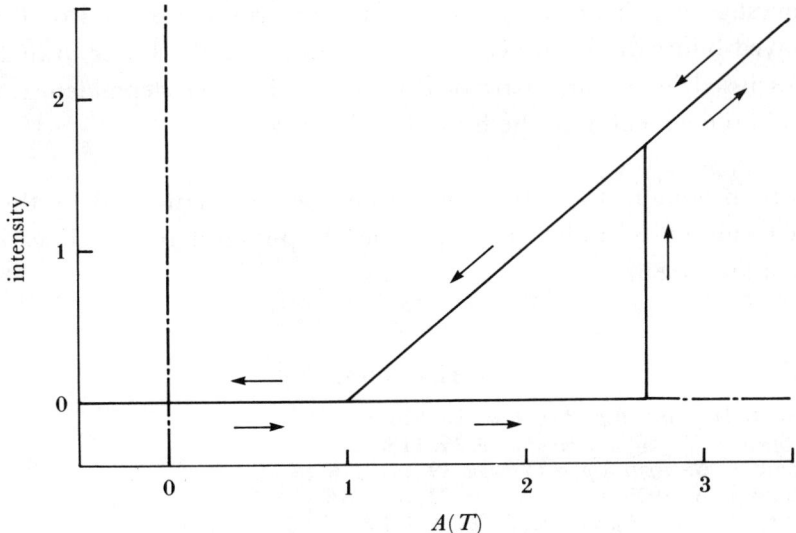

FIGURE 1. Dynamical hysteresis for a laser.

expected from our simple example, the zero intensity state is dynamically stabilized, but there is no significant change of stability for the solution $E^2 = A - 1$. Therefore making a forward sweep followed by a backward sweep will induce two different responses and we create by this mechanism a transient bistable response. Transient o.b. is not new and has been predicted in optics in a different context by Broggi & Lugiato (1984).

(ii) Another example is taken from our study of o.b. in a laser with saturable absorber (l.s.a.) (Erneux & Mandel 1984) and is shown in figure 2. There we have solved the eight semiclassical equations for an l.s.a. and let the pump parameter decrease, starting from the upper branch. For constant A, the linear stability analysis indicates that the upper branch has a Hopf bifurcation at $A = 4.2$ (indicated by a circled bar on the curve), which is reached for $t = 260$. When $A < A_H$ (i.e. $t > 260$) the steady upper branch is unstable. However, owing to the time-dependence of A, the dynamical bifurcation is delayed. In this precise example, the delay is simply larger than the time necessary to reach the endpoint of the upper branch where the jump to the trivial solution occurs.

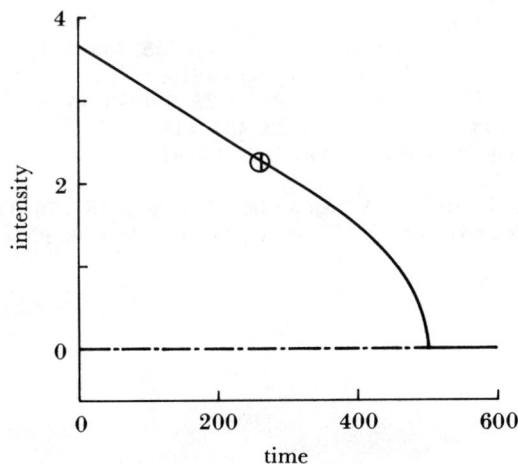

FIGURE 2. Dynamical stabilization: for $t > 260$ the upper branch is unstable with a constant pump parameter and stable for a time-dependent pump parameter.

[99]

We are now investigating the same type of effects on intrinsic o.b., and preliminary results indicate that delayed bifurcations do occur for the upper and the lower branches. Contrary to most effects discussed in the first part of this paper, the time-dependence of the control parameter results in the widening of the hysteresis domain.

This work has been done in the framework of an operation launched by the Commission of the European Community under the experimental phase of the European Community Stimulation Action (1983–85).

REFERENCES

Abraham, E. & Smith, S. D. 1982 *Rep. Prog. Phys.* **15**, 815–885.
Bischofberger, T. & Shen, Y. R. 1979 *Phys. Rev.* A **19**, 1169–1176.
Bonifacio, R. & Lugiato, L. A. 1976 *Optics Commun.* **19**, 172–176.
Bonifacio, R. & Lugiato, L. A. 1976 *Lett. nuovo Cim.* **21**, 510–516.
Bonifacio, R. & Meystre, P. 1978 *Optics Commun.* **27**, 147–150.
Bonifacio, R. & Meystre, P. 1970 *Optics Commun.* **29**, 131–134.
Bowden, C. M., Ciftan, M. & Robl, H. R. (eds) 1981 *Optical bistability.* New York: Plenum Press.
Bowden, C. M., Gibbs, H. M. & McCall, S. L. (eds) 1984 *Optical bistability II.* New York: Plenum Press.
Broggi, G. & Lugiato, L. A. 1984 (In preparation.)
DelleDonne, M., Richter, P. M. & Ross, J. 1981 *Z. Phys.* B**42**, 271–283.
Dorsel, A., McCullen, J. D., Meystre, P., Vignes, E. & Walther, H. 1983 *Phys. Rev. Lett.* **51**, 1550–1553.
Drummond, P. D. 1981 *IEEE J. Quantum Electron.* **QE-17**, 301–306.
Erneux, T. & Mandel, P. 1983 *Phys. Rev.* A **28**, 896–909.
Erneux, T. & Mandel, P. 1984 *Phys. Rev.* A **30** (In the press.)
Firth, W. J. & Wright, E. M. 1982 *Phys. Lett.* **92**, 211–216.
Fork, R. L. 1982 *Phys. Rev.* A **26**, 2049–2064.
Garmire, E., Marburger, J. H., Allen, S. O. & Winful, H. G. 1979 *Appl. Phys. Lett.* **34**, 374–376.
Gibbs, H. M., McCall, S. L. & Venkatesan, T. N. C. 1976 *Phys. Rev. Lett.* **36**, 1135–1139.
Gibbs, H. M., McCall, S. L., Venkatesan, T. N. C., Gossard, A. C., Passner, A. & Wiegman, W. 1979 *Appl. Phys. Lett.* **35**, 451–453.
Goldstone, J. A., Ho, P. T. & Garmire, E. 1981 In Bowden (1981), pp. 187–197.
Harder, Ch., Lau, K. Y. & Yariv, A. 1982 *IEEE J. Quantum Electron.* **QE-18**, 1351–1361.
Hopf, F. A. & Meystre, P. 1970 *Optics Commun.* **29**, 235–238.
Lasher, G. J. 1964 *Solid-State Electron.* **7**, 707–716.
Lugiato, L. A. 1984 In *Progress in optics* (ed. E. Wolf), vol. 21, pp. 71–216. Amsterdam: North-Holland.
Lugiato, L. A. & Milani, M. 1983 *Z. Phys.* B **50**, 171–179.
Mandel, P. & Erneux, T. 1982 *Optics Commun.* **42**, 362–365.
Mandel, P. & Erneux, T. 1984 (In preparation.)
Marburger, J. H. & Felber, F. S. 1978 *Phys. Rev.* A **17**, 335–342.
Miller, D. A. B., Smith, S. D. & Johnston, A. 1979 *Appl. Phys. Lett.* **35**, 658–660.
Moloney, J. V. & Gibbs, H. M. 1982 *Phys. Rev. Lett.* **48**, 1607–1610.
Moloney, J. V., Hopf, F. A. & Gibbs, H. M. 1982 *Phys Rev.* A **25**, 3442–3445.
Rosanov, N. N. & Semenov, V. E. 1981 *Optics Commun.* **38**, 435–438.
Schmidt, H. E., Koch, S. W. & Haug, H. 1983 *Z. Phys.* B **51**, 85–91.
Smith, S. D. 1984 *Nature* **307**, 315–316.
Szöke, A., Daneu, V., Goldhar, J. & Kurnit, N. A. 1969 *Appl. Phys, Lett.* **15**, 376–379.
Weyer, K. G., Wiedenmann, H., Rateike, M., McGillavray, W. R., Meystre, P. & Walther, H. 1981 *Optical Commun.* **37**, 426–430.
Willis, C. R. 1983 *Phys. Rev.* A **27**, 375–380.

Phil. Trans. R. Soc. Lond. A **313**, 291–297 (1984)

Printed in Great Britain

Transverse effects and noise in optical instabilities

By L. A. Lugiato[1], R. J. Horowicz[1]†, G. Strini[1] and L. M. Narducci[2]

[1] *Dipartimento di Fisica dell'Università, Via Celoria 16, 20133 Milano, Italy*

[2] *Department of Physics, Drexel University, Philadelphia, Pennsylvania 19104, U.S.A.*

We analyse the simplest model that describes the dynamics of a two-level, optically bistable system in a ring cavity, and incorporates a Gaussian radial variation of the electric field. We find that the most favourable situation to achieve self-pulsing instabilities is that of atomic and cavity detunings with opposite sign. The results are compared with those of plane-wave theory. We show that noise plays a primary role at the onset of self-oscillatory instabilities, and that its effects are governed by an equation formally identical to the Risken equation for the laser.

1. Introduction

The detailed comparison between theory and experiment in the study of quantum optical systems imposes a thorough analysis of the effects that arise from the radial variation of the electric field in the radiation beam (transverse effects). This is especially true in the case of optical bistability (o.b.), in which the modelling of practical devices requires precise predictions on the behaviour of the system.

From the viewpoint of theory, this implies a renunciation of the comfortable ground of plane-wave theory and coping with the difficulties of two-dimensional theories. From the experimental viewpoint it is necessary to perform detailed and systematic observations of the behaviour of the radial profile, an aspect that is most often disregarded. This is necessary to allow a comparison with theoretical predictions, and to help theory in attaining an adequate description of the physical situation.

In recent years, considerable efforts have been devoted to the study of transverse effects (see, for example, Bowden *et al.* (eds) (1984). A few of these works analyse the transverse effects on instabilities and spontaneous pulsations in o.b. (Moloney *et al.* 1982; Firth & Wright 1982; Lugiato & Milani 1983) or in the laser (Hauck *et al.* 1983). These results suggest that the instability problem including transverse effects is not simply an extension or a generalization of the plane-wave case, but rather an independent piece of physics. In fact, there are examples of plane-wave instabilities that disappear in the Gaussian case (Lugiato & Milani 1983), as well as of instabilities predicted by a full two-dimensional treatment, that are absent in plane-wave theory (Hauck *et al.* 1983).

In this paper we study the transverse effects on a class of instabilities in mixed absorptive and dispersive o.b., previously analysed in the framework of plane-wave theory by Lugiato *et al.* (1982) and Lugiato & Narducci (1984); see also Ikeda & Akimoto (1982). Our treatment is based on the simplest possible model that describes the dynamics of this system and incorporates a radial variation of the electric field. This model, which allows for an analytical

† Permanent address: Departamento de Fisica, Pontifícia Universidade Católica do Rio de Janeiro, Rio de Janeiro, Brazil.

292 L. A. LUGIATO AND OTHERS

study of the stability of the steady-state solutions, is described in §2, in which the stationary solutions are also calculated. The results of the linear stability analysis are illustrated, and compared with those of plane-wave theory, in §3.

Another general class of effects that must be carefully studied for practical purposes are those that arise from noise. Fluctuation phenomena in o.b. have been extensively analysed from a theoretical viewpoint, but almost exclusively in the stationary state (Lugiato 1984). We recently studied some noise effects in the transient or in the stable self-pulsing behaviour, in the critical situations in which these effects are most visible. Another paper in this symposium, (by Broggi & Lugiato), illustrates the role of fluctuations in the critical slowing down, and shows that they give rise to a broad switching-time distribution, as well as an interesting effect of transient noise-induced optical bistability. On the other hand, in §4 of this paper we discuss the effects of noise in the onset of self-pulsing instabilities. The concluding discussion is given in §5.

2. Gaussian one-mode model for optical bistability

We consider a unidirectional ring cavity with spherical mirrors, of total length $\mathscr{L}$. It contains a cylindrical, homogeneously broadened two-level atomic sample of length L and radius d. We assume that the Fresnel number, $w_0^2/\lambda L$, where w_0 is the beam waist and λ the wavelength, is so large that the beam radius is practically constant along the atomic sample. The incident field is assumed to be matched to the fundamental TEM_{00} mode of the cavity.

If we indicate by $\bar{r}$ the radial variable normalized to the beam waist, the one-mode model that we study is given by the dynamical equations (Lugiato & Milani 1983)

$$k^{-1}\frac{d}{dt}f(t) = -f(1+i\theta) + y - 2C\int_0^{d/w_0} d\bar{r}\, 4\bar{r}\exp(-\bar{r}^2)\, P(\bar{r},t), \tag{1a}$$

$$\gamma_\perp^{-1}\frac{\partial}{\partial t}P(\bar{r},t) = D(\bar{r},t)f(t)\exp(-\bar{r}^2) - (1+i\varDelta)\,P(\bar{r},t), \tag{1b}$$

$$\gamma_\parallel^{-1}\frac{\partial}{\partial t}D(\bar{r},t) = -\tfrac{1}{2}\{P(\bar{r},t)f^*(t) + P^*(\bar{r},t)f(t)\}\exp(-\bar{r}^2) - D(\bar{r},t) + 1, \tag{1c}$$

where f and y are the normalized amplitudes of the output and incident field, respectively; $P(\bar{r},t)$ and $D(\bar{r},t)$ are normalized quantities that correspond to the macroscopic atomic polarization and population difference at a distance r from the longitudinal axis. The field damping constant, k, is defined as $cT/\mathscr{L}$, where T is the mirror transmissivity coefficient. The atomic decay rates, $\gamma_\parallel$ and $\gamma_\perp$, are the inverse of the relaxation times T_1 and T_2', respectively. The bistability parameter, C, is usually defined as

$$C = \alpha L/2T, \tag{2}$$

where α is the unsaturated field absorption coefficient. The cavity and atomic detunings θ and $\varDelta$ are given by

$$\theta = (\omega_c - \omega_0)/k, \quad \varDelta = (\omega_a - \omega_0)/\gamma_\perp, \tag{3}$$

where ω_0, ω_c and ω_a are respectively the frequency of the incident field, the frequency of the cavity mode and the atomic transition frequency. The model (1) incorporates the following assumptions: (i) the mean field limit $\alpha L \ll 1$, $T \ll 1$, with $C = \alpha L/2T$ arbitrary; (ii) the time evolution occurs on a timescale much longer than the cavity transit time; and (iii) the transverse profile of the electric field inside the filled cavity corresponds to the Gaussian TEM_{00} mode.

[102]

This one-transverse-mode assumption holds in at least two situations: (1) when all the other modes are detuned enough from the input frequency, ω_0, and (2) when the losses of all the other transverse modes are large enough to maintain their amplitude negligible.

At steady state, one obtains from (1) the equation that links the input and output fields:

$$y^2 = x^2[\{1 + 2Cg(x^2)\} + \{\theta - 2C\varDelta g(x^2)\}^2], \tag{4}$$

where we define

$$x = |f_{st}|$$

and

$$g(x^2) = \frac{1}{x^2} \ln \frac{1 + \varDelta^2 + x^2}{1 + \varDelta^2 + x^2 \exp\{-2(d/w_0)^2\}}. \tag{5}$$

This state equation was derived by Ballagh *et al.* (1981), Arimondo *et al.* (1981), Drummond (1981) and Lugiato & Milani (1983). In the limit $d/w_0 \rightarrow 0$ one recovers the plane-wave theory (Lugiato & Milani 1983). In the following, we shall always consider the opposite case $d/w_0 \rightarrow \infty$.

3. LINEAR STABILITY ANALYSIS

The details of the stability analysis of the steady-state solutions can be found in Lugiato *et al.* 1984*b*). Here we limit ourselves to the discussion of the results.

First, let us recall the picture in the plane-wave theory (Lugiato *et al.* 1982). For suitable choices of the parameters C, $\varDelta$, θ, $\tilde{k} \equiv k/\gamma_\perp$ and $\tilde{\gamma} = \gamma_\parallel/\gamma_\perp$, one finds that a portion of the steady-state curve with positive slope is unstable. A necessary condition is that $\varDelta$ or θ or both are non-zero. With the possible exception of a small range of values of the incident field in correspondence with which the long-term behaviour exhibits precipitation to the lower transmission state, this instability leads to undamped self-pulsing behaviour, with an oscillation period on the order of the cavity build-up time, k^{-1}. When C is very large, the instability range includes a domain of chaotic behaviour. On approaching this domain on either side, one identifies a cascade of period-doubling bifurcations of the usual type.

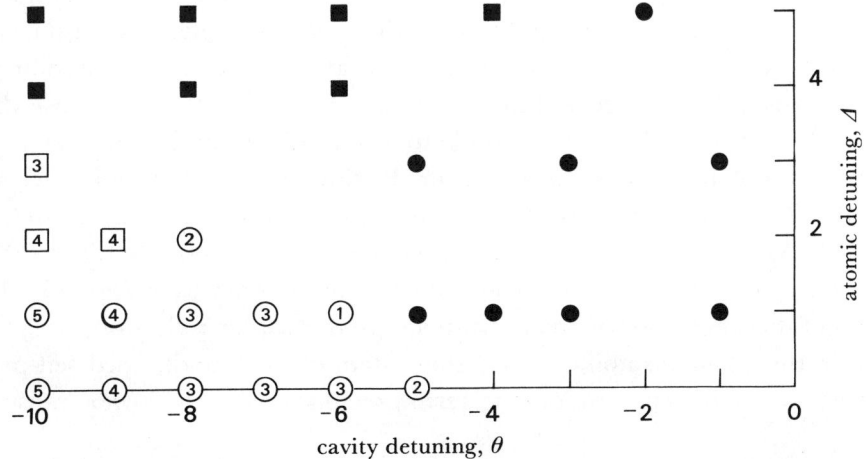

FIGURE 1. Positive-slope instability domain in the $\varDelta$–θ plane, in the plane-wave case. $C = 75$, $\tilde{k} = 0.5$, $\tilde{\gamma} = 2$. Filled circles indicate that the steady-state curve is S-shaped and that no positive-slope instability is there. Filled squares indicate that there is neither bistability nor instability. Open circles indicate that the steady-state curve is S-shaped and that there is a positive slope instability range, the approximate extension of which in the variable x is shown in the circle. Open squares indicate that there is no bistability, but the steady-state curve includes an instability range, the approximate extent of which is indicated in the square.

[103]

We analysed the stability of the steady-state curve for $C = 75$, $\tilde{k} = 0.5$, $\hat{\gamma} = 2$, $-10 \leqslant \Delta$, $\theta \leqslant 10$. Figure 1 shows the extension of the positive-slope instability region in the second quadrant of the Δ–θ plane in the plane-wave case. No instability was found in the first quadrant, provided that Δ was not too small. A simultaneous change of sign in Δ and θ does not change anything with respect to the stability of the stationary states. If we increase the value of $\tilde{k}$ by keeping fixed the other parameters C, Δ, θ and $\hat{\gamma}$, the unstable range in the steady-state curve of transmitted against incident field grows.

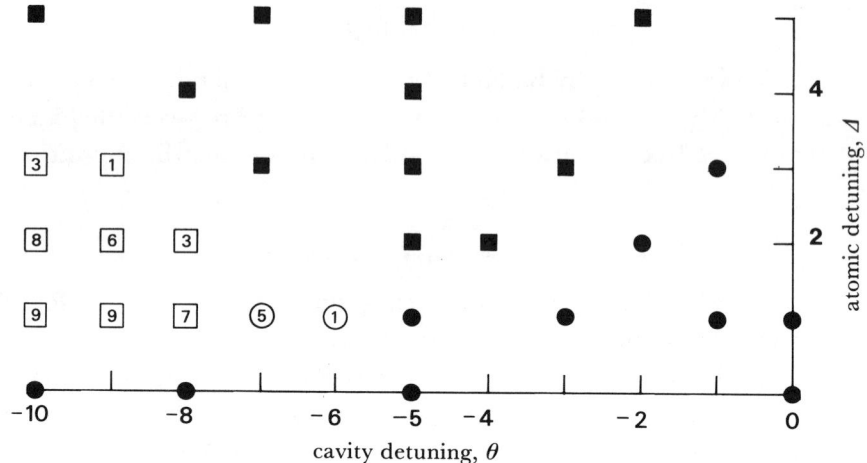

FIGURE 2. As for figure 1, but in the Gaussian case with $d/w_0 = \infty$.

Let us now turn to the Gaussian case. Figure 2 exhibits the extension of the positive slope instability region for the same values of the parameters considered in figure 1. Again, we do not find any positive slope instability in the first quadrant of the Δ–θ plane and in the domain $5 \leqslant \Delta \leqslant 10$, $-10 \leqslant \theta \leqslant 10$. Hence for $C \lesssim 100$ the situation $\Delta\theta < 0$, which is less favourable for bistability than the case $\Delta\theta > 0$, is on the contrary necessary to obtain a self-pulsing instability. The position of the instability domain in the Δ–θ plane is substantially the same as in the plane-wave case; however, we do not find any positive slope instability in the purely absorptive situation $\Delta = 0$. Figure 2 indicates also that in the Gaussian case the situation of S-shaped steady-state curve is less favourable to instabilities than that of a single-valued curve. On comparing the data in figures 1 and 2 for the same values of Δ and θ, we see that, for Δ not too small, the positive-slope instability range in the variable x turns out to be larger in the Gaussian case. As we see from figures 3 and 4, this feature is mainly due to the fact that in the plane-wave case a sizable part of the instability range has negative slope. On the other hand, the situation of the single-valued steady-state curve is ideal for pulsation, because we are sure *a priori* that in the whole instability range the system exhibits undamped self-pulsing. In fact, in this case no competitive process of long-term precipitation to the lower transmission branch is possible.

An interesting quantity is the ratio (instability range)/(instability threshold), where both the instability range and the instability threshold refer to the input field variable y. When the steady-state curve is S-shaped, the instability threshold is the minimum value of y for which the higher transmission steady state is unstable. In figures 3 and 4 the values of this ratio are practically identical and equal to 0.23.

[104]

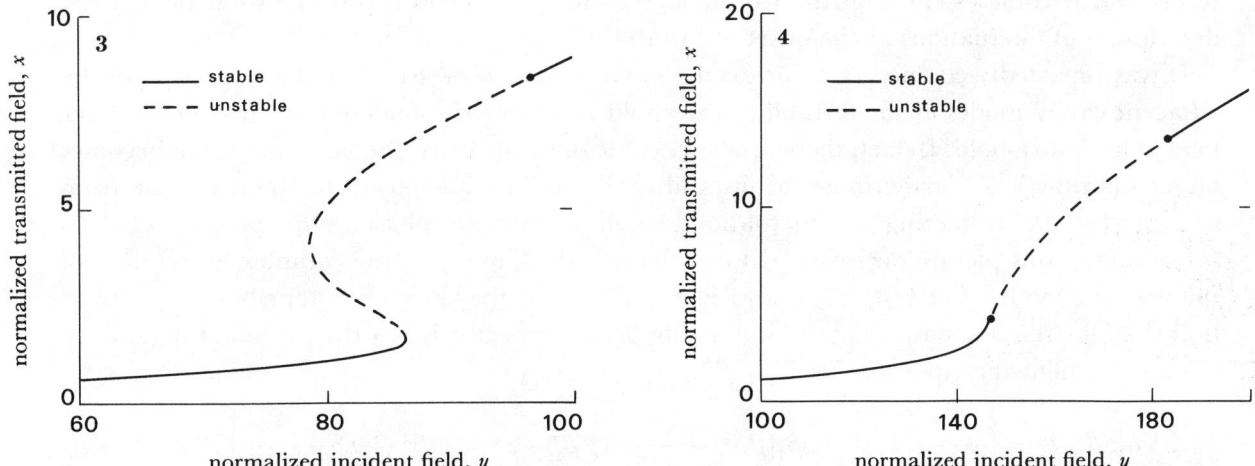

FIGURE 3. Plane-wave case, $C = 75$, $\varDelta = 1$, $\theta = -9$, $\tilde{k} = 0.5$, $\tilde{\gamma} = 2$. The steady-state curve of normalized transmitted field, x, is shown as a function of normalized incident field, y, and the part of the curve that is unstable against fluctuations is indicated.

FIGURE 4. As for figure 3, but in the Gaussian case with $d/w_0 = \infty$.

We are currently analysing the numerical solutions of the full nonlinear time-dependent dynamical equations (1), to see what kind of oscillations arise from this instability in the Gaussian case.

4. NOISE EFFECTS IN THE ONSET OF SELF-PULSING INSTABILITIES

The instability discussed in the previous section arises in the cavity mode nearest to resonance with the incident field (resonant mode), which is the only mode that is taken into account in (1). On the other hand, the first instability that was discovered in the Maxwell–Bloch equations for o.b. arises in the off-resonance longitudinal cavity modes, adjacent to the resonant mode (Bonifacio & Lugiato 1978; Lugiato 1980). Hence in this case the instability must be described by a many-mode theory and leads to spontaneous oscillations with a period on the order of the cavity round-trip time. The slowly varying envelope of the electric field does not oscillate uniformly in the cavity, as in the pulsations arising from the instability described in the previous section, but exhibits a wave profile that propagates and circulates along the cavity.

In Lugiato & Milani (1983) it is shown that in the purely absorptive case this off-resonance mode instability disappears when the electric field has a Gaussian transverse profile. On the other hand, a recent paper (Carmichael *et al.* 1984) demonstrates that in the general mixed absorptive and dispersive case the instability remains, and sometimes is even enhanced with respect to the plane-wave situation.

We analyse the effects of thermal, external and quantum noise on the onset of the off-resonance mode instability in o.b. (Lugiato *et al.* 1984a). For the sake of simplicity, we consider the plane-wave, purely absorptive case. Use of the dressed-mode formalism of o.b. (Benza & Lugiato 1981) turns out to be essential. In Benza & Lugiato (1981) the analysis of noise effects was restricted to the region below threshold, where a simple linearized treatment is possible. Thus, the critical slowing down and the spectral line narrowing effects in the approach to the instability threshold were described. Here we extend this analysis by including the nonlinear terms that become essential in the threshold region. This makes it possible to achieve a full description of fluctuations at the onset of instability.

It was repeatedly claimed (see, for example, Gronchi et al. 1981) that the behaviour of the adjacent cavity modes at the instability threshold is completely analogous to that of the laser field at laser threshold. In fact, these modes become unstable when the side-mode gain becomes larger than the loss. Furthermore, at instability threshold the adjacent modes spring up from zero under the triggering of fluctuations, with a random phase. Our present analysis substantiates this picture decisively. In fact, let us call β_1 and β_{-1} the complex amplitudes of the two side-modes. Let $P(\beta_1, \beta_1^*, t)$ and $P(\beta_{-1}, \beta_{-1}^*, t)$ be the Glauber–Sudarshan P-functions that describe the fluctuations of the side-mode fields, respectively. In the threshold region, we obtain the following equation for $P(\beta_1, \beta_1^*, t)$:

$$\frac{d}{dt} P(\beta_1, \beta_1^*, t) = \left\{ -\frac{\partial}{\partial \beta_1} (\lambda_1 \beta_1 - a\beta_1 |\beta_1|^2) - \frac{\partial}{\partial \beta_1^*} (\lambda_1^* \beta_1^* - a^* \beta_1^* |\beta_1|^2) + D \frac{\partial^2}{\partial \beta_1^* \partial \beta_1} \right\} P(\beta_1, \beta_1^*, t),$$

(6)

where λ_1 is the eigenvalue of side-mode 1; hence $\mathrm{Re}\,\lambda_1$ is negative (positive) below (above) instability threshold. The real part of a and the diffusion coefficient, D, are positive. The imaginary parts of λ and a shift the side-mode frequency with respect to its empty-cavity value. The function $P(\beta_{-1}, \beta_{-1}^*, t)$ obeys a completely similar equation.

The relevant point is that (6) is formally identical to the well known Risken equation (Risken 1966), which describes the statistics of the laser field in the threshold region. This means that we can immediately exploit all the results obtained for the Risken equation, to describe fluctuations at the onset of self-pulsing behaviour. Most importantly, we can conclude that these fluctuations are as remarkable as laser fluctuations at threshold. More precisely, let us distinguish the static and the dynamical aspects.

First, let us consider the asymptotic, long time self-pulsing régime. If we do not observe the whole electric field, but only a sideband in the instability threshold region we find large fluctuations, as we would expect in a critical situation. For instance, it would be interesting to measure the photon statistics, exactly as was done many years ago for the laser. Furthermore, from our knowledge of the laser case we infer that the sideband spectral line not only exhibits narrowing on approaching the instability threshold, but goes on narrowing beyond threshold.

Next, let us consider the approach to the asymptotic, stationary self-pulsing régime. This approach will present anomalous fluctuations, typical of the decay of the unstable state. These fluctuations can in part counteract the critical slowing down that arises at the instability threshold. This effect is similar to that described by Broggi & Lugiato (this symposium).

5. CONCLUDING REMARKS

The matter of instabilities and oscillatory behaviour, both periodic and chaotic, in nonlinear systems is more and more attracting the interest of the scientific community. In the special case of optical bistability, this interest also presents practical aspects that arise from the perspective of realizing all-optical clocks, which can be used for example to drive optical transistors.

We think that both types of instabilities discussed in this paper, namely the resonant-mode and the off-resonance mode instabilities, are good candidates for the construction of an optical clock. They arise in mixed absorptive and dispersive o.b. We have demonstrated that they persist in the presence of a Gaussian transverse profile, and that they arise for values of the parameters that we consider accessible. The pulsation period is different in the two cases: on the order of the cavity build-up time $\mathscr{L}/cT$ in the resonant mode instability, and on the order of the cavity transit time $\mathscr{L}/c$ in the multimode instability.

We have also shown that the appearance of self-oscillations is accompanied by remarkable critical fluctuations. This on the one hand shows the relevance of fluctuations in the dynamical behaviour of the system, and on the other hand suggests new experimental observations of a statistical type on optical systems.

This work has been carried out in the framework of the European Joint Optical Bistability Project (E.J.O.B.) of the Commission of the European Communities.

Work supported by N.A.T.O. Research Grant no. 866/83 and by C.N.R. grant no. CT 83.00029.02.115.14905. R.J.H. is grateful for a grant of the I.C.P.T. Programme for Training and Research in Italian Laboratories.

REFERENCES

Arimondo, E., Gozzini, A., Lovitch, F. & Pistelli, E. 1981 In *Optical bistability* (ed. C. M. Bowden, M. Ciftan & H. R. Robl), pp. 151–172. New York: Plenum.

Ballagh, R. J., Cooper, J., Hamilton, M. W., Sandle, W. J. & Warrington, D. M. 1981 *Optics Commun.* **37**, 143–147.

Benza, V. & Lugiato, L. A. 1981 In *Optical bistability* (ed. C. M. Bowden, M. Ciftan & H. R. Robl), pp. 9–30. New York: Plenum.

Bonifacio, R. & Lugiato, L. A. 1978 *Lett. nuovo Cim.* **21**, 510–515.

Bowden, C. M., Gibbs, H. M. & McCall, S. L. (eds). 1984 *Proceedings of the Topical Meeting on Optical Bistability* New York: Plenum.

Carmichael, H. J., Asquini, L. & Lugiato, L. A. 1984 (In preparation.)

Drummond, P. D. 1981 *IEEE J. Quant. Electron.* **QE-17**, 301–306.

Firth, W. J. & Wright, E. M. 1982 *Phys. Lett.* **92**, 211–216.

Gronchi, M., Benza, V., Lugiato, L. A., Meystre, P. & Sargent, M. III 1981 *Phys. Rev. A* **24**, 1419–1430.

Hauck, R., Hollinger, F. & Weber, H. 1983 *Optics Commun.* **47**, 141–145.

Ikeda, K. & Akimoto, O. 1982 *Phys. Rev. Lett.* **48**, 617–620.

Lugiato, L. A. 1980 *Optics Commun.* **33**, 108–112.

Lugiato, L. A. 1984 In *Progress in optics*, vol. 21 (ed. E. Wolf), pp. 69–216. Amsterdam: North-Holland.

Lugiato, L. A., Casagrande, F. & Horowicz, R. J. 1984a (In Preparation.)

Lugiato, L. A., Horowicz, R. J., Strini, G. & Narducci, L. M. 1984b *Phys. Rev. A.* (In the press.)

Lugiato, L. A. & Milani, M. 1983 *Z. Phys.* B **50**, 171–179.

Lugiato, L. A. & Narducci, L. M. 1984 In *Coherence and quantum optics V* (ed. L. Mandel & E. Wolf), pp. 941–956. New York: Plenum.

Lugiato, L. A., Narducci, L. M., Bandy, D. K. & Pennise, C. A. 1982 *Optics Commun.* **43**, 281–286.

Moloney, J. V., Hopf, F. A. & Gibbs, H. M. 1982 *Phys. Rev.* A**25**, 3442–3445.

Risken, H. 1966 *Z. Phys.* **191**, 302–314.

Phil. Trans. R. Soc. Lond. A **313**, 299–306 (1984)

Printed in Great Britain

Diffusion, diffraction and reflection in semiconductor o.b. devices

By W. J. Firth[1], E. Abraham[1], E. M. Wright[2], I. Galbraith[1]
and B. S. Wherrett[1]

[1] *Department of Physics, Heriot-Watt University, Riccarton, Currie, Edinburgh EH14 4AS, U.K.*
[2] *Max-Planck Institut für Quantenoptik, Postfach 1513, D-8046 Garching, F.R.G.*

For reasons of speed and economy, applications of optical bistability are likely to concentrate on solid-state devices of small volume. In this paper we discuss the modelling and optimization of such devices.

We report calculations of surface reflectivity optimization in the presence of absorption, and show that optical bistability in reflection offers significant advantages over transmission.

Small transverse dimensions, i.e. dense packing, are limited by cross-talk between neighbouring devices. We show that diffraction and diffusion give rise to qualitatively similar effects.

1. Introduction

The macroscopic problem in optical bistability is rather simple, at least at face value. Briefly, given a material with a known behaviour of nonlinear refractive index, one wishes to find the combination of material absorption and end-face reflectivities that will optimize the bistable response for a given distribution of input intensities. This apparently simple requirement hides a formidable mathematical and computational problem that has been tackled in certain cases, but for which no all-embracing model exists. This paper deals with certain aspects of the three phemonema mentioned in the title, but in reverse order. We first argue that the reflected, rather than the transmitted, beam is the more useful output from a bistable device. We then discuss the nature of the 'transverse effects' that arise when physical beams of finite transverse dimensions are considered. We report some recent computer results for this problem for the case of a local nonlinear response. Lastly, we consider the transverse effects of diffusion, when the induced free carriers responsible for the nonlinear refractive index in semiconductors such as indium antimonide are sufficiently long-lived and mobile to migrate significant distances in response to concentration gradients, thereby broadening and smoothing the induced refractive index distribution.

2. Reflection

In this section an analysis of bistable Fabry–Perot action in reflection is outlined. Full details are published elsewhere (Wherrett 1984). Reasons are given why reflection-mode may be preferable to transmission-mode operation in device applications.

Here, and below, our analysis will be framed in terms of a Fabry–Perot cavity with an intensity dependent refractive index, though many of the ideas will be relevant to a wider class of bistable devices. Given a suitable material and laser frequency, the parameters at the designer's disposal are the front (input) and back (transmission) face reflectivities, R_f and R_b (which can be varied by use of coatings), and the material absorption, αD, assumed linear (which can be varied by varying the thickness D of nonlinear material). For any given case, we can define an 'effective reflectivity',

$$R_\alpha = (R_f R_b)^{\frac{1}{2}} \exp(-\alpha D), \tag{1}$$

and a 'finesse factor', $\qquad F = 4R_\alpha/(1-R_\alpha)^2.$

We also define a semi-round-trip phase shift, ϕ, defined as (modulo π)

$$\phi = (3\pi D/\lambda) n_2 \bar{I} - \delta, \tag{2}$$

where δ is the low-intensity cavity detuning, $\bar{I}$ the mean internal intensity, and the refractive index depends on intensity as

$$n(I) = n_0 + n_2 I.$$

It is then possible (Miller 1981; Wherrett 1984) to decouple the microscopic and macroscopic aspects of the bistability problem by scaling intensities in terms of the characteristic intensity of the material, $\lambda\alpha/(3\pi n_2)$. It then turns out, for example, that I_c, the critical intensity to obtain bistability, depends only a universal function of R_f, R_b and $e^{-\alpha D}$. I_c diverges if either R_f or R_b is zero, because there is then no feedback. Equally, as R_f tends to unity I_c diverges as it becomes impossible to get light into the cavity. In contrast, as R_b tends to unity I_c *decreases*, because $\bar{I}$ is enhanced owing to the lower back-face loss. For a 100% reflective back surface there is, of course no transmission whatever and I_c is minimized. One is thus led to examine the properties of the cavity reflectivity, R, given by

$$R = 1 - \frac{1}{(1+F\sin^2\phi)} \frac{(1+R_\alpha^2 - R_f - R_b\,e^{-2\alpha D})}{(1-R_\alpha)^2}. \tag{3}$$

The intensity dependence of R arises entirely from the $\sin\phi$ term: R is thus a minimum when $\sin\phi$ is zero (cavity on resonance) and vice versa, and shows hysteresis if T does. This means that the 'logical value' of R is complementary to that of the transmission, T, which is maximal on resonance. Of more immediate interest is the fact that the range of R between 'off' and 'on' can be large even if αD is substantial, as is shown in figure 1. This could be particularly useful in cascading optical logic devices where, perhaps because of material problems or fabrication problems, it is necessary to cope with substantial absorption. The figure shows clearly that in such a case transmission drops rapidly with αD but that a usable switching range is obtainable in reflection even for $\alpha D = 1$.

These considerations are elaborated elsewhere (Wherrett 1984); the main advantages of the reflection mode (with $R_b = 1$) are identified as (i) lower critical intensity, (ii) better absorption tolerance; (iii) less critical fabrication specifications for R_f, R_b and δ; and (iv) simpler cooling, because the entire back surface can be a heat sink. It should also be noted that the reflection signal is there anyway in Fabry–Perot devices, and would have to be suppressed in cascaded transmission-mode devices. Finally, Wherrett (1984) also demonstrates that optical logic devices and oscillators can be based on reflection-mode as well as on transmission-mode bistable devices.

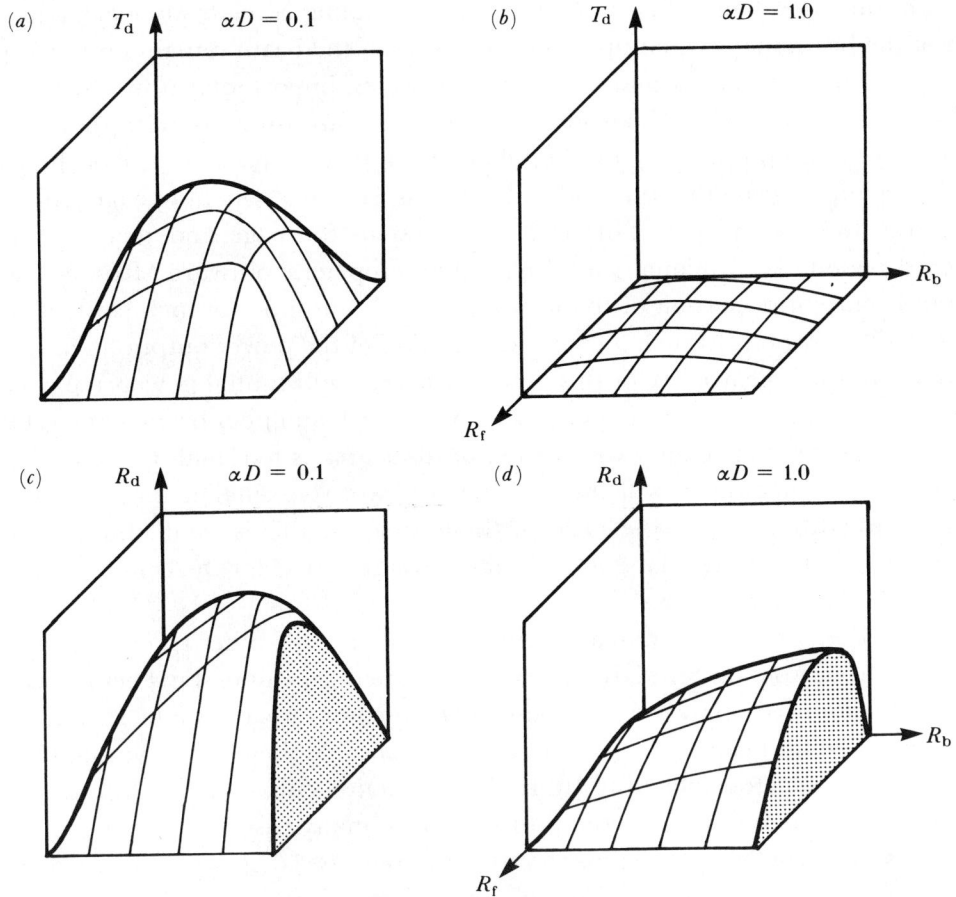

FIGURE 1. Differences between 'on' and 'off' signals: (a, b) the transmission difference, T_d, for $\alpha D = 0.1$ and 1.0; (c, d) the reflection difference, R_d, for $\alpha D = 0.1$ and 1.0.

3. DIFFRACTION

The above analysis is strictly valid only in the plane-wave limit, as indeed is most work in the theory of optical bistability. Consideration of the finite size and non-uniform intensity distribution of the input beam considerably complicates the analysis by introducing a number of 'transverse effects' associated with the intensity and refractive index gradients within the cavity.

Any finite intensity distribution clearly gives rise to a *diffractive* spreading (and phase shifting) of the beams. Because of the complex interplay between amplitude and phase distributions, diffraction is hard to analyse, even numerically, in nonlinear problems. Within the cavity the diffraction properties can be characterized by the *Rayleigh range* $\pi w^2/\lambda$, where w is the beamwidth (for a Gaussian beam, the field amplitude varies as e^{-r^2/w^2}), and $\lambda(= \lambda_0/n_0)$ the wavelength in the medium. The Rayleigh range, z_R, is a measure of the distance over which a light beam remains collimated, and if it is small compared with the cavity length, L, there is strong diffractive coupling between different parts of the beam, and we may expect the beam to switch, if at all, as a unit (whole-beam switching (Moloney & Gibbs 1982)).

The power output will then show hysteresis qualitatively similar to that in plane-wave

[111]

models. On the other hand, if $z_R \gg L$, diffraction coupling is relatively weak and the central portion of the beam may switch up while the wings of the beam remain on the lower branch (part-beam switching). This scenario neglects at least one important feature. Near the switching edge between the 'on' and 'off' portions of the medium, the diffraction length is characterized not by z_R, but more properly by $\pi d^2/\lambda$, where d is a length characteristic of the width of the transition region between the 'on' and 'off' regions. A sharp switching edge will thus diffract strongly into the 'off' region. This will both smooth the edge and tend to 'pull up' the unswitched region. A 'switching wave' may thus propagate outward (Rosanov & Semenov 1981) until that entire portion of the beam for which an upper branch exists is 'on', except for the transition region. In such a case, examination of the power output on a timescale long compared with the time needed for this process will see a substantial power step, roughly equal to the plane-wave intensity step times the area over which an upper branch exists. On lowering the input power, the 'on' region will shrink, until its area is too small to sustain itself against diffraction. The switch-down will thus involve a power step smaller than that on switch-up by a factor roughly equal to the ratio of these areas. If this is small, the power hysteresis characteristic will be rather triangular in shape, whereas that for the central intensity will be much squarer (Gibbs et al. 1982).

Diffraction derives from an optical amplitude gradient: the other transverse effects that we shall discuss derive from the refractive index gradient in the medium. First of these is (nonlinear) refraction; the induced refractive index causes the medium to act as a lens. If $n_2 > 0$, the lens is converging, and the beam may break up into filaments (Moloney, this symposium). If $n_2 < 0$, as in most semiconductors including InSb, the lens action enhances the diffractive spreading: indeed it is much more important than diffraction in many cases.

An important side-effect of the phase gradient is an induced curvature of the internal phase fronts, which clearly leads to loss of finesse, and hence of bistability if large enough. This will be particularly serious in higher orders, and for $n_2 < 0$, whereas for $n_2 > 0$ it can actually be helpful in compensating for the diffractive effects.

If the nonlinear index is due to a mobile excitation, e.g. photocarriers in semiconductors, then these will diffuse in the presence of phase (concentration) gradients. This is discussed in more detail below.

After that fairly extensive preamble, we now set down the field equations describing these phenomena, which must in general be solved numerically. These are, assuming steady-state conditions:

$$\{(\partial/\partial z) - \tfrac{1}{2}\mathrm{i}\mathscr{F}\nabla_\tau^2 + \tfrac{1}{2}\alpha L\}\, F(\boldsymbol{r}) = \mathrm{i}h(\boldsymbol{r})\, F(\boldsymbol{r}), \tag{4}$$

$$\{(-\partial/\partial z) - \tfrac{1}{2}\mathrm{i}\mathscr{F}\nabla_\tau^2 + \tfrac{1}{2}\alpha L\}\, B(\boldsymbol{r}) = \mathrm{i}h(\boldsymbol{r})\, B(\boldsymbol{r}), \tag{5}$$

$$(-l_d^2\nabla_\tau^2 + 1)\, h(\boldsymbol{r}) = -4\mathscr{F}\,\mathrm{sgn}\,(n_2)\,(|F(\boldsymbol{r})|^2 + |B(\boldsymbol{r})|^2). \tag{6}$$

Here the transverse coordinates are scaled to w, ∇_τ^2 is the transverse Laplacian; the longitudinal coordinate is scaled to L, and $\mathscr{F}$ equals $L/2z_R$. The forward and backward field amplitudes, F and B, are scaled so that for example the forward power is

$$P_f(z) = (2P_c/\pi) \iint |F|^2\, \mathrm{d}x\, \mathrm{d}y,$$

$$P_c = \lambda_0^2/2\pi n_0 |n_2|.$$

P_c is the 'critical power for self-focusing' in the case $n_2 > 0$. The third field $h(\mathbf{r})$ is the excitation density, scaled to be the nonlinear phase shift per unit propagation distance. It is assumed to diffuse transversely with a diffusion length l_d (longitudinal diffusion is less significant). In the limit $l_d \to 0$, (6) gives an explicit expression for $h(\mathbf{r})$ that may be inserted into (4) and (5). The resulting equations are valid only if $l_d \gtrsim \lambda$, since for $l_d \lesssim \lambda$ one must include population grating terms, which, in the limit $l_d \to 0$, have the effect of doubling the phase shift imposed by the counter-propagating fields on each other (termed nonlinear non-reciprocity by Kaplan & Meystre 1982). The boundary conditions are

$$B(r, 1) = R_b^{\frac{1}{2}} F(r, 1), \tag{7}$$

$$F(r, 0) = T_f^{\frac{1}{2}} I_{in} + R_f^{\frac{1}{2}} B(r, 0) \, e^{-2i\delta}, \tag{8}$$

essentially as for plane waves.

Only numerical solutions of this system of complex nonlinear partial differential equations have been obtained. Only recently (Moloney, this symposium) has a full two-dimensional problem been examined; most authors have either eliminated one cartesian coordinate or assumed cylindrical symmetry. Methods used have included mode expansions (Ballagh *et al.* 1981; Firth & Wright 1982), which operate in real space but necessarily truncate the mode series; and fast transform techniques (Rosanov & Semenov 1981; Moloney & Gibbs 1982; Moloney 1982 and this symposium) in which the nonlinearity is handled by reducing the medium to a set of slices, and the (linear) propagation between slices is treated by a fast transform algorithm (Siegman 1977).

We report here some new results obtained in the cylindrical geometry using he fast-Hankel transform (f.H.t.) technique for parameters identical to those used by Firth & Wright (1981) with mode expansions. For the chosen parameters, broadly representative of InSb experiments, the plane-wave model gives bistability in first order. Firth & Wright found large bistable loops for low mode numbers, but as more transverse modes were included, these loops narrowed rapidly. By extrapolation beyond the highest mode number used (six), the authors concluded that power hysteresis would probably vanish. Our new results (figure 2), from the use of four slices in the f.H.t. calculation, actually show power hysteresis, but extremely small. With regard

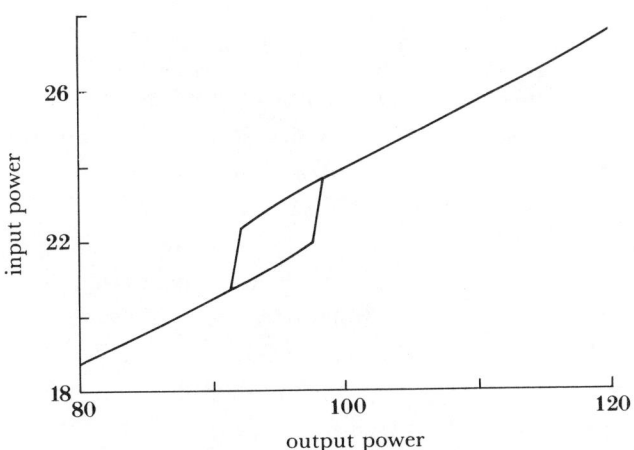

FIGURE 2. Output power as a function of input power, calculated by using the f.H.t. method.
$\mathscr{F} = 1/128$, $Re^{-\alpha D} = 0.218$, $\delta = 0$.

[113]

to intensity profiles, Firth & Wright found smooth quasi-Gaussian profiles on the lower branch, switching to a peak-and-ring profile on the upper branch. The f.H.t. calculations (figure 3) broadly confirm this.

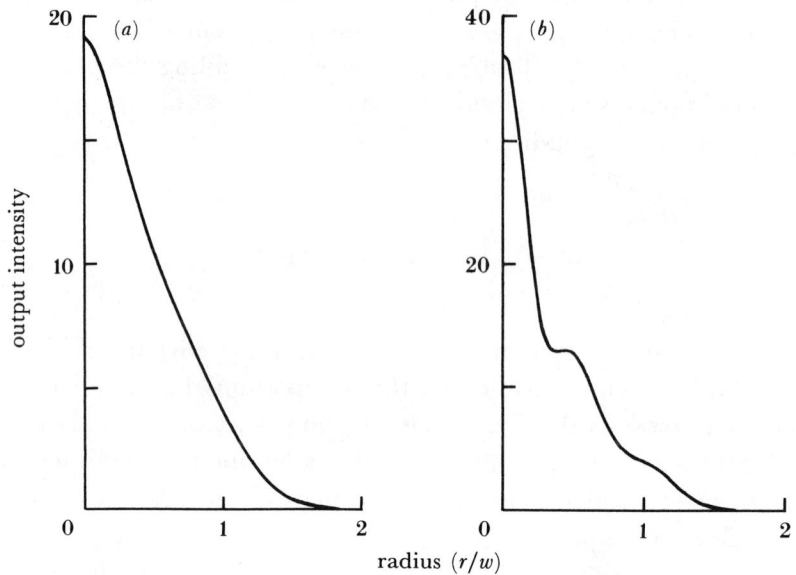

FIGURE 3. Near-field intensity profiles for (a) 'off' and (b) 'on' states. Parameters as in figure 2.

Firth & Wright postulated that this beam reshaping could occur, and show hysteresis, even in the absence of power hysteresis, a phenomenon that they termed 'spatial hysteresis'. Our new results do not show this, though ironically there are two reports (Dagenais & Winful (1984) in CdS, Walker (this symposium) in InSb) in which spatial hysteresis is observed with no apparent power hysteresis. In contrast to the opening discussion in this section, it would seem that the particular parameters used in our computations happen to lie in the transition region between whole-beam switching and part-beam switching, and are thus particularly sensitive to the algorithm and approximations used.

4. DIFFUSION

In this section we present some new analysis and results on transverse effects in optical bistability where diffusion (of the excitation) is the primary mechanism coupling neighbouring regions of the nonlinear medium. Whereas diffraction or nonlinear lensing, or both, will establish a correlation over regions of area $ca.$ λL (see §3), diffusion will be effective over areas of order l_d^2 which, if larger than λL, may lead to whole-beam switching where a purely local theory would predict part-beam switching. The minimum pixel size and power for bistability will then also be determined by l_d^2 rather than λL.

Diffusion has previously been considered mainly with regard to grating effects (see above), but Rosanov (1981) has examined transverse diffusion, though primarily in regard to switching waves analogous to those in diffractive coupling.

There are a number of reasons for believing transverse diffusion to be significant in InSb, not least the fact that material parameters lead to an estimate of $ca.$ 70 μm for l_d at 77 K

(H. A. MacKenzie, private communication); square hysteresis loops are observed despite fairly high Fresnel number, and multiple orders are obtained despite nonlinear lensing and finesse destruction. Diffusion also manifests itself as an increase in the central intensity needed to obtain a given nonlinear phase shift in a (single-pass) defocusing experiment, if the beam area becomes comparable with l_d^2. Preliminary measurements in InSb (F. P. Tooley, private communication) show such an effect for $w \approx 100$ μm. In fact a variational-Gaussian approximation to the carrier profile (6) in a single pass experiment gives a width larger than w by a factor $f^{-\frac{1}{2}}$:

$$f = \{(1 + 32\, l_d^2/w^2)^{\frac{1}{2}} - 1\}/(16\, l_d^2/w^2).$$

It will be seen that $f = 1$ for $l_d = 0$, but $f = \frac{1}{2}$ even for $l_d = \frac{1}{2}w$: the corresponding central phase shift is down by a factor $\frac{4}{9}$ in this case.

Turning now to bistability, we choose the simplest possible model, neglecting diffraction and considering only one cartesian transverse coordinate. We can then integrate the field equations (4, 5), and obtain a simple differential equation for the transverse phase profile:

$$-l_d^2\, \partial^2\phi/\partial x^2 + (\phi + \delta) = \theta\, e^{-x^2/w^2}/(1 + F\sin^2\phi). \tag{9}$$

Here θ is proportional to the input intensity, and can be seen to equal the plane-wave on-resonance nonlinear phase shift (equations (2, 3)). The required boundary conditions are that $\partial\phi/\partial x$ be zero on axis, and that $(\phi + \delta) \approx e^{-x/l_d}$ for $x \gg w$.

Figure 4 shows a hysteresis curve obtained by solving (9) for $F = \frac{1}{2}$ and $w = 2l_d$. Bistability is obtained in second order, and both the on-axis phase shift and the 'power' transmission show the squarish hysteresis loops characteristic of whole-beam switching, which confirms that a diffusion length of *ca.* 50 μm could explain the hysteresis behaviour in InSb.

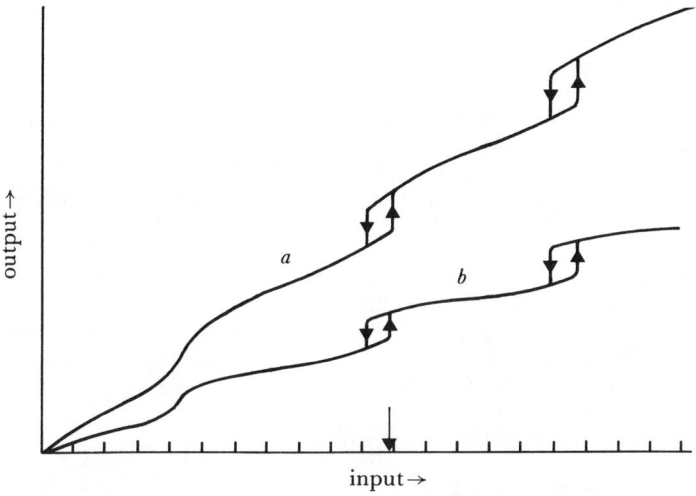

FIGURE 4. Dependence of (*a*) output 'power', and (*b*) on-axis carrier density, on input power: $l_d = \frac{1}{2}w$, $Re^{-\alpha D} = 0.1$, $\delta = 0$.

In computation of figure 4, all phase profiles were smooth. Figure 5 shows intensity profiles corresponding to the lower bistable branch of figure 4, with shoulders arising from the high transmission at $\phi = \pi$. Clearly the neglect of diffraction would be marginal for high-finesse resonators unless $l_d^2 \gg \lambda L$.

[115]

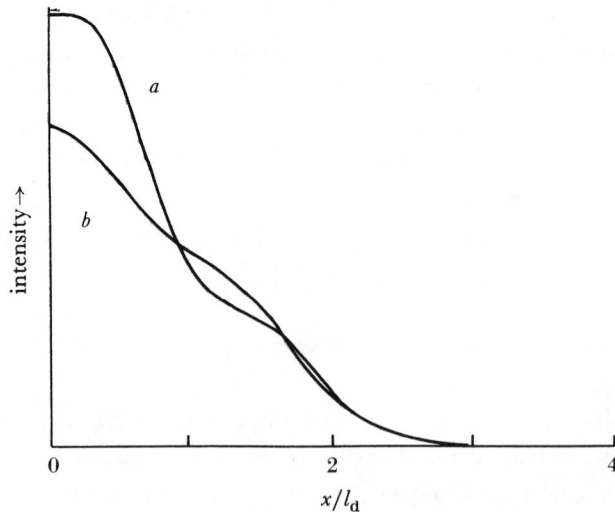

FIGURE 5. Intra-cavity intensity profiles for (*a*) 'on' and (*b*) 'off' states corresponding to the arrowed input in figure 4.

5. CONCLUSIONS

In this paper we have reviewed the salient features of macroscopic bistability theory. We have identified a number of advantages of reflection-mode operation over the conventional transmission mode, discussed the 'transverse effects' and presented some comparative data on diffraction coupling, and, finally, advanced the view that diffusion can explain certain anomalies in the InSb bistability experiments, and supported this with some new numerical results.

REFERENCES

Ballagh, R. J., Cooper, J., Hamilton, M. W., Sandle, W. J. & Warrington, D. M. 1981 *Optics Commun.* **37**, 143.

Dagenais, M. & Winful, H. G. 1984 *Appl. Phys. Lett.* **44**, 574.

Firth, W. J. & Wright, E. M. 1982 *Optics Commun.* **40**, 233.

Gibbs, H. M., Tarng, S. S., Jewell, J. L., Weinburger, D. A., Tai, K., Gossard, A. C., McCall, S. L., Passner, A. & Weigmann, W., 1982 *Appl. Phys. Lett.* **41**, 221.

Ikeda, K., Daido, H. & Akimoto, K. 1980 *Phys. Rev. Lett.* **45**, 709.

Kaplan, A. E. & Meystre, P. 1982 *Optics Lett.* **6**, 590.

Miller, D. A. B. 1981 *IEEE J. Quantum Electron.* **QE-17**, 306.

Miller, D. A. B., Smith, D. S. & Johnston, A. 1979 *Appl. Phys. Lett.* **35**, 658.

Moloney, J. V. 1982 *Optica Acta* **29**, 1503.

Moloney, J. V. & Gibbs, H. M. 1982 *Phys. Rev. Lett.* **48**, 1607.

Rosanov, N. N. 1981 *Soviet Phys JETP* **53**, 47.

Rosanov, N. N. & Semenov, V. E. 1981 *Optics Commun.* **38**, 435.

Siegman, A. E. 1977 *Optics Lett.* **1**, 13.

Wherrett, B. S. 1984 *IEEE J. Quantum Electron.* **QE-20**, 646.

Phil. Trans. R. Soc. Lond. A **313**, 307–310 (1984)

Printed in Great Britain

Generalized multistability and chaos in quantum optics

By F. T. Arecchi

Department of Physics, University of Florence, Florence, Italy, and
Istituto Nazionale di Ottica, Largo Enrico Fermi 6, 50125 Arcetri-Firenze, Italy

Three experimental situations for CO_2 lasers (a laser with modulated losses, a ring laser with competition between forward and backward waves, and a laser with injected signal) are analysed as examples of the onset of chaos in systems with a homogeneous gain line and with a particular timescale imposed by the values of the relaxation constants.

We stress the coexistence of several basins of attraction (generalized multistability) and their coupling by external noise. This coupling induces a low-frequency branch in the power spectrum. Comparison is made between the spectra of noise-induced jumps over independent attractors and the spectrum of deterministic diffusion within subregions of the same attractor. At the borderline between the two classes of phenomena a scaling law holds, relating the control parameter and the external noise in their effect on the mean escape time from a given stability region.

This is a review of a research line at I.N.O. dealing with the role of noise in multistable dynamical systems. The main point is as follows. For some ranges of the control parameters, there can be the simultaneous coexistence of more than one attractor. The basins of attraction are disjoint; hence, provided that the attractors are structurally stable, there is no connection between them. The addition of noise acts as an added dimension to the phase space, which allows a bridging of the attractors. This bridging is evidenced by the appearance of low-frequency tails in the power spectrum, which resemble the so-called $1/f$ noise. Experimental evidence of this effect was given in (*a*) electronic nonlinear devices (Arecchi & Lisi 1982; Arecchi & Califano 1984) such as a Duffing oscillator driven by external modulation (figure 1); (*b*) a CO_2 laser with modulated losses (Arecchi *et al.* 1982) (figure 2); (*c*) a CO_2 laser in a ring configuration implying a competition between the two counter-propagating fields via scattering through the nonlinear inversion grating (Tredicce *et al.* 1984 and this symposium); (*d*) a CO_2 laser with an injected signal at a frequency different from the free-running laser frequency (Arecchi *et al.* 1984*c*). The theory of noise-induced jumps among many attractors was given at two levels: (*e*) with reference to a one-dimensional cubic iteration map, which is the first generalization of the logistic map allowing for more than one attractor (Arecchi *et al.* 1984*a*); (*f*) by a numerical study of the parameter space of a Duffing equation (Arecchi *et al.* 1984*b*) with particular emphasis on the transition region where the régime of two separate attractors undergoes crisis (Grebogi *et al.* 1982). This case is particularly relevant in that it allows comparison with a completely different phenomenon, i.e. the deterministic diffusion within subregions of the same attractor (Geisel & Nierwetberg 1982; Grossmann & Fujisaka 1982) (see figures 3 and 4).

Around the crisis (transition from noise-induced jumps to deterministic diffusion) a universal scaling law relates the control parameter and the external noise in their effect on the mean

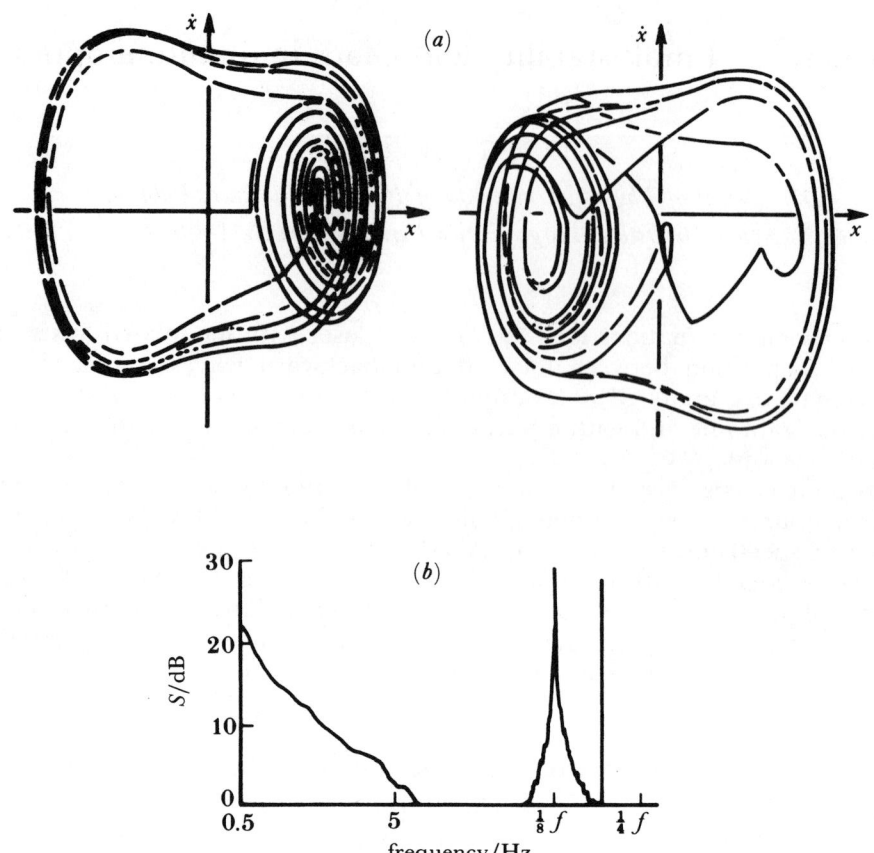

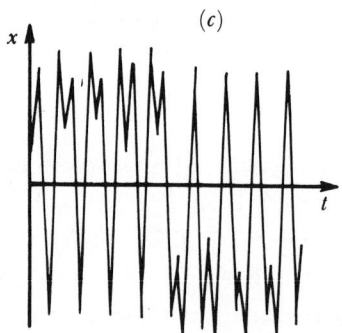

FIGURE 1. Experimental generalized multistability in an analogue system with an electronic nonlinearity studied by Arecchi & Lisi (1982): (a) phase plane $(x, \dot{x})$; (b) power spectrum; (c) signal x against time.

escape time from a given stability region. The main experimental outcome of these phenomena is the appearance of low-frequency tails in the power spectra, which display a $f^{-\alpha}$ behaviour. The slope α in a double-logarithmic plot has been measured (Arecchi & Lisi 1982; Arecchi & Califano 1984; Arecchi et al. 1982; Tredicce et al. 1984 and this symposium) to vary between 0.6 and 1.7 and has been proved theoretically (Arecchi et al. 1984 a) to depend on the number of attractors among which jumps occur and on the Lyapunov exponents of the attractors.

[118]

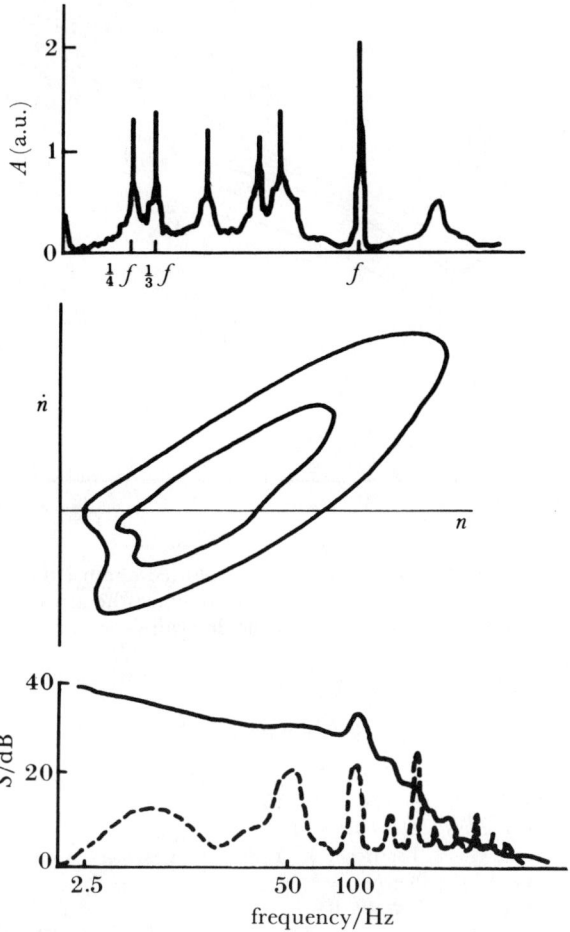

FIGURE 2. Experimental generalized multistability in a CO_2 laser with modulated losses studied by Arecchi *et al.* (1982): (*a*) two superposed ($\frac{1}{3}f$ and $\frac{1}{4}f$) power spectra; (*b*) two coexisting attractors; (*c*) low-frequency part of the spectrum with noise (solid line) and without noise (broken line).

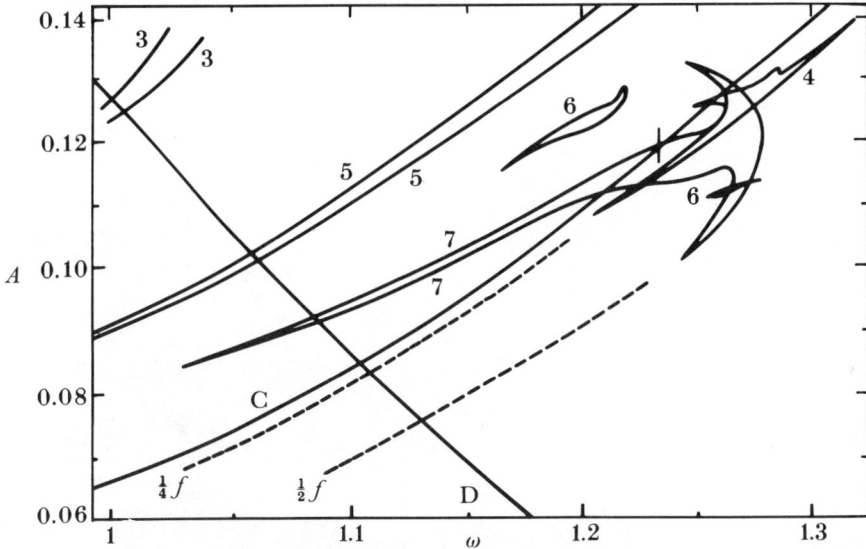

FIGURE 3. Coexistence of many basins of attraction from numerical studies of the Duffing oscillator equations (Arecchi *et al.* 1984*b*) in the phase space of the amplitude (A) and the frequency (ω) of the modulation. Numbers denote the periodicity of the attractors. The vertical line in the parameter space indicates the range over which A is changed in figure 4.

[119]

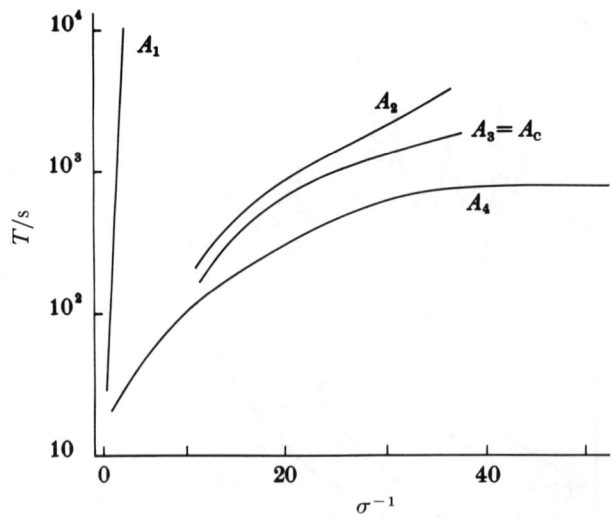

FIGURE 4. The mean escape time from the period-7 attractor plotted against the amplitude of the applied noise (Arecchi *et al.* 1984b) A_1 and A_2 correspond to conditions in which the attractor is stable against infinitesimal perturbations. $A_3 \equiv A_c$ corresponds to the frontier of the deterministic destabilization; A_4 is beyond the crisis.

REFERENCES

Arecchi, F. T., Badii, R. & Politi, A. 1984a *Phys. Rev.* A, **29**, 1006.
Arecchi, F. T., Badii, R. & Politi, A. 1984b *Phys. Lett.* **103** A, 3.
Arecchi, F. T. & Califano, A. 1984 *Phys. Lett.* **101** A, 443.
Arecchi, F. T., Lippi, G., Puccioni, G. & Tredicce, J. 1984c In *Coherence in quantum optics V* (ed. L. Mandel & E. Wolf) pp. 1227–1231. New York: Plenum.
Arecchi, F. T. & Lisi, F. 1982 *Phys. Rev. Lett.* **49**, 94.
Arecchi, F. T., Meucci, R., Puccioni, G. & Tredicce, J. 1982 *Phys. Rev. Lett.* **49**, 1217.
Geisel, T. & Nierwetberg, J. 1982 *Phys. Rev. Lett.* **48**, 7.
Grebogi, C., Ott, E. & Yorke, J. A. 1982 *Phys. Rev. Lett.* **49**, 1507.
Grossmann, S. & Fujisaka, H. 1982 *Phys. Rev.* A **26**, 1779.
Tredicce, J., Lippi, G. L., Abraham, N. B. & Arecchi, F. T. 1984 *Infrared Phys.* (In the press.)

Phil. Trans. R. Soc. Lond. A **313**, 311–319 (1984)

Printed in Great Britain

All-optical logic in optical waveguides

By H. A. Haus and N. A. Whitaker, Jr

Department of Electrical Engineering and Computer Science and Research Laboratory of Electronics, Massachusetts Institute of Technology, Cambridge, Massachusetts 02139, U.S.A.

The motivation for the development of all-optical logic elements in integrated optical waveguides is reviewed. The difficulties inherent in the 'all-optical' approach are outlined. The all-optical, universal logic for pipeline operation based on the Mach–Zehnder interferometer is described and experiments on a prototype all-optical modulator fabricated in $LiNbO_3$ are presented. The paper concludes with a discussion of a device design for operation with shorter lengths and compatible with semiconductor laser drives.

1. Introduction

The advances made in the last few years in reducing the loss and increasing the bandwidth of optical fibres are unprecedented. According to the latest reports, a fibre link 161.5 km in length was operated at 420 Mbit s^{-1} at a wavelength of 1.55 μm (Kasper *et al.* 1983). Shorter fibre links could accommodate higher bit rates, and further progress will no doubt be made in reducing the dispersion in all-optical fibres. The stage is set for optical communication links whose bit rates are enormous compared with contemporary usage. Thus far, however, the optical signals are detected, processed and recorded electronically, not optically.

All-optical signal-processing devices will compete successfully with the current electronic devices by offering unique processing capabilities. Some of these capabilities are listed below.

(*a*) For signals that are already optical, an all-optical signal-processing stage has a natural advantage, provided that it is not of much greater complexity than its electronic counterpart.

(*b*) The absolute bandwidths that are achievable in optical circuits are much larger than those of electronic circuits. It is not an accident that the shortest pulses measured (Fujimoto *et al.* 1984) are optical, not electronic.

(*c*) The introduction of single-mode fibres has advanced the fabrication technology of optical waveguide signal-processing devices because such devices are compatible only with single-mode fibres, not with their more common multimode counterpart.

The circumstances cited above give all-optical, single-mode waveguide signal-processing devices a competitive edge over their electronic counterparts. We are not addressing here two-dimensional signal-processing devices which by their two-dimensionality offer certain unique characteristics. It is well to keep in mind that the handicaps of all-optical single-mode waveguide devices are severe in the race against their electronic competitors, as shown below.

(i) Optical waveguides have large dimensions. Their cross-sectional widths are at least several wavelengths and their interaction lengths are often of the order of millimetres. This must be compared with a microwave FET of 1 μm 'length' and submillimetre 'widths'.

(ii) The topology of optical waveguides is cumbersome. They cannot be 'bent' as easily as a transmission line, and waveguides cannot easily be made to cross.

(iii) Electronic devices need to operate with switching energies that are large compared with kT to prevent random switching (k = Boltzmann's constant, T = thermodynamic temperature). Optical devices must have switching energies large compared with the photon energy $h\nu$ to avoid photon noise.

(iv) Optical nonlinearities are weak compared with electronic nonlinearities.

(v) Optical devices require a continuous optical input, whereas maintenance of a switching state in CMOS is achieved with virtually no supply of power.

The great single advantage offered by all-optical signal processing is its speed; if speed of response is an overriding concern, optical signal processing will always maintain a competitive edge over the electronic alternative.

In this paper we concentrate on a proposed travelling-wave all-optical waveguide device that can be adapted to operate as an AND gate, OR gate, XOR gate, or as an inverter. The device operates in the 'pipeline mode', i.e. a stream of pulses is continuously processed, interacting within a time interval τ and emerging at the output after a delay equal to the interaction time. Since the required pulse energy is proportional to $1/\tau$, lower pulse energies can be achieved by increasing τ, thereby increasing the delay. The speed of the device is solely a function of the energy relaxation time in the material, but not the interaction time. This should be contrasted with the standing-wave Fabry–Perot type devices (Gibbs *et al.* 1979), in which the interaction time also determines the response time.

In §2 we describe the operation of the device and experimental results obtained with a device fabricated in $LiNbO_3$ that performed as a picosecond all-optical waveguide modulator. Section 3 discusses a proposed version of the device that can operate, in principle, with 'multiple quantum well' (m.q.w.) structures at power levels compatible with semiconductor laser diodes. The response time of the device is sharpened (relative to the 20 ns relaxation time in a m.q.w. structure) by means of lateral carrier diffusion, as outlined in §4.

2. The all-optical Mach–Zehnder interferometer

One requirement for an all-optical logic device is that it should operate with mutually incoherent optical excitations because maintenance of phase coherence is too severe a requirement in any practical system. It is also desirable to use the same frequency for the interacting pulses. The proposed all-optical logic gate based on the Mach–Zehnder interferometer satisfies these requirements (Lattes *et al.* 1983). This device uses single-mode optical waveguides as shown in figure 1. A pulse enters the central guide in a TM mode. The two 'control' guides are excited by pulses in the TE mode. The interferometer performs different functions depending on the d.c. bias voltage applied to the electrodes.

If the field is such that the phase difference in the two arms of the interferometer is π and no pulse enters the 'control' guides, then an antisymmetric field pattern is produced at the output waveguide 'Y'. Because the output waveguide is a single-mode guide this field (which corresponds to a higher-order antisymmetric mode) is not confined to the waveguide and escapes into the substrate: no output occurs. Next consider the case when one TE pulse enters either one of the two control guides. If the intensity of the control pulse is adjusted so as to produce an index change Δn such that $(\omega/c)\,\Delta nl = \pi$, the bias π phase shift is cancelled and the TM pulse emerges at the output.

Of course the π phase shift occurs only at the peak intensity, and cancellation over the entire

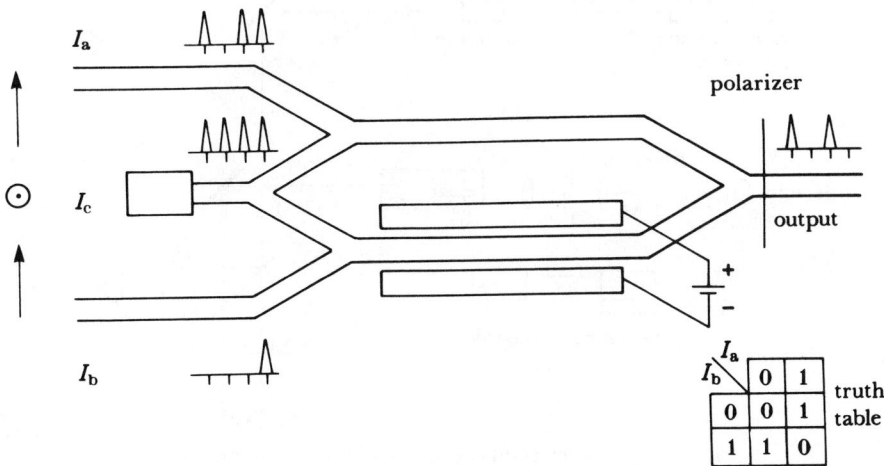

FIGURE 1. Schematic diagram of the optically modulated Mach–Zehnder waveguide interferometer.

pulse duration is not perfect, as some distortion occurs. This distortion can be overcome if there is birefringence such that the pulses 'slip' through one another, effectively broadening the time window within which the change Δn occurs as seen by the TM pulse. The device operates as an XOR gate, as indicated by the inset in figure 1. Figure 2 shows other possible functions of the interferometer, corresponding to different choices of the bias and different choices of inputs and outputs.

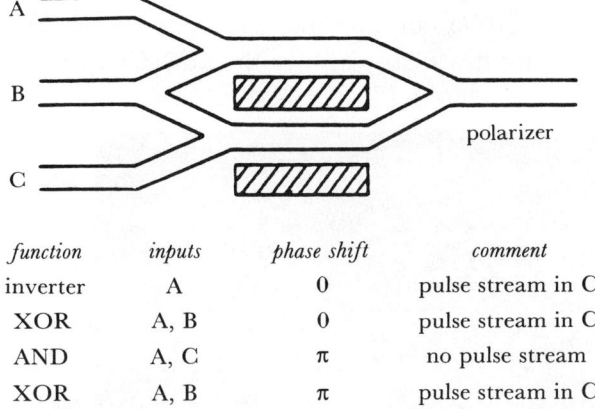

function	inputs	phase shift	comment
inverter	A	0	pulse stream in C
XOR	A, B	0	pulse stream in C
AND	A, C	π	no pulse stream
XOR	A, B	π	pulse stream in C

FIGURE 2. Possible functions of interferometer for different biases and different input ports.

Such a device has been fabricated in Z-cut Y-propagating LiNbO$_3$ to demonstrate some of its features (Lattes *et al.* 1983). The power required to operate an XOR gate of total interaction length $L = 2$ cm was estimated by Miller's rule to be of the order of 150 W. In fact the operation of the device was the first measurement of the coefficient $\chi^{(3)}_{xxzz}$ expressed as n_2 and found to be $n_2 = 3 \times 10^{-9}$ cm^2 MW^{-1}. This called for a drive power roughly 10 times greater than predicted. Because only about 2 W peak power was provided in the interaction region, only *modulation* of one optical pulse by the other could be demonstrated. The control pulse was shifted relative to the controlled pulse by an interferometer arrangement, as shown in figure 3. The electrode voltage was switched to give phase shifts of $-\frac{1}{2}\pi$ and $\frac{1}{2}\pi$ alternately to provide maximum positive and negative response sensitivity and the response was measured on a dual trace scope. The

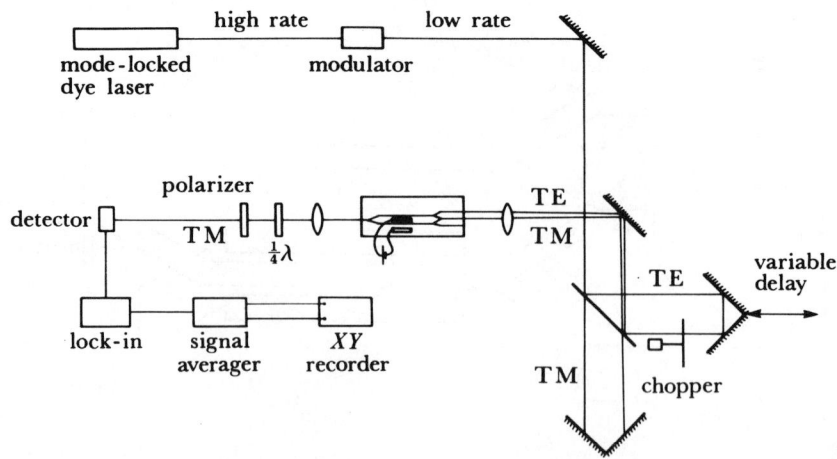

FIGURE 3. Experimental set-up for testing a modulator.

measured full-width half maximum of the pulses shown in figure 4 is 10 ps. The pulses appear to be 'extended' because the pulses slip through one another owing to the birefringence of LiNbO$_3$. Other pertinent experimental parameters are summarized in table 1.

This experiment demonstrated, in effect, the first all-optical modulator. A practical device, operated from a semiconductor laser and of shorter length, requires a material of much larger nonlinearity. The measured n_2 in GaAs is two orders of magnitude larger than the n_2 found in LiNbO$_3$ (Chen & Carter 1982).

We are currently fabricating such an interferometer in GaAs. This structure is not yet practical owing to its excessive length (1 cm) and the required peak power (20 W). The device becomes practical only if one could drastically reduce the power requirement. The very large

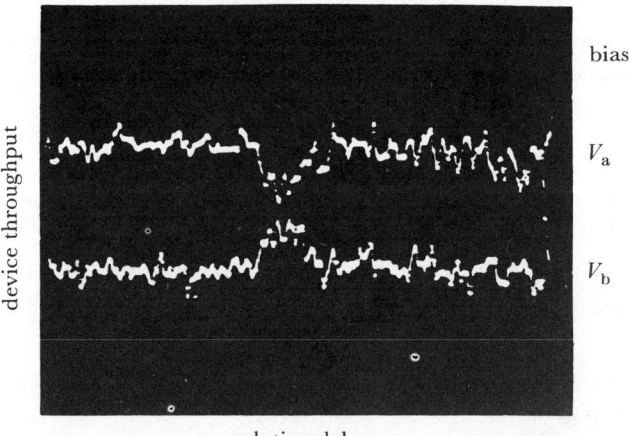

FIGURE 4. Response of modulator. The two traces are obtained by switching the bias voltage to produce $\frac{1}{2}\pi$ and $-\frac{1}{2}\pi$ phase shifts respectively and displaying the interferometer response on a dual-trace oscilloscope.

TABLE 1. EXPERIMENTAL CONDITIONS

source	synchronously pumped Kr$^+$ ion laser
dye	oxazine 750/pc and ethylene glycol
tuning range	720–900 nm
pulse duration	5 ps
repetition rate	80 MHz
this work	peak power, 200 W; λ, 850 nm

nonlinearities found in multiple quantum well GaAs would require a much lower drive power. The response time of room-temperature excitons in m.q.w. GaAs is of the order of 20 ns and additional modifications are therefore necessary to achieve a fast all-optical gate. In the last section we discuss some of these issues as well as a design that operates with 1 W peak power and less than 300 ps response time. Faster operation at low power levels still awaits new material developments of a new device design.

3. A WAVEGUIDE INTERFEROMETER WITH M.Q.W. STRUCTURE

Operation of a multiple quantum well device faster than 50 MHz (*ca.* $1/(20$ ns)) requires a speeding-up of the carrier relaxation rate. Fast traps to aid carrier recombination are not appropriate because they fill up under steady-state excitation. Carrier sweepout by using either drift or diffusion are possibilities. To achieve acceptable nonlinear index changes, it is necessary to produce carrier densities greater than 10^{17} cm^{-3}. Under these conditions carrier transport can be adversely affected by ambipolar effects.

For the carrier drift not to be ambipolar (and unacceptably slow) it is necessary for the applied field not to be reduced appreciably when the positive and negative charge carriers are pulled to the opposite electrodes. For a density of $N = 10^{17}$ cm^{-3} photocarrier pairs to be swept out over a distance of $d = 1$ μm, the applied field, E, would have to be greater than eNd/ϵ, where e is the charge on the electron and ϵ the dielectric constant. This corresponds to a field of 1.4×10^6 V cm^{-1} for a relative dielectric constant of 13. This field is large compared with a field, E_d, that destroys the excitons.

This field may be estimated from $eE_d d = B$, where d is the exciton diameter and B is the two-dimensional exciton binding energy. This gives $E_d = 2 \times 10^4$ V cm^{-1}. Carrier removal by drift only can be accomplished over distances of the order of 0.1 μm. This distance is small compared with a wavelength and entails operation of the nonlinear medium in the fringing field of the optical waveguide. Whereas sweepout by drift may not be impossible, the device proposed here uses sweepout by diffusion.

Figure 5 shows a schematic diagram of such a device, designed to operate by carrier diffusion. The vertical and horizontal scales are widely different. The inset shows the cross section through the waveguide with the layered m.q.w. material. The barrier and well dimensions are chosen to be approximately 10 nm, which is a workable compromise for good exciton confinement without excessive inhomogeneous broadening (Chemla *et al.* 1984). The optical index is changed in the electro-optic tuning section by means of the applied electrodes on top of the waveguides.

The excitonic resonance is dissipative at the centre of the exciton line with a reactive part that peaks and then diminishes as one detunes from line centre. Detuning is also accompanied by a decrease of the nonlinearity, so that minimization of drive power and minimization of device throughput loss are at odds. The response I_0/I_i of the device, equal to the ratio of the output power to the input power of the controlled pulse, is controlled by a change of absorption coefficient ($\Delta\alpha$) and a phase change ($\Delta\phi$) away from the applied π phase shift in one arm. This can be easily determined from the symmetric part of the resulting mode pattern,

$$I_0/I_i = e^{-\alpha_0 L}\{\sin^2\left(\tfrac{1}{2}\Delta\phi\right) e^{-\frac{1}{2}\Delta\alpha L} + \tfrac{1}{4}(1 - e^{-\frac{1}{2}\Delta\alpha L})^2\},\tag{1}$$

where α_0 is the linear (constant) loss.

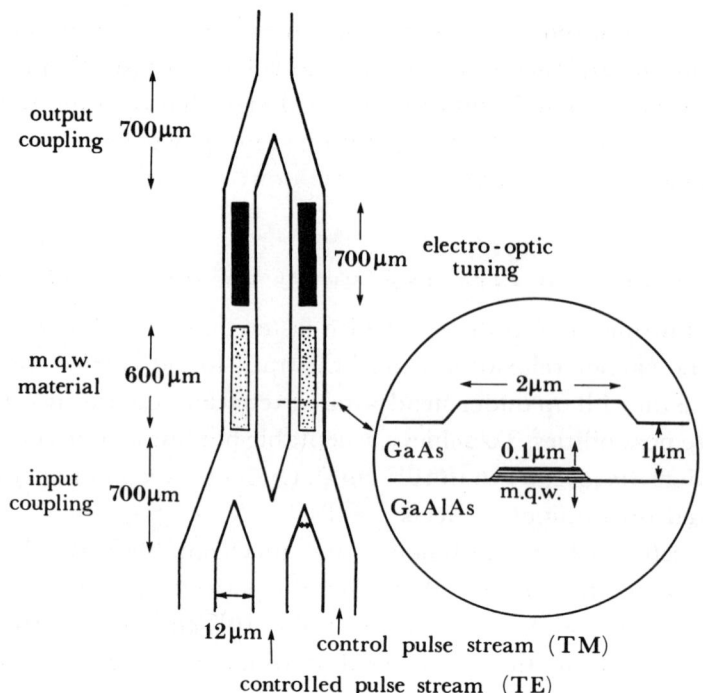

FIGURE 5. Schematic diagram of interferometer design incorporating m.q.w.

The effective phase shift, $\Delta\phi$, and loss saturation can be evaluated from the assumed simple saturation formula

$$\alpha(I) = \alpha_0/(1+I/I_s),\tag{2}$$

where

$$I_s = \hbar\omega/2L_z A_x \alpha_0 \tau_p.$$

Here L_z is the width of the quantum well, A_x is the exciton area and τ_p is the pulse length. The susceptibility has the same intensity dependence. No measurements are available of the m.q.w. absorption spectrum with the polarization perpendicular to the layers. The control mode, assumed to be polarized perpendicularly to the layers, produces carrier density changes that affect the absorption, as indicated by (2). We assume that the absorption of the control mode is given by the same formula as that of a mode polarized parallel to the layers. The controlled mode is assumed to be weaker than the control mode so that its saturation effects can be ignored.

From (2) we find that α changes according to the law

$$\Delta\alpha = \alpha_0 - \frac{\alpha_0}{1+(I/I_s)} = -\frac{\alpha_0(I/I_s)}{1+(I/I_s)}.$$

The change of index follows the same saturation law except that the proportionality constant is n_0, rather than α_0; n_0 can be related to α_0 by Kramers–Kronig relations and is a function of the frequency deviation from the heavy-hole exciton absorption-line centre. The intensity decays with distance due to the absorption saturation of photons and the creation of charged-carrier pairs. Ignoring the absorption saturation ($I(x) \approx I_i e^{-\alpha_0 x}$), $\Delta\phi$ is obtained from the integral

$$\Delta\phi = \frac{2\pi}{\lambda} n_0 \int_0^\infty dx \frac{I_i e^{-\alpha_0 x}/I_s}{1 + (I_i e^{-\alpha_0 x}/I_s)} = \frac{2\pi}{\lambda} \frac{n_0}{\alpha_0} \ln\left(\frac{1 + (I_i/I_s)}{1 + (I_i/I_s) e^{-\alpha_0 L}}\right). \tag{3}$$

In a similar way the effective change in absorption due to saturation is obtained:

$$\Delta\alpha L = -\ln\left(\frac{1 + (I_i/I_s)}{1 + (I_i/I_s) e^{-\alpha_0 L}}\right). \tag{4}$$

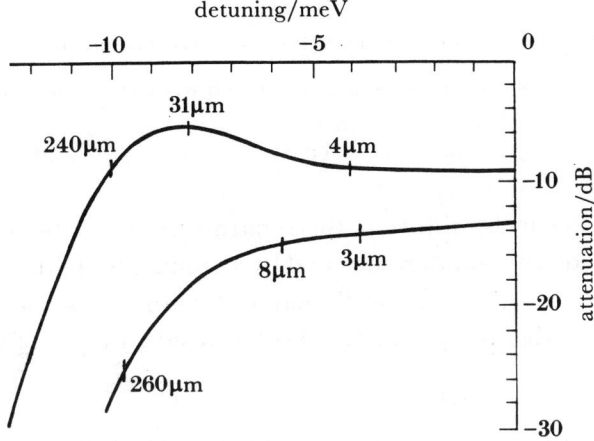

FIGURE 6. Throughput of a waveguide interferometer as a function of detuning from heavy-hole exciton absorption peak. A Gaussian profile for $\alpha_0(\omega)$ has been assumed. Pulse width is 15 ps. Curves are for 1 W (upper) and 0.1 W (lower). Length L of m.q.w. is indicated. A filling factor of 5 % is assumed.

Figure 6 shows a plot of the waveguide interferometer throughput in the 'ON' state (symmetry assures a 0 in the 'OFF' state) as a function of frequency detuning from line centre of the heavy-hole exciton. The curves in this graph show the optimum throughput attainable as the frequency is varied, and the input power and pulse length are held fixed. As the detuning is increased, the loss gets lower but the saturation intensity increases in turn. At a detuning of approximately 9 meV, a device 30 μm in length can be operated with 15 ps pulses and about 6 dB throughput loss. Notice that the actual nonlinear interaction length in the waveguide is approximately 20 times greater, because the entire waveguide cross section is not filled with m.q.w. material.

4. SWEEPOUT BY DIFFUSION

If the device is to be speeded up by sweepout, the carriers have to be removed sufficiently rapidly that no carrier buildup occurs in c.w. operation. Suppose each optical pulse produces N_0 carriers. Suppose further that the input digital signal can be represented by the random sequence $f(n)$, where f takes on the values 0 and 1 with equal probability. Each pulse is assumed to produce the number of carriers N_0 needed to produce a π phase shift, and pulses are separated by a time T. Immediately after the $n = 0$ pulse has arrived, the total number of carriers is given by

$$N(nT)|_{n=1} = N_0 f(0) + N_0 \sum_{m=1}^{\infty} f(m) a^m, \tag{5}$$

where $a(<1)$ is the fraction of the carriers that remain after time T. If the device works reactively, the average response becomes (neglecting background absorption)

$$\frac{I_0}{I_i} = 1 - \tfrac{1}{8}\pi^2 \frac{a^2}{(1-a^2)(1-a)}, \tag{6}$$

where the second term represents a bias due to the long-term accumulation of carriers in the quantum well material. The root mean square deviation of the throughput (σ) is computed similarly and tabulated in table 2. In order to achieve a root-mean-square deviation of less than 5% we need to sweep out at least 80% of the carriers between pulses.

TABLE 2. ROOT-MEAN-SQUARE FLUCTUATIONS ON NORMALIZED OUTPUT INTENSITY

(The variable a is the fraction of carriers remaining in the m.q.w. material after one inter-pulse period.)

a	0.1	0.2	0.3	0.4
σ	0.01	0.06	0.17	0.40

If the carrier sweepout occurs by diffusion, two separate limits can be considered in estimating the device speed. The speed can be underestimated by assuming that diffusion is one-dimensional both inside and outside the quantum well material with the same diffusion constant D (ambipolar). For this case, the number of carriers left inside a layer of width W is given by

$$N(x, t) = N_0 \, e^{-x^2/4Dt'}/\sqrt{(4Dt')}, \tag{7}$$

where $t' = t + W_0^2/2D$ and W_0 is determined from the width of the initial distribution. The number of carriers inside the layer as a function of time is easily expressed in terms of the error function.

Since carrier diffusion outside the layers is essentially two-dimensional and therefore much more rapid, the other limit is obtained by assuming that the carrier distribution vanishes at the layer edges. For this case, the carrier distribution is given by

$$N(x, t) = N_0 \, W \tfrac{1}{2}\pi \cos(\pi x/W) \, e^{-t/\tau}, \tag{8}$$

where $\tau = W^2/(\pi^2 D)$. Notice that the carrier distribution decays much more rapidly in this case owing to the exponential time dependence. If we assume that the curvatures of the carrier distributions are the same at the origin, these two cases are illustrated in table 3. For ambipolar

TABLE 3. FRACTION OF CARRIERS (PERCENTAGES) REMAINING IN THE M.Q.W. MATERIAL AT VARIOUS TIMES

(Case 1 is the Gaussian distribution (equation (7)); case 2 is the cosine distribution (equation (8)). The diffusion constant is 20 cm² s⁻¹.)

time/ps	case 1	case 2
0	100	100
50	45	37
150	28	5
300	20	0.2

diffusion $D \approx 2D_h$ and is taken to be 20 cm² s⁻¹ for this example. Although the actual device recovery time lies somewhere between these two cases, 300 ps should be adequate to sweep out 80% of the carriers.

5. CONCLUSIONS

We have described an experiment on an all-optical waveguide modulator that showed a response on a picosecond timescale. Full modulation or operation of the device as a logic gate called for an exorbitant peak power in the $LiNbO_3$ realization. A GaAlAs waveguide system would permit operation at greatly reduced, and thus realizable peak powers. To reduce power requirements to levels available from semiconductor laser diodes, greater optical nonlinearities than those available in bulk GaAs are required.

An m.q.w. structure was investigated with this objective, with the natural relaxation time speeded up by carrier diffusion. An optical gate design was carried out that predicts pulse separation times less than 300 ps and peak powers of 1 W with pulse widths of 15 ps. Note that the m.q.w. model gave a net loss. This is characteristic of a two-level system with an appreciable linear loss coefficient α_0 that decreases, roughly, with the square of the frequency deviation from line centre, and a nonlinear index that decreases with the inverse of the cube of the frequency deviation. The detuning cannot decrease the linear loss $e^{-\alpha_0 L}$ sufficiently if a net nonlinear phase shift of π is requred within the length L. A minimum insertion loss is found, as in figure 5. Net loss is also found when the controlled mode is made more intense than the control mode. The throughput is defined as the intensity of the output pulse of the controlled mode divided by the intensity of the input pulse of the control mode.

The set of numbers found in this investigation is not yet as good as one would need for practical all-optical signal-processing devices, but improvement of these numbers by an order of magnitude through the use of a higher optical nonlinearity could make these devices of practical importance. The intensive work on nonlinear optical phenomena currently pursued in several laboratories may arrive at improved materials. We are not aware of any fundamental physical constraints that would prevent such hoped-for improvements.

This work was supported in part by the National Science Foundation under grant no. ECS83-10718.

REFERENCES

Chemla, D. S., Miller, D. A. B., Smith, P. W., Gossard, A. C. & Weigmann, W. 1984 *IEEE Jl Quantum Electron.* (In the press.)

Chen, Y. J. & Carter, G. M. 1982 *Appl. Phys. Lett.* **41**, 307.

Fujimoto, J. G., Weiner, A. M. & Ippen, E. P. 1984 Submitted to *Appl. Phys. Lett.*

Gibbs, H. M., McCall, S. L., Venkatesan, T. N. C., Gossard, A. C., Passner, A. & Wiegmann, W. 1979 *Appl. Phys. Lett.* **35**, 451.

Kasper, B. L., Link, R. A., Campbell, J. C., Dentai, A. G., Vadhanek, R. S., Henry, P. S., Kaminow, I. P. & Ko, J. S. 1983 In *Digest of the 9th European Conference on Optical Communications, Geneva*, post-deadline paper.

Lattes, A., Haus, H. A., Leonberger, F. J. & Ippen, E. P. 1983 *IEEE Jl Quantum Electron.* **QE-19**, 1718.

Phil. Trans. R. Soc. Lond. A **313**, 321–326 (1984)

Printed in Great Britain

Applications of guided waves to nonlinear optics

By G. I. Stegeman[1], C. T. Seaton[1] and H. G. Winful[2]

[1] *Optical Sciences Center, University of Arizona, Tucson, Arizona 85721, U.S.A.*
[2] *GTE Laboratories Inc., 40 Sylvan Road, Waltham, Massachusetts 02254, U.S.A.*

Electromagnetic waves guided by single or multiple interfaces offer exciting possibilities for nonlinear optics because of the high power densities that can be achieved with small total powers. Progress in this area with the use of second-order and third-order nonlinearities is reviewed, including the current status of second harmonic generation by codirectional and contradirectional guided waves. A number of third-order nonlinear phenomena have recently been observed and are described. In particular, new nonlinear modes unique to guided waves have been verified to show optical limiter characteristics. Coherent anti-Stokes Raman scattering has been demonstrated experimentally in thin-film waveguides with unparalleled conversion efficiencies.

Introduction

Nonlinear interactions occur whenever the optical fields associated with one or more laser beams are large enough to produce polarization fields proportional to the product of two or more fields. These nonlinear polarization fields radiate with the generated field under optimum conditions (phase-matching), growing linearly with propagation distance; hence the key to obtaining efficient nonlinear optical interactions is to maintain high optical intensities over as long a distance as possible.

Optical beams can be confined to an optical wavelength in one dimension by total internal reflection at the boundary of a film whose refractive index is higher than its surroundings. Diffractionless propagation (in the confined dimension) occurs down the film for centimetre distances, limited by absorption or scattering or both. Nonlinear interactions can take place either in the film, or in the neighbouring media via the evanescent fields that accompany the guided wave. Rectangular channels with cross-sectional dimensions of optical wavelengths can also be used to confine optical beams optimally, provided that the channel region has a higher index than its surroundings.

Most nonlinear guided wave phenomena reported to date are analogues of similar interactions previously studied with plane waves. The nonlinear polarization is usually written as

$$P_{nl} = \epsilon_0 \chi^{(2)} : EE + \epsilon_0 \chi^{(3)} : EEE + \dots, \tag{1}$$

where $\chi^{(2)}$ and $\chi^{(3)}$ are the second-order and third-order susceptibilities respectively, and E is the total field. For second-order interactions, waves at the sum and difference frequencies of the input waves can be generated. In the third-order case for frequency inputs at ω_a, ω_b and ω_c, the nonlinear polarization and hence the radiated fields can have frequency components $\omega_a \pm \omega_b \pm \omega_c$, which leads to a large range of phenomena. We discuss these various applications based on guided waves in this paper.

SECOND-ORDER NONLINEAR PHENOMENA

The principal application of second-order guided-wave nonlinearities is to the generation of second-harmonic radiation. For efficient conversion of codirectional waves, the effective refractive index of both the fundamental and harmonic waves must be equal (phase-matching), just as in the plane-wave case. Since a waveguide is inherently a dispersive medium owing to geometry alone, and because multiple solutions to the dispersion relations exist at a fixed frequency, waveguide phase-matching constraints are considerably less restrictive than those for plane waves.

The best results so far (Sohler & Suche 1983) have been obtained in channel in-diffused $LiNbO_3$ waveguides. Sohler & Suche (1983) have obtained a conversion efficiency of 10^{-3} with a 1 mW He–Ne laser. By making a matched resonator out of the channel waveguide those authors predict that this value can be increased by an additional order of magnitude. Further increases in efficiency will probably require the use of organic thin films (Stegeman & Liao 1983).

Generation of non-phase-matched second harmonics normal to the waveguide surface can be obtained by mixing two oppositely propagating guided waves (Normandin & Stegeman 1982), for example in Ti:in-diffused $LiNbO_3$ waveguides. As illustrated schematically in figure 1, this process can be used to take the autoconvolution of optical pulses in the picosecond

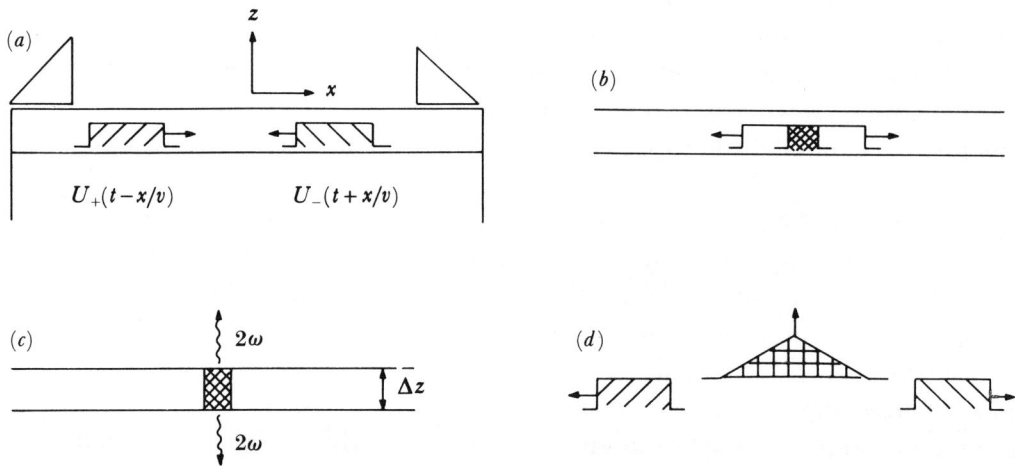

FIGURE 1. Schematic of the nonlinear mixing of two input waveforms that produce a convolution signal at the harmonic frequency: (a) two approaching pulse forms in a waveguide; (b) overlap of the two waveforms at an instant in time; (c) radiation by the polarization field; (d) convolution waveform generated by the nonlinear mixing of the (now departing) input pulses.

time domain. The harmonic convolution signal then falls on a charge carrier (c.c.d.) (which is a linear array of detectors) placed parallel to the surface. Hence the temporal autoconvolution of the waveform now appears as a spatially distributed, digitized signal along the c.c.d. array. This device has applications as a picosecond transient digitizer for capturing single events, provided that new highly nonlinear organic materials can be used (Liao et al. 1983).

THIRD-ORDER NONLINEAR PHENOMENA

The latest advances in the application of guided waves to nonlinear optics utilize third-order nonlinearities. Most of the reported experiments involve a field-dependent refractive index (or, equivalently, dielectric constant). In addition coherent anti-Stokes Raman scattering has been observed in waveguides.

Nonlinear guided waves

Many of the applications of third-order nonlinearities stem from a field-dependent refractive index. Rewriting (1),

$$P = P_1 + P_{nl} = \epsilon_0[n_0^2 + \chi^{(3)} : EE] E, \tag{2}$$

where clearly the second term in the square brackets can be interpreted as a field-dependent refractive index. For plane waves the nonlinear term reduces to $2n_0 n_{2,E}|E|^2$ where $n_{2,E}$ is called the intensity-dependent refractive index. The situation for guided waves is more complex (Stegeman 1982). For isotropic waveguide media

$$P_{nl,i} = 2n_0 n_{2,E}[\tfrac{2}{3}E_i E_j E_j^* + \tfrac{1}{3}E_i^* E_j E_j], \tag{3}$$

which differs from the plane-wave case when evanescent fields are involved, as is always true for guided waves.

Analytical dispersion relations can be obtained for nonlinear thin-film waveguides (Akhmediev 1982; Boardman & Egan 1984; Stegemen et al. 1984a). For TE-polarized waves (E parallel to the surface), one solves the nonlinear wave equation $\nabla^2 E - k_0^2[n_0^2 + \alpha|E|^2] E = 0$ in all of the waveguiding media and matches tangential boundary conditions across the film interfaces. For example for a film of thickness h and refractive index n_f bounded on one side by a linear medium of refractive index n_s and on the other by a nonlinear medium characterized by $n = n_0 + \alpha|E|^2$, the dispersion relation is

$$\tan(\kappa k_0 h) = \kappa[s\tan(s\kappa z_1) + p]/[\kappa^2 - sp\tan(sk_0 z_1)],$$

where $k_0 = \omega/c$, $\kappa^2 = n_f^2 - \beta^2$, $p^2 = \beta^2 - n_s^2$, β is the effective mode index and z_1 is related to the guided wave power. A numerical example is shown in figure 2, where curves a and b

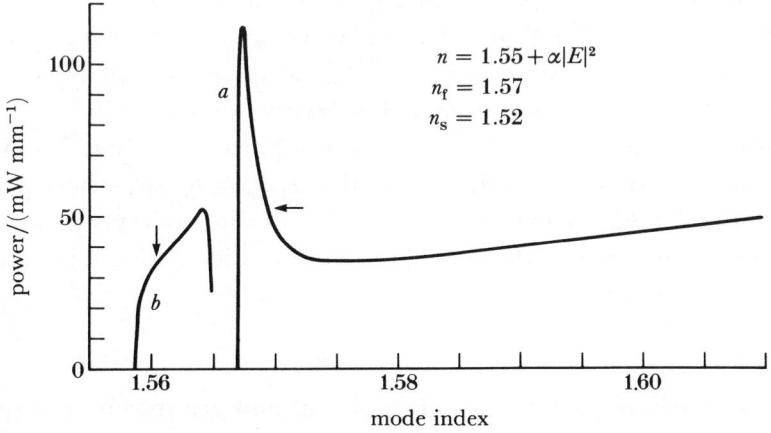

FIGURE 2. The mode index, β, versus the guided wave power for $TE_0(a)$ and $TE_1(b)$ modes guided by a film of thickness 2 μm.

correspond to the TE_0 and TE_1 modes. Note the maximum in the power that can be transmitted in both cases and that $\beta > n_f$ is possible. If both bounding media are nonlinear, or if the film is, multiple new modes are also predicted.

There is experimental evidence (Vach *et al.* 1984) for such power-dependent modes. Shown in figure 3 is the transmitted against incident TE_1 power for a glass waveguide with a liquid-crystal medium on top. The limiting action is clearly shown, as well as the two modal branches that coexist at the same input power.

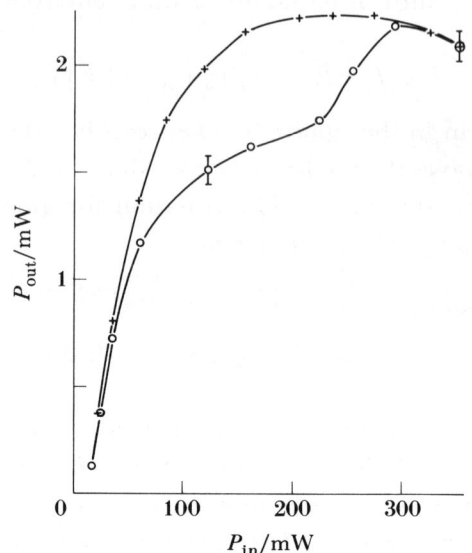

FIGURE 3. The power transmitted in the TE_1 mode for both increasing ($+$) and decreasing ($\circ$) incident powers.

Optical bistability

One of the important potential applications of a field-dependent refractive index is to optical bistability (Miller *et al.* 1979). The feedback required can be obtained by using gratings, either as mirrors or for distributed feedback (Winful *et al.* 1979; Winful 1981). Assuming a $\frac{1}{2}\pi$ phase shift is required (which is the maximum) and a highly nonlinear material such as InSb ($\lambda \approx 5.5\ \mu m$) in thin film form, the power required for bistability can be as low as $15\ \mu W\ mm^{-1}$ (Stegeman 1982) or even $10\ nW$ in channel waveguides (Stegeman *et al.* 1984*b*). A collaborative programme is currently under way to test these concepts, involving researchers at the University of Arizona and Heriot-Watt University.

Nonlinear gratings can be used for a variety of all-optical functions. The Bragg condition at which maximum reflection occurs from a grating is $\boldsymbol{\beta}_s = \boldsymbol{\beta}_i + \boldsymbol{\kappa}$, where $\boldsymbol{\beta}_i$ and $\boldsymbol{\beta}_s$ are the incident and scattered guided wavevectors, and $\boldsymbol{\kappa}$ is the grating wavevector. Since the β can be controlled optically, the grating characteristics can be controlled optically. This makes switching, logic, etc., all possible.

Nonlinear couplers

The operation of distributed couplers such as prisms and gratings is changed for nonlinear waveguides. This has been pointed out by Carter & Chen (1983), who predicted switching, and by Liao & Stegeman (1984), who showed that the coupling efficiency can be dramatically reduced. In a prism of refractive index n_p, for example, optimum coupling is usually obtained when light is incident through the prism at an angle θ to the surface normal such that $\beta = n_p k_0 \sin \theta$. If β depends on guided wave power, this synchronous coupling condition is lost and coupling efficiency falls.

Coherent couplers

A coherent coupler consists of two almost identical parallel channel waveguides in close enough proximity for the guided wave fields of one to overlap the other. Light introduced into one channel is transferred to the second after a characteristic distance, which varies inversely with $\Delta\beta$, the difference in propagation constants of the two waveguides (Jensen 1982). For an intensity-dependent wavevector the coupling condition becomes power-dependent. This effect has been observed with $LiNbO_3$ channel waveguides (Lattes *et al.* 1984) and is discussed by Hans & Whitaker (this symposium).

Coherent anti-Stokes Raman scattering

Coherent anti-Stokes Raman scattering is the nonlinear mixing of two photons at frequency ω_1 with one photon at frequency ω_2 to produce a nonlinear polarization field at frequency $2\omega_1 - \omega_2$. The key is to arrange the propagation wavevectors so that $2\boldsymbol{\beta}_1 - \boldsymbol{\beta}_2 = \boldsymbol{\beta}_3$, where $\boldsymbol{\beta}_3$ is a guided-wave wavevector of frequency $2\omega_1 - \omega_2$, i.e. the process is phase-matched. Under these conditions, the signal power, P_3, is (Stegeman *et al.* 1983)

$$P_3 \propto L^2 \left| \chi_b^{(3)} + \sum_r \frac{\chi_r^{(3)}}{(\omega_1 - \omega_2 - \omega_r) + i\Gamma_r} \right|^2 P_1^2 P_2, \qquad (4)$$

where L is the propagation distance, and $\chi_b^{(3)}$ and $\chi_r^{(3)}$ are the background and resonance terms respectively. As the difference frequency, $\omega_1 - \omega_2$, is tuned through a Raman transition (frequency ω_r and lifetime Γ_r^{-1}), the signal is resonantly enhanced and both the frequency, ω_r, and linewidth, Γ_r, can be determined.

This phenomenon has been observed in thin-film polystyrene waveguides (Hetherington *et al.* 1984). For the characteristic $992\ cm^{-1}$ vibration of a ring structure, a peak signal corresponding to 0.2% conversion of the incident into the spectroscopic signal has been measured, demonstrating the very high efficiency of this technique.

SUMMARY

Nonlinear integrated optics is, with the exception of second-harmonic generation, a young emerging technology. Guided waves are the optimum means of providing the high power densities and long interaction distances necessary for practical applications of nonlinear optics. Within the last year, nonlinear guided waves, nonlinear coherent couplers, intensity-dependent coupling and coherent anti-Stokes Raman scattering have all been observed. One can confidently expect that both degenerate four-wave mixing and bistability will both be reported before the end of 1984.

This research was supported by the NSF (ECS-8117483 and ECS-8304749) and the Joint Services Optics Program of ARO and AFOSR.

References

Akhmediev, N. N. 1982 *Soviet Phys. JETP* **26**, 299–303.

Boardman, A. D. & Egan, P. 1984 In *Proc. Int. Conf. on the Dynamics of Interfaces* (*J. Physique, Paris, Colloq.*) (In the press.)

Carter, G. M. & Chen, Y. J. 1983 *Appl. Phys. Lett.* **42**, 643–644.

Hetherington, W. M. III, Van Wyck, N. E., Koenig, E. W., Stegeman, G. I. & Fortenberry, R. M. 1984 *Optics Lett.* **9**, 88–89.

Jensen, S. M. 1982 *IEEE Jl Quantum Electron.* **QE-18**, 1580–1585.

Lattes, A., Haus, H. A., Leonberger, F. J. & Ippen, E. P. 1983 *IEEE Jl Quantum Electron.* **QE-19**, 1718–1723.

Liao, C., Bundman, P. & Stegeman, G. I. 1983 *J. appl. Phys.* **54**, 6213–6217.

Liao, C. & Stegeman, G. I. 1984 *Appl. Phys. Lett.* **44**, 164–166.

Miller, D. A. B., Smith, S. D. & Johnston, A. 1979 *Appl. Phys. Lett.* **35**, 658–660.

Normandin, R. & Stegeman, G. I. 1982 *Appl. Phys. Lett.* **40**, 759–761.

Sohler, W. & Suche, H. 1983 In *Proc. SPIE Conf. Integrated Optics III*, vol. 48 (ed. L. D. Hutcheson & D. G. Hall), pp. 163–171. Washington: SPIE Press.

Stegeman, G. I. 1982 *IEEE Jl Quantum Electron.* **QE-18**, 1610–1619.

Stegeman, G. I. & Liao, C. 1983 *Appl. Optics* **22**, 2518.

Stegeman, G. I., Fortenberry, R., Karaguleff, C., Moshrefzadeh, R., Hetherington, W. M. III & Van Wyck, N. E. 1983 *Optics Lett.* **8**, 295–297.

Stegeman, G. I., Seaton, C. T., Chilwell, J. & Smith, S. D. 1984 *Appl. Phys. Lett.* (In the press.)

Stegeman, G. I., Liao, C. & Winful, H. G. 1984 In *Proceedings of the Topical Meeting on Optical Bistability* (ed. C. Bowden & S. L. McCall). New York: Plenum Press. (In the press.)

Vach, H., Seaton, C. T., Stegeman, G. I. & Khoo, I. C. 1984 *Optics Lett.* (In the press.)

Winful, H. G., Marburger, J. H. & Garmire, E. 1979 *Appl. Phys. Lett.* **35**, 379–381.

Winful, H. G. 1981 In *Proc. Int. Conf. Excited States and Multiresonant Nonlinear Optical Processes in Solids, Aussois, France*, pp. 55–58.

Phil. Trans. R. Soc. Lond. A **313**, 327–332 (1984)
Printed in Great Britain

Optical nonlinearity and bistability in liquid crystals

By Y. R. Shen

Department of Physics, University of California, Berkeley, California 94720, U.S.A., and
Max-Planck-Institut für Quantenoptik, 8046 Garching bei Munchen, Federal Republic of Germany

Molecular reorientation and laser heating induced by an optical field can yield significant changes in the refractive indices in a nematic liquid crystal. A c.w. laser beam is intense enough to induce a phase retardation much larger than 2π in a nematic film less than 100 µm thick. Optical bistability in such a film sandwiched between mirrors can be readily observed. Coupling between the two mechanisms for induced refractive indices can lead to interesting results in the bistable operation.

Liquid crystalline materials are known to be composed of highly anisotropic molecules that can be easily reoriented by external fields (see, for example, Sheng 1975). This is particularly true for the mesophase owing to the very strong correlation among molecules. The situation is a close analogy to spins in a ferromagnetic phase. Typically, a d.c. field of $E \approx 100$ V cm^{-1} or $H \approx 0.1$ T is sufficient to induce a significant molecular reorientation and lead to a refractive index change, Δn, as large as 0.01 to 0.1. Since, for molecular reorientation, an optical field is equivalent to a d.c. field as long as there is no strong permanent dipole on the molecules (Herman & Serinko 1979), the same Δn can be induced by a laser intensity of *ca.* 100 W cm^{-2}, readily obtainable from a c.w. laser beam. It immediately suggests that liquid crystals should be an ideal medium for studies of highly nonlinear optical phenomena resulting from the optical-field-induced Δn.

An example is shown in figure 1 (Durbin *et al.* 1981), where we have presented both the experimental data and the theoretical calculation of the induced phase shift

$$\Delta\phi = \int_{-\frac{1}{2}d}^{\frac{1}{2}d} (\omega/c)\,\Delta n\,\mathrm{d}z$$

as a function of the laser intensity, I, in a thin film of homeotropically aligned 4-cyano-4'-pentylbiphenyl (5CB) of thickness $d = 250$ µm. Because of the strong anchoring force on the molecules at the boundary surfaces, Δn is not uniform across the film, but is nearly zero at $z \approx \pm\frac{1}{2}d$ and maximum around $z = 0$. The theoretical curves are calculated from the field-induced molecular reorientation, which is derived from minimization of the free energy of the system with appropriate boundary conditions. With a normally incident laser beam, the homeotropically aligned molecules cannot be reoriented unless the field intensity is above a certain threshold value. Such critical behaviour is known as the Freedericksz transition in liquid crystals (see Sheng 1975). As seen in figure 1, a small change in the laser intensity can induce a rather appreciable change of the phase shift in some region, especially near the Freedericksz transition. The operating point for differential operation can be selected by the application of a bias field, which can be either d.c. or optical because for molecular reorientation the two are equivalent. Thus, for example, if we use a d.c. magnetic field parallel to the liquid crystal

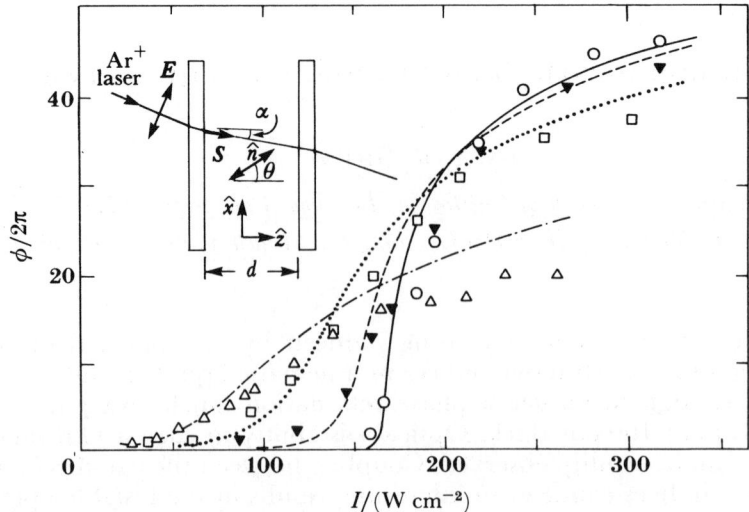

FIGURE 1. Experimental data and theoretical curves for the phase shift $\Delta\phi$ induced in a 250 μm homeotropically aligned 5CB film by an Ar$^+$ laser beam at different angles α: circles and solid curve, $\alpha = 0°$; solid triangles and broken curve, $\alpha = 3°$; squares and dotted curve, $\alpha = 11°$; open triangles and dot-dashed curve, $\alpha = 30°$. The inset shows the experimental geometry.

film to bias the operating point on the $\alpha = 0$ curve near the transition, then a laser intensity of *ca.* 1 W cm^{-2} will be able to induce a π phase shift. This means that a 1 mW He–Ne laser beam will be intense enough to induce optical bistability in such a film sandwiched between two mirrors.

In addition to molecular reorientation, laser heating also contributes to the change in refractive index. Although 5CB has no absorption band in the visible region, laser heating through residual absorption, presumably due to defects or impurities, can still be noticeable. A 350 W cm^{-2} Ar$^+$ laser can raise the local temperature of the medium by *ca.* 2 K. Figure 2 shows how the refractive indices of 5CB change with temperature (Chu *et al.* 1980). It is seen that even well into the nematic phase, we find values of $\Delta n_\parallel \approx -3.2 \times 10^{-3}$ K^{-1} (or -2×10^{-5} cm^2 W^{-1}) and $\Delta n_\perp \approx 7.6 \times 10^{-4}$ K^{-1} (or *ca.* 4×10^{-6} cm^2 W^{-1}). The corresponding

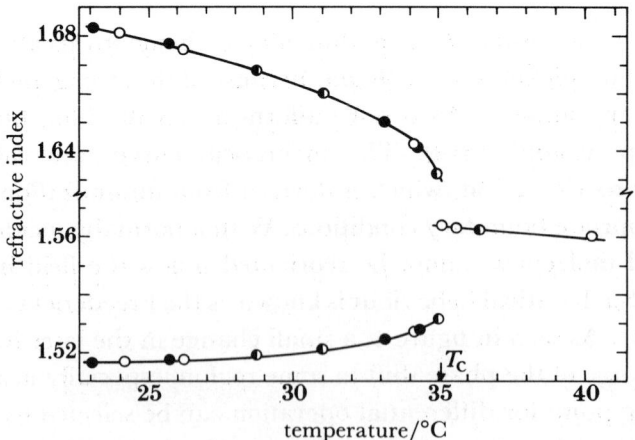

FIGURE 2. Refractive indices of 5CB against temperature, measured by the surface plasmon technique (o) and by the critical angle method (●).

[138]

induced phase shifts in a $250\,\mu\text{m}$ film are $\Delta\phi_\| \approx (10^{-2})2\pi\ \text{rad}\ \text{W}^{-1}\ \text{cm}^2$ and $\Delta\phi_\perp \approx (2\times10^{-3})2\pi\ \text{rad}\ \text{W}^{-1}\ \text{cm}^2$, respectively. This nonlinearity is not as large as that from molecular reorientation, but is certainly large enough to be observed even with a c.w. laser beam. Furthermore, the two mechanisms have very different response times, and hence their effects can be easily separated.

The dynamic response of a nematic substance to an applied laser field is actually quite complicated. The molecular reorientation angle, θ, should obey a diffusion-type equation, which, after some simplification, takes the form (Van Doorn 1975)

$$\gamma\,\partial\theta/\partial t - K\nabla^2\theta = T_H + T_{\text{opt}}, \tag{1}$$

where γ is an effective viscosity coefficient, K is an effective elastic constant, and T_H and T_{opt} are the torques due to the applied magnetic and optical fields, respectively. By knowing that the spatial dependence of θ is usually dominated by the Fourier component $\cos(\pi z/d)$, and assuming that the optical-field-induced orientation is sufficiently small, an integration of the above equation with a suitable proportional constant should yield the following approximate dynamic equation for the induced phase shift $\Delta\phi_\theta$ (Durbin et al. 1983):

$$\left(\frac{\partial}{\partial t} + \frac{1}{\tau_\theta}\right)\Delta\phi_\theta = \frac{\alpha_\theta}{\tau_\theta} I. \tag{2}$$

This is in the form of a Debye relaxation equation, with τ_θ being the relaxation time and α_θ representing the reorienting strength. We can show that $\tau_\theta \approx \gamma/[\pi^2 K/d^2 - GH^2]$, which is of the order of a few seconds to a few tens of seconds for a 5CB sample of ca. 100 μm thick, where G is a constant depending on the initial orientation induced by the d.c. magnetic field (Hsiung et al. 1984). The constant α_θ also depends on the initial orientation and can be of the order of 6 rad W^{-1} cm^2. The temperature rise in a medium due to laser heating of course obeys the thermal diffusion equation. A similar simplification procedure as in the reorientation case leads to the dynamic equation for the thermally induced phase shift $\Delta\phi_T$ (Durbin et al. 1983):

$$\left(\frac{\partial}{\partial t} + \frac{1}{\tau_T}\right)\Delta\phi_T = -\frac{\alpha_T}{\tau_T} I, \tag{3}$$

where $\tau_T \approx d^2/\pi^2 D$ is of the order of 0.1 s for a 100 μm sample (D being the heat diffusion constant), and α_T is directly proportional to the absorption coefficient of the medium.

The above values of Δn (or $\Delta\phi$) and their dynamic characteristics allow us to estimate the strengths of various nonlinear optical effects in liquid crystals. We consider here only optical bistability. It is well known that if a laser beam is intense enough to induce a phase shift of $\Delta\phi = \pi$ in a nonlinear medium filling a Fabry–Perot interferometer, one should be able to observe optical bistability. With a 100 μm 5CB film in an appropriate bias magnetic field, it requires a beam intensity of only a few watts per square centimetre, assuming that molecular reorientation is the dominant mechanism for nonlinearity. Indeed, as shown in figure 3, optical bistability can be easily observed in an 83 μm 5CB film sandwiched between two mirrors (Durbin et al. 1983). The first bistable loop is complete at $I_{\text{in}} \approx 6\ \text{W}\ \text{cm}^{-2}$. With increasing laser intensity, $\Delta\phi$ soon becomes larger than multiples of π, and hence multiple bistable loops are readily generated.

In the above case I have assumed that the laser polarization is parallel to the biasing magnetic field. The laser heating contribution to Δn is then negligible in comparison with the molecular

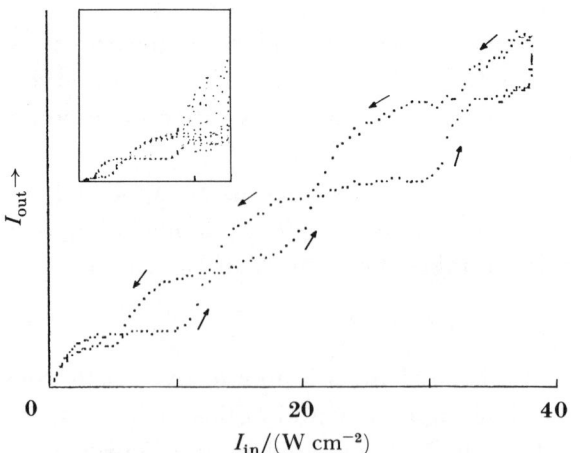

FIGURE 3. Multiple bistable loops obtained from a Fabry–Perot interferometer consisting of an 83 μm 5CB nematic film sandwiched between two mirrors of reflectivities 55 and 75 % at 514.5 nm. A bias magnetic field of 103 kA m⁻¹ parallel to the laser polarization was applied to the system. The inset shows bistable loops and the onset of oscillation.

reorientation effect in the steady-state operation. Laser heating also induces a $\Delta n_\perp$ that is even smaller than $\Delta n_\parallel$. It is, however, possible to increase the laser heating contribution by dissolving some absorbing molecules in the liquid crystal so that $\Delta\phi_T$ and $\Delta\phi_\theta$ become comparable. We can then have an interesting case of a nonlinear Fabry–Perot interferometer with two input beams or orthogonal polarizations, which are coupled through nonlinearity in the cavity (Kaplan & Meystre 1982), i.e.

$$\Delta\phi_\parallel = aI_\parallel\, T_\parallel + bI_\perp\, T_\perp, \quad \Delta\phi_\perp = cI_\parallel\, T_\parallel + dI_\perp\, T_\perp, \tag{4}$$

where a, b, c, d are constant coefficients and $T_\parallel$ and $T_\perp$ are the transmission coefficients of the interferometer for the two polarizations.

The different dynamic responses of the two mechanisms can make the bistable operation even more interesting. Since τ_T is nearly two orders of magnitude smaller than τ_θ, the bistable operation will actually be dominated by the laser heating effect if the operating time for light to go around the bistable loops is in the range $\tau_T \lesssim t_{\mathrm{op}} \ll \tau_\theta$. As t_{op} becomes comparable with or larger than τ_θ, the bistable operation is then dominated by molecular reorientation. The two mechanisms also induce phase shifts of opposite signs ($\Delta n_\parallel > 0$ from molecular orientation and $\Delta n_\parallel < 0$ from laser heating). This, together with the different response times, can cause the output from the interferometer to self-oscillate. The phenomenon can be understood physically from a graphic construction (Durbin *et al.* 1983).

Figure 4a shows that as the laser intensity increases, the output (gauged by $\Delta\phi$) of the Fabry–Perot interferometer rises in steps in a stable fashion. When the operating point, O, is reached, however, one finds that the intensity is now high enough for the interferometer to go into bistable operation with the laser heating mechanism alone, as indicated by the broken line. The operating point, O, is then no longer stable against thermal fluctuations, and in a time short compared with τ_θ it shifts to either A or B. Yet neither A nor B is an operating point in the steady state. As time goes on, molecular reorientation begins to respond to the intensity change in the cavity. The operating point should then move from A (or B) along the Fabry–Perot transmission curve towards O, but as it reaches C (or C′) the system again becomes unstable

[140]

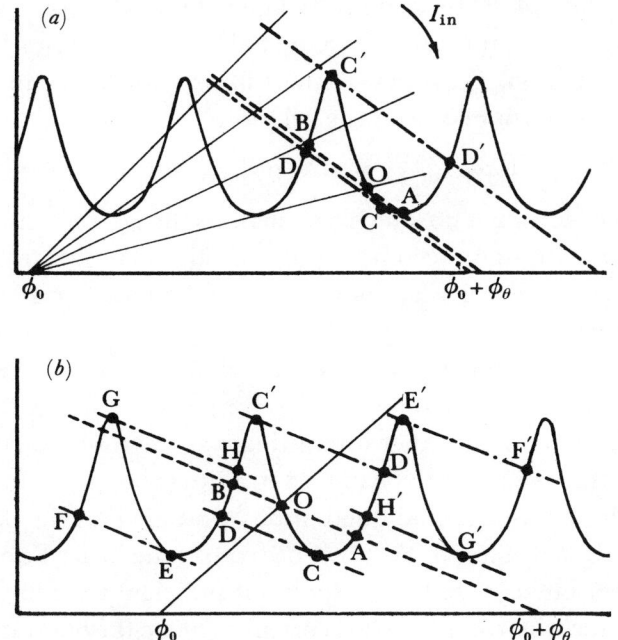

FIGURE 4. Graphic construction showing how the interplay between laser-induced molecular reorientation and thermal effect can lead to indefinite oscillation in the output of a nonlinear Fabry–Perot interferometer. Intersection of the Fabry–Perot transmission curve and solid line gives the steady-state operating point, O. The broken line describes the phase shift due to laser heating alone. In (a) the operating cycle is ACDBC'D'A; in (b) the operating cycle is ACDEFGHBC'D'E'F'G'H'A.

under thermal fluctuations, and is quickly switched to D (or D'). Afterwards, the operating point is shifted to C' (or C) by molecular reorientation, switched to D' (or D) by thermal action, and shifted back to A (or B). It then repeats the cycle ACDBC'D'A, and the output appears in the form of an indefinite oscillation as shown in figure 5. The period of oscillation is of the order of τ_θ.

The picture here predicts that oscillation can occur only when the laser intensity has increased

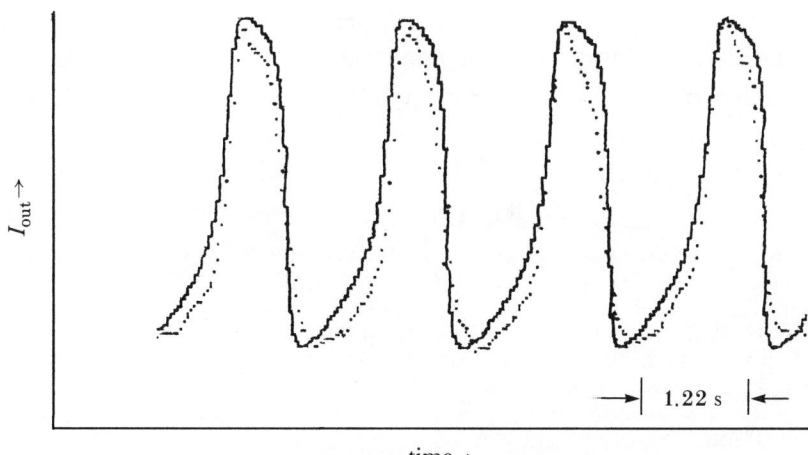

FIGURE 5. Oscillatory output (dotted line) from the Fabry–Perot interferometer described in figure 3 observed with an input laser intensity $I_{in} = 70$ W cm^{-2} and a bias magnetic field $B = 123$ kA m^{-1}. The solid curve is a theoretical fit.

to such a value that the slope of the broken line in figure 4 is less negative than the most negative tangent that one can draw from the point O to the Fabry–Perot transmission curve. The lowest possible intensity for oscillation, I_{th}, arises when O happens to be at an inflection point of the Fabry–Perot curve. It is determined from the relation

$$|C\alpha_T I_{th}|^{-1} = |d\mathscr{I}/d\phi|_{max} \approx F/\pi$$

at which the oscillation condition is no longer satisfied. As the laser intensity is increased further, spatial self-phase modulation may become significant and complicate the oscillation pattern of the output. The effect of the intensity distribution of the transverse beam on oscillation has not yet been worked out.

From the graphical construction, it can be seen that if the laser heating mechanism is stronger, the output will break into oscillation at a lower laser intensity. The oscillation pattern will also change with the laser intensity in a rather interesting way. At a sufficiently high intensity with $\alpha_\theta + \alpha_T < \alpha_\theta$, the operating point O can be moved to a situation shown in figure 4 b. Here again, the point O is not stable against thermal fluctuations. Assume that the operation is first shifted to A, moved to C, and shifted again to D. Now, because of the lower Fabry–Perot transmission coefficient at D, the laser intensity in the cavity is not sufficient to maintain the corresponding $\Delta\phi_\theta$ present. The molecules therefore have to reorient to move the operating point further down the curve towards E, at which it switches to F. The movement down the curve would continue if the laser intensity in the cavity at F is still too low. Otherwise, the operating point will move up the curve from F to G, H, C', D', E', F', G', H', and back to A. Clearly, in this case, the oscillation pattern involving six switching actions in a cycle is much more complicated than the one shown in figure 5. At even higher intensities, more switching actions should occur in a cycle, and the oscillation pattern could become hard to analyse and look somewhat like noise.

Whether liquid crystals can be used as the nonlinear medium to study bifurcation and chaos in an all-optical system has yet to be explored. Interesting results of such study, however, have already been obtained on hybrid systems (Song *et al.* 1983; Zhang *et al.* 1983). In general, liquid crystals as nonlinear media in bistable devices offer a lot for our imagination. The many, easily controllable parameters of the material should allow us to study the various physical phenomena connected with bistability in great detail.

This work was supported by the National Science Foundation under grant number DMR81-17366. The author acknowledges the major contributions of S. D. Durbin to this work and the wonderful hospitality of Professor H. Walther at the Max-Planck-Institut für Quantenoptik.

REFERENCES

Chu, K. C., Chen, C. K. & Shen, Y. R. 1980 *Molec. Cryst. liq. Cryst.* **59**, 97–108.
Durbin, S. D., Arakelian, S. M. & Shen, Y. R. 1981 *Phys. Rev. Lett.* **47**, 1411–1414.
Durbin, S. D., Arakelian, S. M. & Shen, Y. R. 1983 *Optics Lett.* **8**, 39–41.
Herman, R. M. & Serinko, R. J. 1979 *Phys. Ref.* A **19**, 1757–1769.
Hsiung, H., Shi, L. P. & Shen, Y. R. 1984 (In preparation.)
Kaplan, A. & Meystre, P. 1982 *Optics Commun.* **40**, 229–232.
Sheng, P. 1975 In *Introduction to liquid crystals* (ed. E. B. Priestley, P. J. Wojtowicz & P. Sheng), pp. 115–120. New York and London: Plenum Press.
Song, J. W., Lee, H. Y., Shin, S. Y. & Kwon, Y. S. 1982 *Appl. Phys. Lett.* **43**, 14–16.
Van Doorn, C. Z. 1975 *J. Phys., Paris* **36**, C1–C261.
Zhang, H.-J., Dai, J.-H., Peng-Ye, W. & Chaio-Ding, J. 1983 In *Laser spectroscopy VI* (ed. H. P. Weber & W. Luthy), p. 322. Berlin: Springer-Verlag.

Phil. Trans. R. Soc. Lond. A **313**, 333–340 (1984)

Printed in Great Britain

Optical bistability in semiconductor lasers

By J. E. Carroll and I. H. White

Department of Engineering, University of Cambridge, Trumpington Street, Cambridge CB2 1PZ, U.K.

Absorptive, dispersive and modal bistabilities are considered in semiconductor injection lasers. Previous work is briefly reviewed and the discussion is concerned with experimental work with transverse modal bistability in twin stripe injection lasers. A simple circuit analogy indicates how modal bistability can arise and how it is possible to have stronger light output on the side where the current drive is weakest. Experimental results are reviewed for an optical flip-flop with the use of twin-stripe lasers with three such lasers coupled together in an optical logic circuit. Various mode patterns that have been observed by using different geometries of twin-stripe lasers are discussed. Use is foreseen for active bistable devices in front-end optical signal processing.

1. Introduction

It has long been recognized that the semiconductor injection laser could form an active element for information processing (Lasher 1964; Basov 1968; Basov *et al.* 1972). Segmented lasers were considered that could switch between two states, with the bistable action arising because of saturable absorption. Initial work on such devices was often concerned with puslations between the two states (Nathan *et al.* 1965; Lee *et al.* 1970) and was used to study the undesirable effects of inhomogeneities within lasers for optical communications.

In communications, the need for stable injection lasers naturally led to experiments with external optical resonators, which can help frequency stability. However, such optical feedback can sometimes cause pulsations and hysteresis in the current–light characteristics (Broom *et al.* 1970; Lang *et al.* 1980) especially in lasers containing defects or saturable absorption. Hysteresis can indicate bistability, which with optical feedback can be caused by a complex range of mechanisms. Bistability with saturable absorption has been demonstrated by Lau *et al.* (1982). Dispersive bistability has been demonstrated by using frequency-selective optical feedback (Glas *et al.* 1982; Bazhenov *et al.* 1982).

In passive optical bistable devices the optical frequency is fixed, but in lasers the lasing frequency and mode can change. Modal bistability appears to be an important additional mechanism available in active devices. Mode hopping between longitudinal modes in lasers is well known and is a form of bistability or even multistability, when it is possible to switch to one of several modes. At present, stable longitudinal mode hopping is fraught with difficulties that need careful investigation. Transverse mode hopping looks more promising.

In a search for good structures that encourage stable transverse modes it is inevitable that structures are found that encourage modal instability. For example symmetrical twin-stripe lasers can encourage a high-power zero-order mode (Ripper *et al.* 1970), but asymmetry can lead to beam steering (Scifres *et al.* 1978) and instabilities (Kirkby 1978). Shore recognized the potential bistable action of twin-stripe lasers, and carried out extensive theoretical work on these structures (Shore *et al.* 1981, 1983; Shore 1982 *a*). Bistable operation triggered by an external optical source was indicated (Shore 1982 *b*). Later in this paper, experimental work (White *et al.* 1982, 1983) will be outlined demonstrating an optical flip-flop using twin-stripe lasers, along with new bistable results.

Injection lasers, then, exhibit absorptive bistability, dispersive bistability and modal bistability. Only a brief review is given of the first two effects.

2. ABSORPTIVE BISTABILITY

Figure 1 shows a schematic construction for an injection laser that will exhibit absorptive bistability (Lasher 1964; Basov 1968). Segment 1 is an active region on forward bias carrying a current *I*, and segment 2 acts as a saturable absorber. A simplified account of the bistable action is as follows: the total loss in both segments is too high to permit lasing action for the

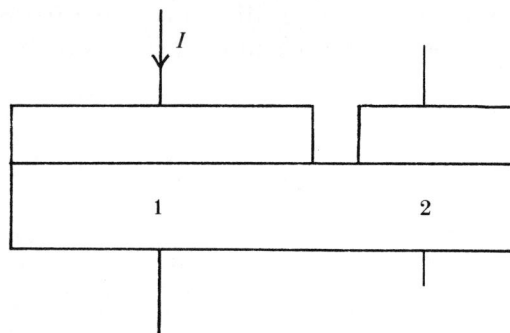

FIGURE 1. Segmented bistable injection laser (schematic). Segment 1 is the active region carrying current I; segment 2 is the saturable absorber.

combined cavities, with the current *I*. Only spontaneous emission is then generated by *I*, as in a light-emitting diode. The device is 'off'. To switch into the 'on' state, *I* is momentarily increased, or light is injected, so as to increase the light intensity sufficiently to saturate the loss in segment 2. The total incremental loss in the two segments is then low enough to permit lasing action with the current *I* generating sufficient stimulated emission to keep the loss saturated in segment 2. Lasing action is maintained and the device is 'on'.

Early experiments often exhibited pulsations rather than bistability. This may have been because of electrical feedback between the two segments. Harder *et al.* (1981, 1982) have shown that good bistable action requires an adequately high electrical resistance between segments 1 and 2 to prevent electrical feedback. Alternatively, it may have been that the nonlinearities were inadequate in the early work. Kawaguchi (1982 *a, b*) has shown that bistability can occur with multiple unpumped segments and that it is possible to switch between bistable states by using external trigger pulses.

Different configurations have been demonstrated. Coupled cleaved cavities (Dutta *et al.* 1984) have exhibited a bistable frequency change. Lau *et al.* (1982) have found that feedback with an external mirror can help to enhance the bistability. However, the detailed mechanisms at work in these experiments may well be a mixture of dispersive as well as absorptive bistability.

3. Dispersive bistability

The refractive index of the material in an injection laser is affected by the electron density. Typically one finds for GaAs at the frequencies of laser action that a density of 10^{24} m^{-3} changes the refractive index by about 0.7%. The electron density can be changed in nanoseconds through local current injection combined with diffusion and recombination. The electron density is also changed by the optical intensity stimulating recombination within a picosecond timescale. A laser that was 100 µm long would then require an optical input energy of about 2J per square metre of cross section to change the cavity's optical length by $\frac{1}{4}\lambda$. If the recombination time, τ, for the electrons were 1 ns then this optical energy, spread over τ, would require an input power around 2 mW per square micrometre of cross section of active region to effect a useful change in the laser's optical length.

Dispersive optical bistability can be demonstrated with an external feedback system, as shown in figure 2. There have been many experiments with optical feedback, but to establish *dispersive*

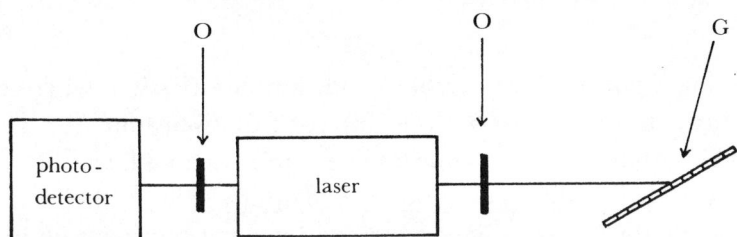

FIGURE 2. Schematic arrangement for observation of dispersive bistability in injection lasers. O, Lens or microscope objectives; G, grating or holographic element giving frequency-selective feedback into laser.

bistability it is useful to make the feedback at a single frequency (Bazhenov *et al.* 1982; Glas *et al.* 1982). Bazhenov, for example, used a holographic grating to demonstrate such bistability. At low power levels the optical length of the laser was such that the feedback at the selected frequency had a phase that prevented oscillation. Changing the optical power level then altered the optical length of the laser so that laser action was possible. Optical bistability could be observed at the 1 mW level, in rough agreement with previous estimates.

4. Modal bistability and twin-stripe lasers

To set the scene, we consider a simple model of coupled resonators (figure 3) with equal positive resistive loading but possibly asymmetrically distributed negative resistance. The circuit parameters are adjusted so that there are two modes that exhibit steady state oscillations with the same negative resistances but slightly different frequencies (see appendix). The mesh currents, i_1 and i_2, can differ markedly in amplitude so that in the power, $|i_1^2| R$ and $|i_2^2| R$, transferred to the positive resistance loads in each arm, can be asymmetrical and need not have the same sense of asymmetry as the negative resistance. The net power transferred to the positive resistances can be shown to be equal to the power given out by the negative resistances if and only if the system is oscillating in one or the other of the two permitted modes. Thus linear circuit theory, along with power conservation, shows that there is modal bistability; one mode

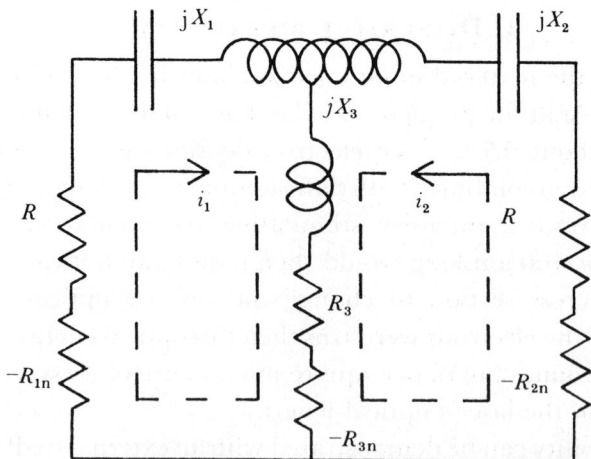

FIGURE 3. Equivalent circuit to demonstrate modal bistability. It is possible to have $|i_1| > |i_2|$ independent of whether $R_{1n} \lessgtr R_{2n}$. Only one mode at a time to give stable oscillation with the same negative resistances but different frequencies for the two modes: see Appendix.

or the other will oscillate but not both. Other conditions are possible where both modes may oscillate simultaneously, but these are not relevant for this discussion.

Transverse modal bistability of twin-stripe lasers exhibits various forms. The two laser stripes may be regarded as coupled oscillators as above. The optical gain in the medium provides the negative resistance while the facets radiate power and present positive radiation resistances to the circuit, with the resonator providing the reactance. A linear model combined with power conservation gives a helpful start to understand mode competition and the fact that the asymmetry need not correspond with the drive asymmetry (Mukai *et al.* 1983), but the model is not adequate for all cases because of the changes, within the resonator, of refractive index with electron density, which can in turn be changed by the optical intensity. Progress is being made elsewhere with theoretical models (Shore 1982 *a*, *b*; Shore *et al.* 1983), but here we report experimental results.

Figure 4*a* shows a section for twin-stripe lasers using GaAs/GaAlAs heterojunctions made by Dr R. G. Plumb of Standard Telecommunication Laboratories. Three types of these lasers have so far been investigated. Closely coupled twin-stripe lasers ($s = d = 3\ \mu m$) exhibit index guiding when the currents in the two stripes have similar values. A dip in the electron concentration between the two stripes gives sufficient increase in the refractive index to provide waveguiding. If the current in one of the stripes dominates over the other then the light–current characteristics are similar to the gain guided output from a single $3\ \mu m$ stripe laser (White *et al.* 1982). At a critical ratio and level of the two current drives it is possible to have an index-guided mode that sits centrally between the two stripes, or alternatively to have a mode that is gain guided (figure 4*b*). This latter mode permits the peak optical intensity to cross from beneath one stripe at one facet to the other stripe at the other facet (White & Carroll 1983; Shore 1983), indicating beam steering, which could be useful for changing coupling to a fibre. At present the fastest time to switch between these two modes has been 800 ps, but this has not been optimized.

With these lasers both electrical bistability (applying an electrical pulse to trigger the laser from one mode to the other) and optical bistability have been demonstrated (White *et al.* 1983). Figure 5*a* gives a schematic layout of the latter experiment. The twin-stripe lasers exhibit

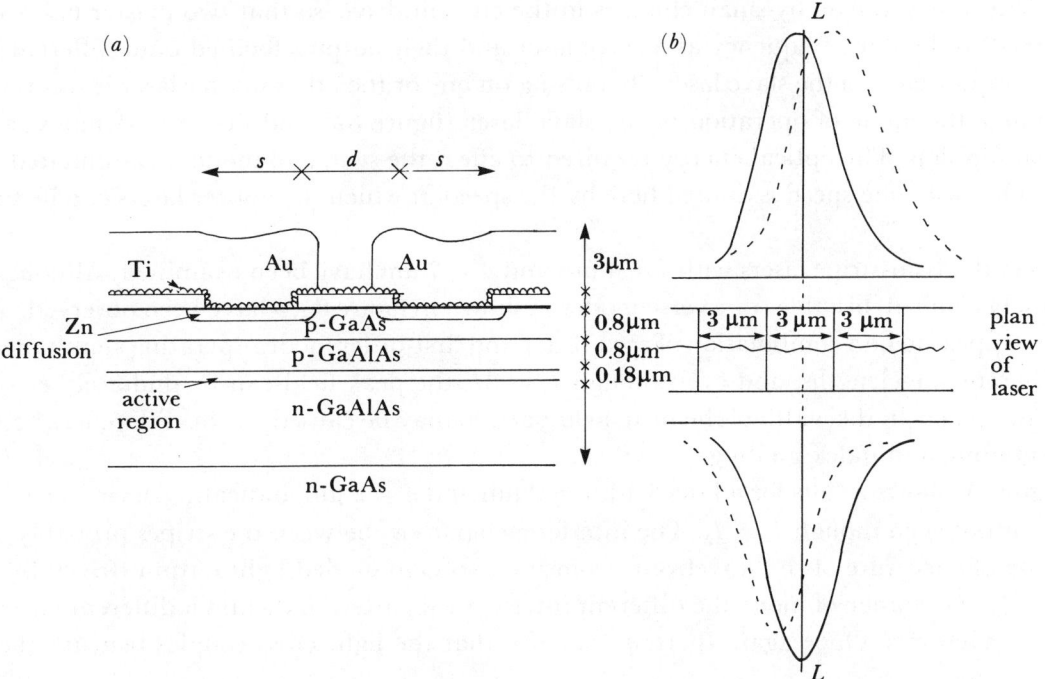

FIGURE 4. Twin-stripe laser: (a) schematic cross-section (s, stripe width, d, separation); (b) optical near-field distributions (s = d = 3 μm for bistable states A (——) and B (----) measured at each facet with $I_1 = I_2$, 10% above threshold.

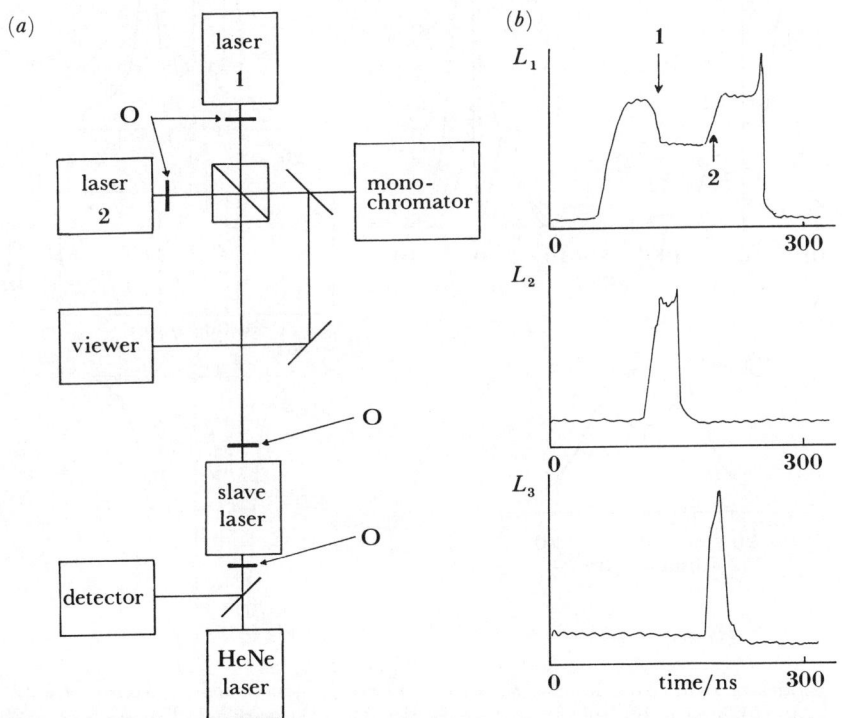

FIGURE 5. Optical flip-flop: (a) experimental layout (O, microscope (×50) objectives); (b) timing of optical outputs–inputs. L_1, Light output (relative units) at a point on the near field from bistable slave laser; L_2 and L_3, light outputs (relative units) from master lasers into slave laser to trigger bistable switching at 1 and 2 respectively.

[147]

sufficient tuning range, by small changes in the current drive, so that two master lasers could be tuned to the same frequency as a slave laser and their outputs focused onto different parts of the 'input' facet of the slave laser. By pulsing on one or the other master laser it was possible to change the mode of operation of the slave laser (figure 5b) and hence to demonstrate an optical flip-flop. The optical energy required to effect the switch of mode was estimated to be 1 pJ. The switching speed is limited here by the speed at which the master lasers can be turned on.

Recently, twin-stripe lasers with $s = 4$ μm and $d = 7$ μm have been examined. Although not so closely coupled, bistable transverse modes as shown in figure 6a, b have been observed. These modes appear to be a combination of zero-order and first-order modes operating simultaneously at different wavelengths, and even though $I_1 = 2I_2$ the peak field can be under either stripe. The differences in the width of the near-field pattern may be caused by the differences between gain guiding and index guiding.

Figure 6c shows results for a laser with $s = 2$ μm and $d = 2$ μm, indicating asymmetry in the light output even though $I_1 = I_2$. The interference pattern between the stripes probably arises because of curvature of the wavefronts from the two gain guided light output driven by each stripe. At the change of mode the different interference pattern indicates a different curvature of the wavefronts. Once again there is evidence that the light cross-couples beneath the two

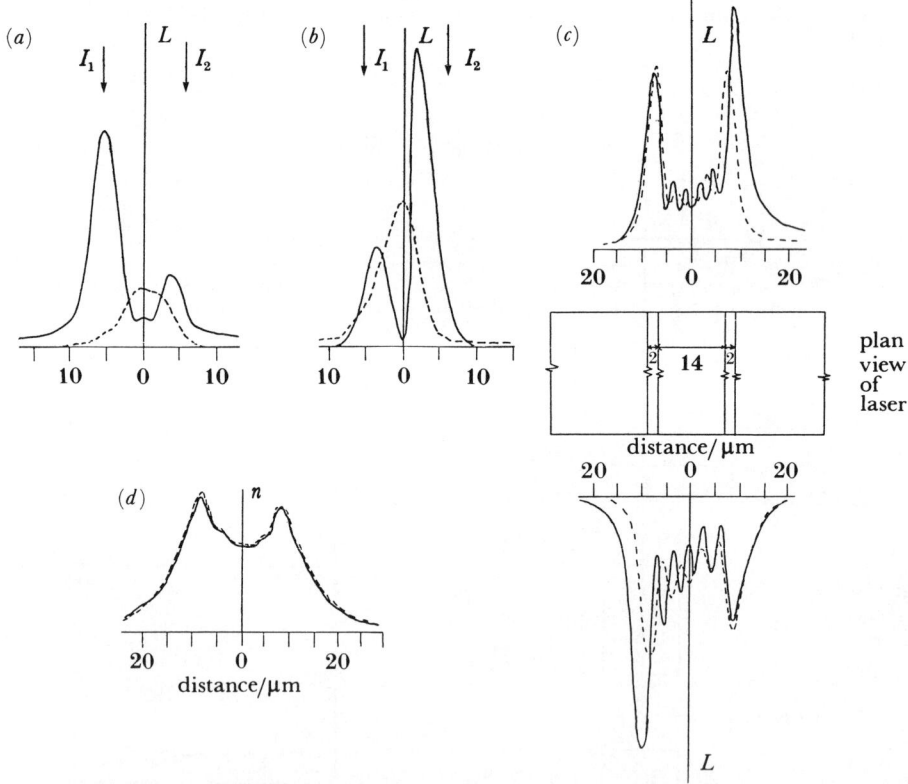

FIGURE 6. Bistable modes of twin-stripe lasers. (a) Spectrally resolved near-field patterns of a twin-stripe laser ($s = 4$ μm; $d = 7$ μm) biased to bistable state A ($I_1 = 2I_2$, 10 % above threshold): ——, $\lambda = 865.7$ nm; ----, $\lambda = 863.8$ nm. (b) As above but for state B: ——, $\lambda = 860.4$ nm; ----, $\lambda = 865.2$ nm. (c) Spectrally resolved near-field patterns of a twin-stripe laser ($s = 2$ μm; $d = 14$ μm) biased to bistable state A (——) and state B (----). ($I_1 = I_2$, 10 % above threshold.) (d) Charge carrier concentrations with conditions as in (c).

stripes. Within experimental error the electron density (figure 6d) has the identical transverse pattern for the two modes and it is suggested that the optical patterns are an 'in-phase' and an 'anti-phase' mode with bistability. Shore & Rozzi (1983) have suggested that it may be possible to achieve rapid switching between suitable transverse modes. On physical grounds one would expect that to achieve fast switching it is desirable to avoid changes in the electron density or the optical energy. A modal pattern such as that of figure 6c may then be close to the optimum for fast switching. Further experiments are under way to test this hypothesis.

5. CONCLUSIONS

Injection lasers have been shown to exhibit absorptive, dispersive and modal bistability. Modal bistability can be understood with a linear circuit model combined with power conservation. Different transverse near-field patterns for twin-stripe lasers have been measured indicating changes that could alter the coupling into an optical fibre or alter the coupling into another bistable laser. An all-optical flip-flop has been demonstrated. The full potential of switching between bistable modes has yet to be explored but there is hope that, with minimal change of either electron densities or optical energy, this switching can achieve picosecond speeds and require little energy to activate.

In active bistability, the element is biased into stimulated emission by an electrical current, though optical pumping should not be ruled out. The 'passive' device can be biased optically into a nonlinear régime by a 'floodlight'. It is therefore easier to envisage a two-dimensional 'passive' array of bistable switches. True two-dimensional active integrated arrays are probably further away from development and this remains a question for the future. However, some lasers have threshold currents in the milliampere range so that the bias power could be low enough for a modest amount of integration to form an all-optical signal-processing circuit at the sending or receiving end of a fibre. For such 'front-end' optical processing, active bistable elements may be a real competitor to other devices.

The authors are deeply indebted to Dr R. G. Plumb for his continued encouragement and help. The S.E.R.C. is thanked for a grant towards this work and a fellowship for one of us (I.H.W.).

APPENDIX

With the notation of figure 3, write

$$R - R_{1n} = (a+b)\,R; \quad R - R_{2n} = (a-b)\,R; \quad R_3 - R_{3n} = cR;$$

$$jX_1 = j(A+B)\,R; \quad jX_2 = j(A-B)\,R; \quad jX_3 = jCR;$$

where a, b, c, A, B and C may take positive or negative values and $j = \sqrt{-1}$. Then from mesh analysis, eliminating the currents i_1 and i_2, the condition for steady oscillation is

$$\begin{bmatrix} (a+c)+j(A+C)+(b+jB) & c+jC \\ c+jC & (a+c)+j(A+C)-(b+jB) \end{bmatrix} = 0.$$

Hence $\qquad (a+c)+j(A+C) = \pm\sqrt{\{b^2+c^2-B^2-C^2+2j(Bb+Cc)\}}.$

[149]

The condition that both modes have the same negative resistances assuming the same loads (same positive resistances) but different frequencies (different reactances) is satisfied if

$$a+c = 0 \quad \text{and} \quad \mathrm{j}(A+C) = \pm \mathrm{j}K, \quad \text{where} \quad K^2 = B^2 + C^2 - b^2 - c^2,$$

with
$$i_1/i_2 = -(c+\mathrm{j}C)/(b+\mathrm{j}B\pm\mathrm{j}K) \quad \text{and} \quad bB+cC = 0.$$

Hence
$$|i_1/i_2|^2 = G \quad \text{or} \quad 1/G,$$

where $G = (c^2+C^2)/(2B^2+C^2-c^2+2KB)$ is not, in general, unity. It may be checked that the power given out equals the power in,

$$|i_1|^2 R_{1\mathrm{n}} + |i_2|^2 R_{2\mathrm{n}} + |(i_1+i_2)|^2 R_{3\mathrm{n}} = |i_1|^2 R + |i_2|^2 R + |(i_1+i_2)|^2 R_3,$$

for only one or other of the eigenmodes but not a mixture of both modes. In the twin-stripe laser, the asymmetry and level of the bias currents determines, through the asymmetry of the equivalent negative resistances, the asymmetry and level of total output power.

REFERENCES

Basov, N. G. 1968 *IEEE Jl Quantum Electron.* **QE-4**, 855–864.

Basov, N. G., Culver, W. H. & Shah, B. 1972 In *Laser handbook* (ed. F. T. Arecchi & E. O. Schulz-DuBois), pp. 1650–1693. Amsterdam: North-Holland.

Bazhenov, V. U., Bogatov, A. P., Eliseev, P. G., Okhotnikov, O. G., Pak, G. T., Rakvalsky, M. P., Soskin, M. S., Taranenko, V. B. & Khairetdinov, K. A. 1982 *Proc. Instn elect. Engrs* **129** (I), 77–82.

Broom, R. F., Mohn, E., Risch, C. & Salathe, R. 1970 *IEEE Jl Quantum Electron.* **QE-6**, 328–334.

Dutta, N. K., Agrawal, G. P. & Focht, M. W. 1984 *Appl. Phys. Lett.* **44**, 30–32.

Glas, P. & Muller, R. 1982 *Optics Quantum Electron.* **14**, 375–389.

Harder, C., Lau, K. Y. & Yariv, A. 1981 *Appl. Phys. Lett.* **39**, 382–384.

Harder, C., Lau, K. Y. & Yariv, A. 1982 *Appl. Phys. Lett.* **40**, 124–126.

Kapon, E., Lindsey, C., Katz, J., Margalit, S. & Yariv, A. 1984 *Appl. Phys. Lett.* **44**, 389–391.

Kirkby, P. A. 1978 Ph.D. thesis, University of Southampton.

Kawaguchi, H. 1982 *Proc. Instn elect. Engrs* **129** (I), 141–148.

Kawaguchi, H. 1982 *Appl. Phys. Lett.* **41**, 702–704.

Lang, R. & Kobayashi, K. 1980 *IEEE Jl Quantum Electron.* **QE-16**, 347–355.

Lasher, G. J. 1964 *Solid State Electron.* **7**, 707–716.

Lau, K. Y., Harder, C. & Yariv, A. 1982 *Appl. Phys. Lett.* **40**, 369–371.

Lee, T. P. & Roldan, R. H. R. 1970 *IEEE Jl Quantum Electron.* **QE-6**, 339–352.

Mukai, S., Yakima, H., Uekusa, S. & Sone, A. 1983 *Appl. Phys. Lett.* **43**, 432–434.

Nathan, M. I., Marinace, J. C., Rutz, R. F., Michel, A. E. & Lasher, G. J. 1965 *J. appl. Phys.* **36**, 473–480.

Ripper, J. E. & Paoli, T. L. 1970 *Appl. Phys. Lett.* **17**, 371–373.

Scifres, D. R., Streifer, W. & Burnham, R. D. 1978 *Appl. Phys. Lett.* **33**, 702–704.

Shore, K. A. 1982*a* *Quantum Electron.* **14**, 177–181.

Shore, K. A. 1982*b* *Quantum Electron.* **14**, 321–326.

Shore, K. A. 1983 *Elect. Lett.* **19**, 874–875.

Shore, K. A., Davies, N. G. & Hunt, K. 1983 *Optics Quantum Electron.* **15**, 547–548.

Shore, K. A. & Rozzi, T. E. 1981 *IEEE Jl Quantum Electron.* **17**, 723–731.

Shore, K. A. & Rozzi, T. E. 1983 *Optics Quantum Electron.* **15**, 497–506.

White, I. H. & Carroll, J. E. 1983 *Elect. Lett.* **19**, 337–339.

White, I. H., Carroll, J. E. & Plumb, R. G. 1982 *Proc. Instn elect. Engrs* **129** (I), 121–126.

White, I. H., Carroll, J. E. & Plumb, R. G. 1983 *Elect. Lett.* **19**, 558–560.

Phil. Trans. R. Soc. Lond. A **313**, 341–347 (1984)

Printed in Great Britain

Optical resonators driven by radiation pressure

By A. Dorsel[1], J. D. McCullen[2], P. Meystre[3], H. Walther[1,3]
and E. M. Wright[3]

[1] *Sektion Physik, Universität München, D-8046 Garching, Federal Republic of Germany*
[2] *Department of Physics, University of Arizona, Tucson, Arizona 89721, U.S.A.*
[3] *Max-Planck-Institut für Quantenoptik, D-8046 Garching, Federal Republic of Germany*

The combined effects of radiation pressure and gravitational force can be used to stabilize a moving mirror to a high degree of accuracy. A noise analysis shows that, under typical conditions, a three-mirror configuration can lead to a mirror confinement within about 1.5 nm, for incident laser powers of 0.5 W.

1. Introduction

Consider a plane Fabry–Perot interferometer in which one mirror is fixed and the other is very light, and suspended to swing as a pendulum. The light pressure produced by the intensity W_{in} in the cavity drives it towards equilibrium with the gravitational and inertial forces. At steady state, the mirror displacement x from its rest position in the absence of light is then proportional to W_{in}. Since x determines the interferometer spacing, there is a one-to-one correspondence between this situation and usual optical bistability in the presence of a Kerr medium, except that we now have a change of the physical cavity length instead of its optical length. In a previous paper (Dorsel *et al.* 1983) we showed experimentally that this system can display a bistable response (radiation-pressure optical bistability). In addition, we showed that when the incident light intensity is sufficient, the movable mirror becomes extremely stable, and the motion at the mechanical resonance is suppressed. We call this behaviour 'mirror confinement'. In view of its potential applications, efforts to analyse this control mechanism, and to exploit it further have proceeded. In particular, we found that a three-mirror device (the movable mirror is suspended between two fixed mirrors) is capable of far superior confinement and stability (McCullen *et al.* 1984). The improved stability is due to the fact that radiation pressure now acts on both sides of the movable mirror, the greater force coming from the side closest to resonance. By suitable choice of the round-trip cavity phase shift, these competing forces can be used to produce a high level of mirror confinement and stability. In §2 a review of the two-mirror system is given, and in §3 the three-mirror arrangement is discussed. A white-noise analysis of the systems is presented in §4. In §5 we discuss the effects of ground noise, which are relevant for applications, for example as a narrow-band seismometer. Finally, §6 is a summary and conclusion.

2. Two-mirror resonator

The interferometer constructed for our experimental studies has been described elsewhere (Dorsel *et al.* 1983). Bistability was obtained by slowly varying the input power across the bistable region, from feeble power to high power, and back. Scanning times ranged from 2 to 5 min, times long compared with the damping time of the mirror.

In addition to optical bistability, we also observed an effect we call mirror confinement, whereby for sufficient fixed input power the movable mirror became extremely stable against mechanically resonant oscillations at the pendulum frequency. It was found that the effective resonant frequency Ω_{eff} for the system could become significantly different from the mechanical frequency Ω of the pendulum for input power sufficiently beyond the bistability threshold.

Mirror confinement can be simply understood in terms of the potential describing the motion of the moving mirror. This potential can be derived from the dynamical equation for the moving mirror

$$\ddot{x} + \gamma\dot{x} + x = P(x), \tag{1}$$

where x measures the displacement of the mirror from its rest position in the absence of light, γ is the pendulum damping constant, and $P(x)$ is the radiation pressure force per unit mass. For scaling convenience, x is measured in units of half-wavelengths of the incident light (assumed monochromatic) and time is measured in units of the inverse of the pendulum frequency Ω. For the two-mirror system $P(x)$ is given by

$$P_2(x) = \frac{K_2}{1 + F_2 \sin^2 \frac{1}{2}\phi}, \tag{2}$$

where $\phi = 2\pi x - \phi_0$, ϕ_0 being the detuning of the resonator in the absence of light. The constants K_2 and F_2 are given by

$$K_2 = \frac{4R'(1-R)\,W}{\{1-\surd(RR')\}^2\,mc\lambda\Omega^2}; \quad F_2 = \frac{4\surd(RR')}{\{1-\surd(RR')\}^2}, \tag{3}$$

where R and R' are the intensity reflectivities of the fixed and movable mirrors, respectively, W is the input intensity, and m is the movable mirror mass.

The potential corresponding to the dynamical equation (1) (we set $\gamma = 0$) is given by

$$V(x) = \tfrac{1}{2}x^2 - \int_0^x \mathrm{d}y\, P(y) \tag{4}$$

and is plotted in figure 1 a for $R = 0.99$, $R' = 0.95$, $K_2 = 109$ (we use these values for examples throughout). This value for K_2 corresponds to an applied light power of 0.5 W, and $\Omega = 7\ \mathrm{s}^{-1}$, with a mirror mass 60 mg. The potential minima correspond to possible stable states of the system. It is clear that if the mirror is captured into one of these wells, it thereafter responds significantly only to driving forces whose frequencies are near the oscillation frequency in the well. As the input power, and thus the internal power, increases, the depth and curvature of the well increase, and the effective resonant frequency of the system departs from the pendulum frequency. For sufficient input powers the system can therefore be made extremely stable against mechanically resonant oscillations.

3. Three-mirror system

The chief drawback presented by the two-mirror system in confining the mirror is that the radiation pressure force is exerted only in one direction. Thus on one side of the equilibrium point the force changes rapidly with position (on a scale determined by the cavity finesse),

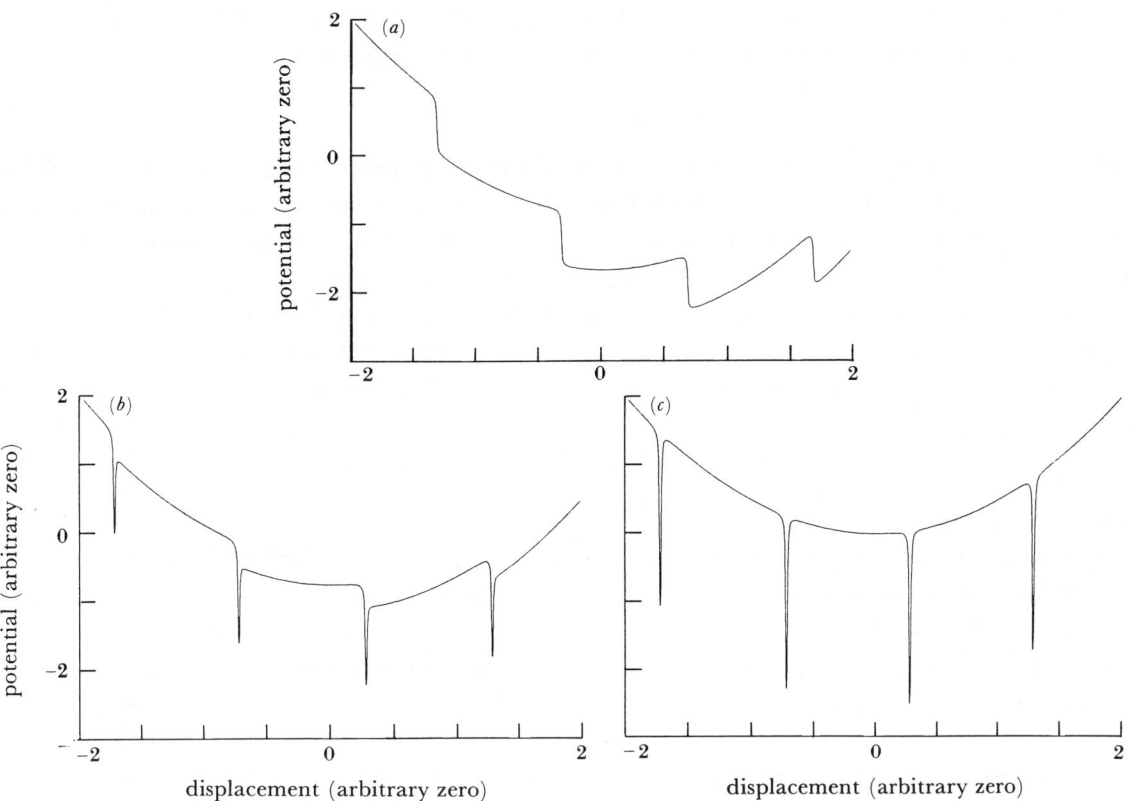

FIGURE 1. Potentials for (a) the two-mirror case; (b) the three-mirror case with one-sided illumination; (c) the three-mirror case with symmetric illumination, for $R = 0.99$, $R' = 0.95$, $K_2 = 109$.

whereas on the other side the restoring force comes from the usual pendulum forces, and varies much more slowly with position. The situation can be significantly improved by suspending the movable mirror inside a fixed tuned Fabry–Perot cavity, thereby creating two coupled interferometers. The radiation pressure force can now be in either direction, depending upon which of the cavities is closer to resonance, with a concomitant sharpening of its positional dependence. By appropriately tuning the main cavity, reversal of the direction of the radiation pressure occurs when the overall transmission is large, providing maximal radiation forces to confine the pendulum.

Equation (1) remains valid for the three-mirror system, but with

$$P_3 = \frac{K_3\{f(1+R) - 2\cos(\phi - \phi_1)\}}{1 + F_3\{f\cos(\phi - \tfrac{1}{2}\phi_1) - \cos\tfrac{1}{2}\phi_1\}\{fR\cos(\phi - \tfrac{1}{2}\phi_1) - \cos\tfrac{1}{2}\phi_1\}}, \tag{5}$$

where we have taken the laser field to be incident on one side of the fixed Fabry–Perot only. In this equation ϕ_1 is the (round-trip) phase detuning of the main cavity, $f = \surd(R'/R)$, and F_3 and K_3 are given by

$$F_3 = \frac{4R}{(1-R)^2}; \quad K_3 = \frac{1}{f}\left(\frac{1-fR}{1-R}\right)^2 K_2. \tag{6}$$

The denominator in (5) has a sharp minimum, similar to the denominator in (2). The value for ϕ at which it minimizes, however, is $\sin(\phi - \frac{1}{2}\phi_1) = 0$, rather than $\sin\frac{1}{2}\phi = 0$. The value for ϕ_1 that makes this minimum the smallest is given by the equation

$$\cos\tfrac{1}{2}\phi_1 = \tfrac{1}{2}f(1+R). \tag{7}$$

With this choice of ϕ_1, the numerator of (4) passes through zero at the same value for ϕ for which the denominator minimizes. The force thus changes sign in a region of maximum transmission. Roughly speaking, this causes the equilibrium points of the mirror to occur at nodes of the standing wave of the light resonant in the main cavity.

Potential curves for the three-mirror system are plotted in figure $1\,b, c$. Curve $1\,c$ is obtained with equal intensities incident on both sides of the fixed Fabry–Perot and assuming that the light beams are mutually incoherent; the force then consists of the sum of two terms, each of the form of (5):

$$P_s = 2K_3 \sin\tfrac{1}{2}\phi_1 \sin(\phi - \tfrac{1}{2}\phi_1)/D, \tag{8}$$

where D is the denominator in (5).

The curves for the three-mirror system show potential wells around equilibrium points of the movable mirror that are much narrower than their two-mirror counterparts. The examples in figure 1 are for the choice of cavity phase given by (7), which results in the narrowest wells. With ϕ_1 greater than this, the wells become deeper and the spacing between the steep walls increases, making the potential more or less flat on the bottom. For a given choice of ϕ_1, the depth of the well depends upon the intensity of the incident light. The shape, however, depends on the cavity parameters. For the case shown in figure 1, the width at half maximum is about 0.017 of the incident wavelength, or 8.5 nm with 500 nm incident light. The effective resonant frequencies for the three-mirror system can be estimated by a linearized analysis (see §5) of the dynamics represented in (1).

4. White-noise analysis

The effect of white noise on (1) can be investigated by solving the corresponding Fokker–Planck equation

$$\frac{\partial \mathcal{W}}{\partial t} = \left\{ -u\frac{\partial}{\partial x} + \frac{\partial}{\partial u}\left(\gamma u + \frac{\partial V}{\partial x}\right) + \mathcal{D}\frac{\partial^2}{\partial u^2}\right\}\mathcal{W},$$

where $\mathcal{W}(x, u, t)$ is the probability density in position–velocity space and u is the movable mirror velocity in units of $u_0 = \frac{1}{2}\lambda\Omega$. $\mathcal{D} \equiv \gamma(u_{\mathrm{th}}/u_0)^2$ is the usual diffusion coefficient.

In the steady state ($\partial\mathcal{W}/\partial t = 0$) the solution of this equation is given by Risken (1984)

$$\mathcal{W}(x, u) = N\exp\left[-(u_0/u_{\mathrm{th}})^2\{\tfrac{1}{2}u^2 + V(x)\}\right], \tag{9}$$

where N is a normalization constant.

Integrating over velocities yields the probability density $Q(x)$ of finding the particle between x and $x + \mathrm{d}x$. We find

$$Q(x) = \exp\{-(u_0/u_{\mathrm{th}})^2\,V(x)\}\bigg/ \int_{-\infty}^{\infty} \mathrm{d}y\,\exp\{-(u_0/u_{\mathrm{th}})^2\,V(y)\}, \tag{10}$$

which has the form of a partition function. We are interested in the confinement of the movable mirror when it is initially prepared in one of the potential minima. If this well corresponds to essentially the same potential value as one of its close neighbours, then, according to (10), the steady-state probabilities for the mirror to be in either well will be equal. This arises because in the limit $t \to \infty$, tunnelling produces the distribution (10). However, on more realistic time scales (say hours) tunnelling will be negligible if $Q(x)$ for the initial well is narrow compared with the well width. We therefore consider individual wells and treat the remainder of the potential curve as that of the gravitational potential. In figure 2 we have plotted $Q(x)$ as a

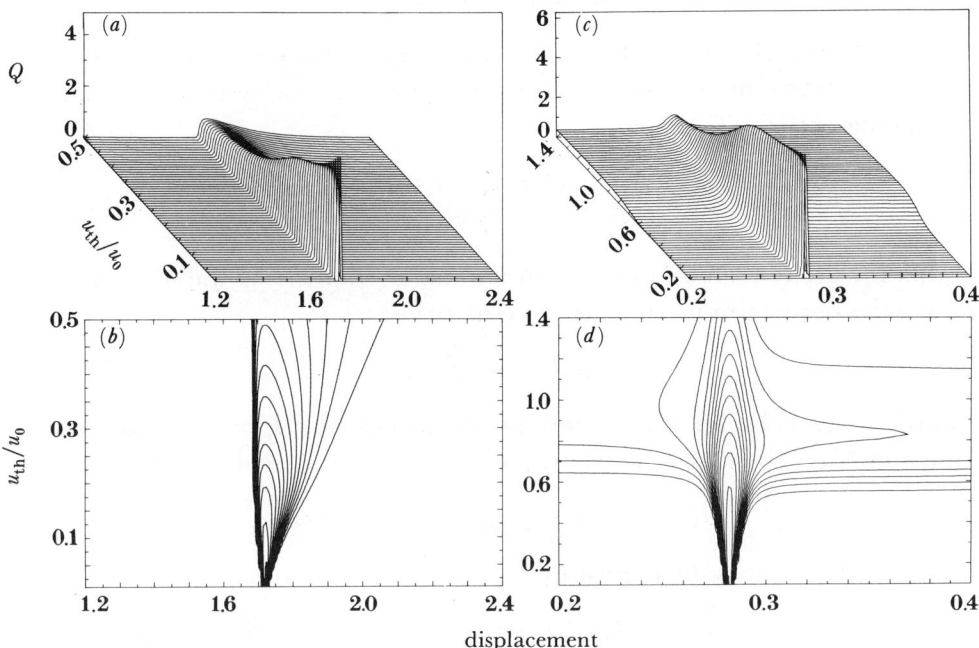

FIGURE 2. Probability density $Q(x)$ plotted against x and u_{th}/u_0 together with the corresponding contour plot for the two-mirror (a, b) and three-mirror (c, d) systems. These results correspond to the deepest potential wells in figure 1 a, b.

function of x and (u_{th}/u_0), along with a corresponding contour plot, for both the two-mirror (figure 2 a, b) and three-mirror (figure 2 c, d) cases. In each case we choose the deepest potential well (see figure 1 a, b). Figure 2 a shows that the stability of the two-mirror system to noise is somewhat limited owing to the slowly varying gravitational potential. This feature leads to fast tunnelling from the initial well to the neighbouring well on the slow side of the potential for relatively low values of u_{th}/u_0. In contrast the three-mirror system shows far superior stability to noise. Figure 2 c shows clearly that the probability density remains essentially constant in width, right up to a critical point at which the profile broadens abruptly and stability is lost. This feature is more clearly displayed in figure 2 d, where the contours diverge abruptly at $u_{th}/u_0 \approx 0.6$. For the case of two-way driving of the three-mirror system, the results are similar to those in figure 2 c, d, except that the curves are more symmetrical.

5. EFFECTS OF GROUND NOISE

The white-noise analysis of the previous section, though useful for comparing the relative stability and confinement properties of the various systems, is inappropriate to the description of ground noise, which is the main noise source as far as applications are concerned. To evaluate

[155]

its effects, we use a linearized form of (1), in which $P(x)$ is expanded to first order in x around an equilibrium point $x_0 = 0$. In the presence of ground noise source, $G(t)$, this equation becomes (x is now the displacement from the chosen equilibrium point)

$$\ddot{x} + \gamma \dot{x} + \Omega_e^2 x = G(t), \tag{11}$$

where the effective frequency Ω_e is found from

$$\Omega_e^2 = 1 + (\partial P/\partial x)_{x_0}, \tag{12}$$

which is around $\Omega_e^2 \approx 1.147 \times 10^6 \, W$ for the parameters considered. This harmonic oscillator approximation is clearly most appropriate for the three-mirror systems (see figure 1 b, c). If two-way pumping occurs, Ω_e^2 becomes

$$\Omega_e^2 \approx \frac{4\pi \sqrt{\{4 - f^2(1 + R^2)\}}}{fT^2 T'} \, W. \tag{13}$$

Assuming that the ground noise causes a translational motion $z(t)$ of the whole system through stationary air, the noise term in (11) can be written

$$G(t) = -\gamma \dot{z} - \ddot{z}. \tag{14}$$

Fourier transforming (11), for a unit impulse displacement $z(t) = \delta(t)$ yields the frequency transfer function $H(\omega)$ (Bendat & Piersol 1966),

$$H(\omega) = (\omega^2 - i\gamma\omega)/(\Omega_e^2 - \omega^2 + i\gamma\omega). \tag{15}$$

The spectral density $S_x(\omega)$ is then given by

$$S_x(\omega) = |H(\omega)|^2 S_z(\omega), \tag{16}$$

where S_z is the spectral density of the ground noise, and the mean squared displacement of the moving mirror by

$$\langle x^2 \rangle = \int S_x(\omega) \, d\omega. \tag{17}$$

Ground noise is characterized by a spectral density of the general form $S_z(\omega) \approx C/\omega^4$, C being a constant dependent on location. For the Munich area $C \approx 6 \times 10^{-11}/\lambda^2 \Omega^3$ (Billing *et al.* 1983), the λ and Ω dependence arising from scaling. For $\lambda = 500$ nm we obtain

$$\langle x^2 \rangle \approx \frac{250}{\Omega^3} \int_{\omega_c}^{\omega_u} \frac{1}{\omega^4} \left\{ \frac{\omega^4 + \gamma^2 \omega^2}{(\Omega_e^2 - \omega^2)^2 + \gamma^2 \omega^2} \right\} d\omega, \tag{18}$$

where ω_c is taken as a lower bound of the earth vibration frequencies, which has a value *ca.* 1 min^{-1}. In the scaled units we are using this gives $\omega_c \approx 0.1/\Omega$. The upper value, ω_u, is taken as a few times Ω_{eff}, say 10, then $\omega_u \approx 10^4 \sqrt{C}$.

These values yield an x_{rms} displacement of the moving mirror of $x_{rms} \approx 1.5$ nm, for $\gamma = 0.5$. With higher values of damping, x_{rms} slowly increases, because the effective resonator Q is decreased. This shows, however, that confinement to a small range of motion within the confines of the potential well is possible.

6. SUMMARY AND CONCLUSIONS

Our results show that control of the position of a movable mirror to within a range of a few tenths of a nanometre by using radiation pressure, should be readily achievable. This opens the way to a number of applications, since such a system represents a very sensitive, narrow-banded tunable transducer between mechanical and optical signals. Specifically, its frequency can be tuned by varying the input laser intensity and its bandwidth by changing the phase detuning, ϕ_1, of the main cavity. For the optimal detuning ϕ_1 given by (7), a bandwidth of a few hertz can readily be achieved. Since the stabilization of the mirror is under optimum conditions about a maximum in transmission, a second, weak laser slightly detuned from the stabilizing laser should be used for detection.

Possible basic physical applications include acousto-optical studies, very accurate photoelectric effect measurements, and atomic and molecular beam diagnostics. On the more applied side, these systems can be developed into narrow-banded seismometers. An even more intriguing idea is the geometrical stabilization of a very large thin-pellicle space telescope (Labeyrie 1979).

This work was carried out partly in the framework of an operation launched by the Commission of the European Communities under the experimental phase of the European Community Stimulation Action (1983–85).

REFERENCES

Bendat, J. S. & Piersol, A. G. 1966 *Measurement and analysis of random data.* New York: Wiley.

Billing, H., Winkler, W., Schilling, R., Rüdiger, A., Maischberger, K. & Schnupp, L. 1983 In *Quantum optics, experimental gravitation, and measurement theory* (ed. P. Meystre & M. O. Scully), p. 525. Plenum.

Dorsel, A., McCullen, J. D., Meystre, P., Vignes, E. & Walther, H. 1983 *Phys. Rev. Lett.* **51**, 1550.

Labeyrie, A. 1979 *Astron. Astrophys.* **77**, L1.

McCullen, J. D., Meystre, P. & Wright, E. M. 1984 *Optics Lett.* **9**, 193.

Risken, H. 1984 *The Fokker–Planck equation, methods of solutions and applications* (Springer Series in Synergetics).

Phil. Trans. R. Soc. Lond. A **313**, 349–355 (1984)
Printed in Great Britain

Applications of all-optical switching and logic

By P. W. Smith

Bell Communications Research, Holmdel, New Jersey 07733, U.S.A.

In this paper I review the present status of all-optical switching and logic elements. I then discuss their future potential, taking account of limitations imposed by materials, considerations of system architecture, and fundamental physical mechanisms. I conclude by describing two areas in which all-optical signal-processing systems are likely to have a major impact.

1. Introduction

One of the major driving forces behind the current interest in optical bistability and all-optical switching and logic is certainly the hope that important device applications will arise. How realistic are these hopes, and how close are we to these applications? In this paper I summarize the present status of research in these areas, and offer some speculations on the future potential of photonic switching and logic.

2. Present status: overview

There are at present a large variety of bistable optical devices that have been demonstrated. Although many are based on the nonlinear Fabry–Perot resonator, a number of non-resonant devices, including the nonlinear interface and the self-focusing bistable device, have also been studied (Bowden 1983). At this meeting we have heard of new ideas for devices based on bistability due to increasing absorption (D. A. B. Miller, H. Haug & S. Schmitt-Rink, M. Dagenais), self-induced electro-optic effect (SEED) (D. A. B. Miller), radiation pressure (P. Meystre & H. Walther), and novel guided-wave structures (G. I. Stegeman & H. G. Winful).

A wide range of optical materials has been used. Most materials have limitations based on wavelength and power requirements. At this meeting, we have heard of novel results from the use of multiple quantum well (m.q.w.) material (D. A. B. Miller; H. M. Gibbs), and exciton effects in semiconductors (R. Levy *et al.*; H. M. Gibbs; M. Dagenais; D. A. B. Miller).

Clearly, photonic devices are still in the active research stage. There are few current practical applications. There are, however, many exciting prospects.

3. Prospects

A number of recent developments have increased the interest in digital optical signal-processing devices and techniques. Laser technology has now advanced to the point that lasers are being used in consumer electronics. Optical fibre communication systems are being widely installed. Integrated optics spectrum analysers are being marketed.

At this meeting, we have already heard some ideas for optical computing (A. C. Huang), and optical switching and logic (A. C. Walker *et al.*; H. A. Haus). Later in this paper I shall

23-2

touch on some possibilities for optical parallel processing and mode-locking of semiconductor lasers. Can we make some general statements about future applications?

Figure 1 shows a 'plot' of switching time benchmarks for a variety of mechanisms and devices. The longest time point is the cycle time of the fastest currently available electronic computer: the CRAY-2. The fastest time point is a visible optical cycle. Note that all time points below *ca.* 10 ps are optical. I shall comment further on the potential of photonic devices for ultrafast switching later in this paper.

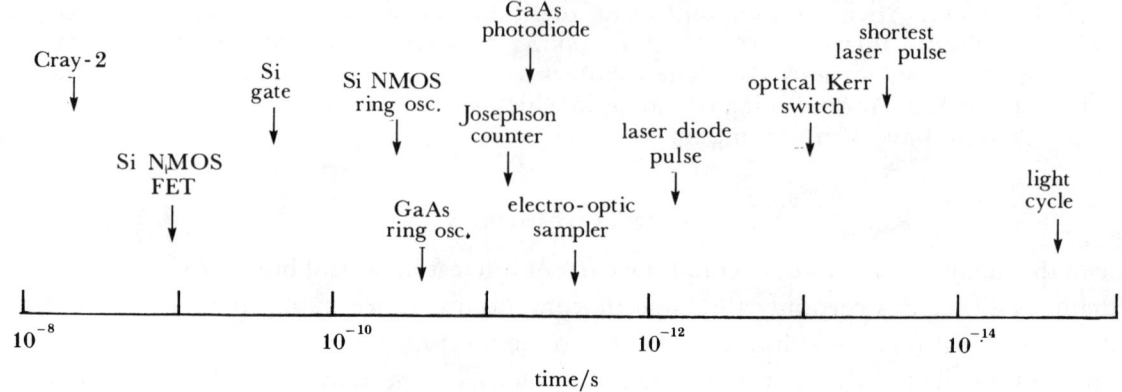

FIGURE 1. 'Time line' showing time benchmarks that are pertinent for several switching and signal-processing technologies.

Figure 2 shows an extrapolation of a plot of the capacity of experimental optical fibre communications systems as a function of year. If the present trend towards higher-capacity systems continues, there will soon be a severe incompatibility between the bit-rate capacity and the limits imposed by the electronics used for switching and signal handling. Clearly there is a need for fast, high-capacity optical systems to 'push back' the boundary between the optical

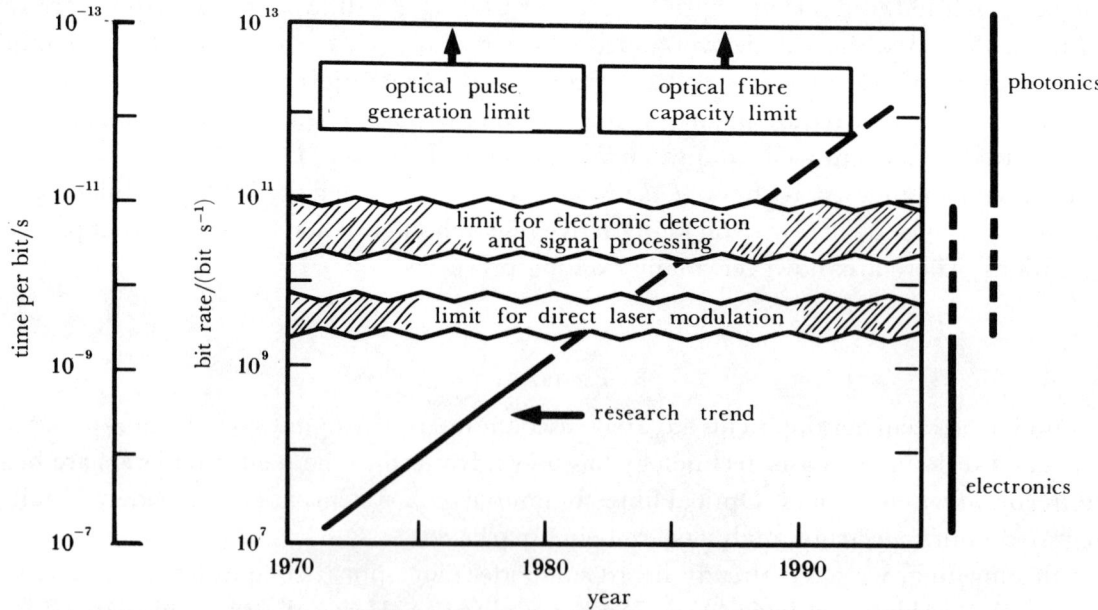

FIGURE 2. An extrapolation of current progress in optical communications systems capacity over the next decade.

[160]

(transmission) part of the system and the electronic (signal processing and switching) part. To focus more precisely on the role that optical signal-processing elements can play, it is necessary to examine their fundamental limitations.

4. Towards fundamental limits

Much work has already been done in the search for materials with the right combination of properties for photonic switching applications. In general we can state that the ideal material should have a large optical nonlinearity at the wavelength of interest, fast response time, high resistance to optical damage, and adequate transparency. In addition, because in many systems there will be an interface with electronics, it would be desirable to use a semiconductor optical material that would facilitate the fabrication of the optics–electronics interface. So far, multiple quantum well (m.q.w.) material comes the closest to meeting all these requirements.

It is important to design systems that can take advantage of the special properties of optical switching devices. A. C. Huang (this symposium) has described novel designs for optical computers based on this philosophy. The high degree of parallelism possible with optical devices (see §5) permits computer and switching system architectures very different from those currently used for electronic systems.

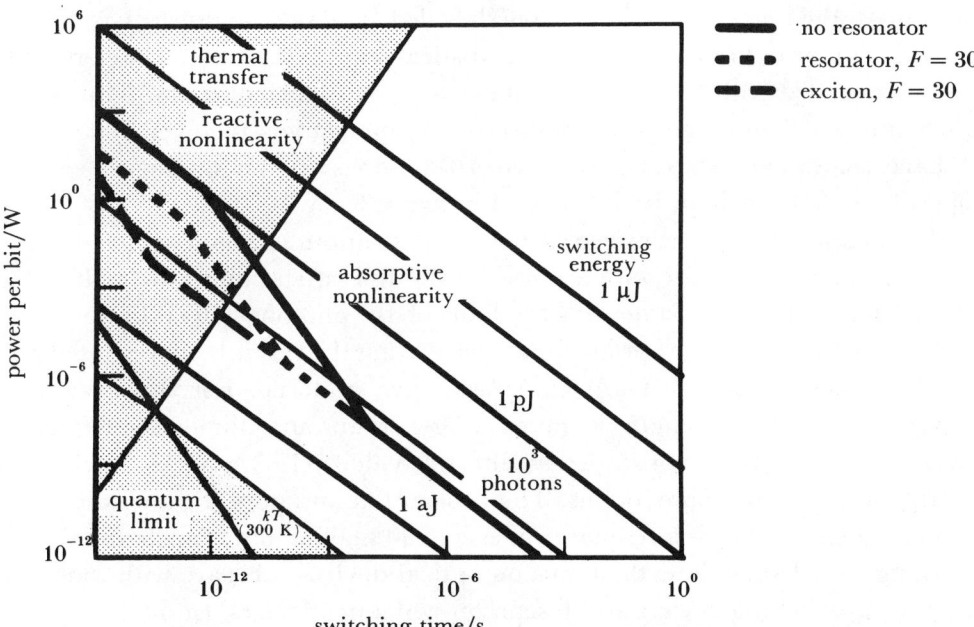

FIGURE 3. Limits for optical switching devices, with λ set to 0.85 μm. The ordinate is the power that must be applied for a switching operation, and the abscissa is the length of time that the power must be applied. The lines at 45° are lines of constant switching energy.

Figure 3 shows some fundamental limits for optical switching devices (Smith 1982). Certain general considerations will apply to all switching elements. Many years ago, J. von Neumann pointed out from thermodynamic arguments that a single (non-reversible) 'yes–no' switching operation must dissipate a minimum of about kT of energy (here k is Boltzmann's constant and T is the absolute temperature). Quantum mechanical considerations lead to the assertion that a quantum switching operation must use at least h/τ of energy, where h is Planck's constant and τ is the switching time. Because these mechanisms represent a 'noise' level for the signal of interest, we should expect that a practical device would require a signal level of many times the kT or $h\nu$ limits.

Keyes (1970) at I.B.M. has examined in depth the physical limits on a number of different types of switching devices (see also Keyes 1975). He shows that for a repetitive operation the heat dissipated per switching operation sets an upper limit to the achievable switching rate (a higher rate would result in an unacceptable temperature rise in the device). The region affected by this consideration is also shown in figure 1 and is labelled 'thermal transfer'. (Note that this area represents power *dissipated* per switching operation. The actual power used may be higher if it is not absorbed in the switching element.) A higher power can also be used if the device is operated at less than the maximum repetition rate.

If an optical switching device operates by absorbing light and saturating an optical transition in some material, certain general relations between power absorbed and switching time can be derived. It can be shown that, for the case where the response time, τ, of the switching element is taken to be the shortest time compatible with the material response, the switching power can be represented by the line labelled 'absorptive nonlinearity'.

A similar argument can be made for a material exhibiting an optical Kerr effect, i.e. the refractive index has a term proportional to the light intensity. For a material with the largest-known value of electronic nonlinearity, the polydiacetylene PTS ($n_2 = 6 \times 10^{-12}$ cm^2 W^{-1}), we obtain the limits shown by the line labelled 'reactive nonlinearity'.

A third type of limit is imposed by statistical considerations. To have reliable switching, the noise associated with 'on' and 'off' states must be sufficiently low. This noise will depend on the number of light photons, or number of absorbing atoms, involved in the switching operation. I have somewhat arbitrarily selected 10^3 photons as the number necessary for low-noise operation, and this limit is also plotted in figure 3.

These limits have been derived without any assumptions about the use of an optical resonator. If we assume a resonator with a finesse of 30, the optical limits are reduced as shown by the dotted line in figure 3. Note that the limit of 10^3 photons will still apply. I have also shown in figure 1 the limits appropriate for a device using the recently investigated room-temperature excitonic nonlinearity in GaAs–GaAlAs m.q.w. material. For this excitonic resonance, the effective oscillator strength is much larger than an atomic oscillator strength. Recent measurements have demonstrated an effective value of $|\mu|^2 \approx 50(ea_0)^2$, where e is the electronic charge and a_0 is the Bohr radius. The broken line in figure 3 shows the limits for an optical switching element with a resonator finesse of 30 making use of such an excitonic nonlinearity.

In figure 4 I show how the limits on optical devices compare with those on semiconductor electronic switching devices and Josephson switching devices. In the 10^{-7} to 10^{-11} s region, the limits for optical devices are slightly below the best current semiconductor device performance. At switching speeds of 10^{-12} to 10^{-14} s, however, optical devices appear to have no competition. This unique capability for subpicosecond switching is one of the most exciting aspects of this new technology.

The switching power required in this short-time region puts the operating point within the 'thermal transfer' region discussed earlier. For this reason it does not appear feasible to design a *general purpose* high-speed, digital optical computer. However, for many applications these thermal limits will not present severe problems.

There are many factors that relate to the choice of a switching technology that cannot be

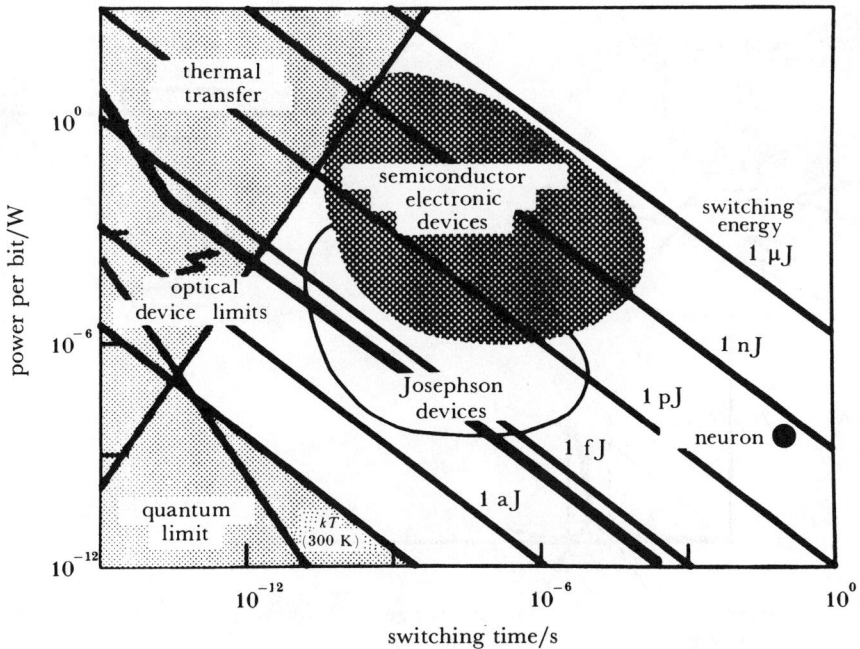

FIGURE 4. Comparison of optical switching device limits with present results for other technologies.

shown on a power–time plot. In many cases it is desirable to perform some signal-processing operation on a light signal, either because the incoming signal is in the form of light or because freedom from electromagnetic interference is desired. Optical switching devices typically operate at room temperature. In many cases, they have extremely large bandwidths and can be adapted for many special functions such as rapid parallel processing of information. For these reasons there will be cases where optical switching systems will be used, even in an area of figure 4 in which other technologies show a switching-energy advantage.

5. PREDICTIONS: TWO AREAS OF MAJOR IMPACT

There are many applications of optical switching devices that can be predicted. I select here two areas that appear to have great promise.

(a) High-speed switching

In optical communications systems the characteristics of large bandwidth, high speed and the ability to process signals already in the form of light should be particularly useful. It would be possible, for example, to make an optical multiplexer that would take several optical data streams, each with the maximum data rate compatible with electronic devices, and time-division multiplex them onto a single fibre. At the other end a similar all-optical device could demultiplex to get back to data rates that can be handled with electronic components.

A multiplexer can be made from a number of triggerable switching elements, as shown in figure 5a. A trigger pulse with the proper time synchronization is required to multiplex pulses as shown. Each element could be made from a properly designed bistable optical device, as shown in figure 5b. This bistable device consists of a suitable nonlinear optical material in a

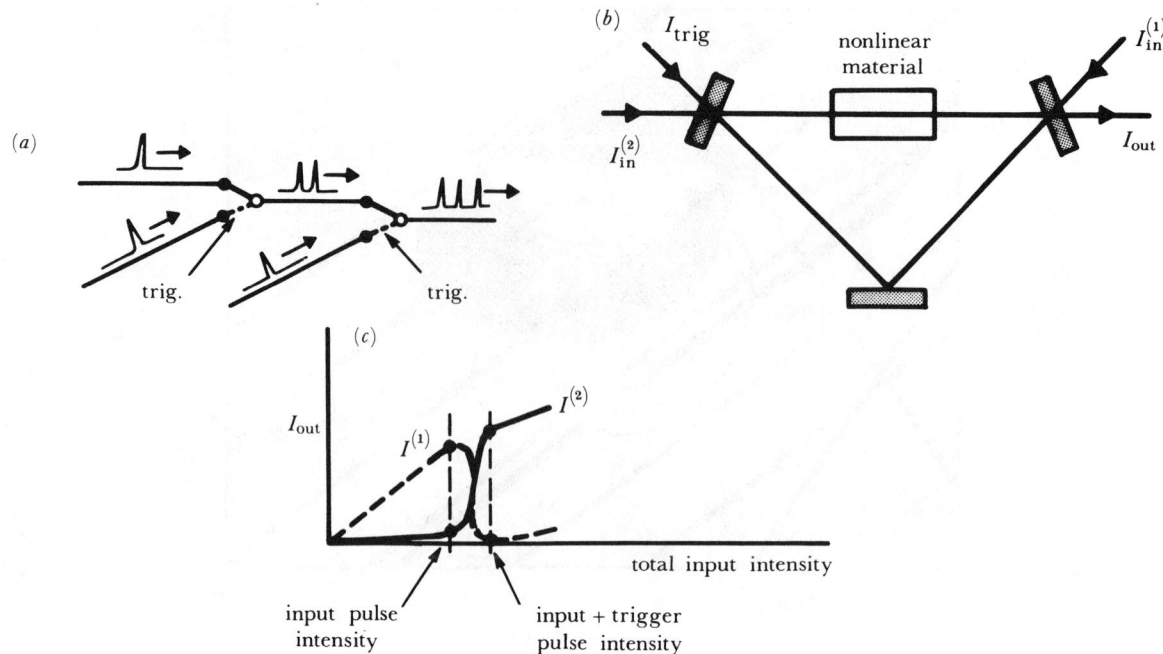

FIGURE 5. Optical time-division multiplexer: (*a*) overall schematic diagram; (*b*) ring triggerable bistable element; (*c*) output characteristic of ring bistable element.

ring resonator. The ring geometry allows separation of the inputs and outputs. However, some polarization selectivity may have to be employed to avoid interference effects between the two input beams $I_{in}^{(1)}$ and $I_{in}^{(2)}$. Let us assume here that pulses in these two beams are never present simultaneously in the element. The output intensity depends on the total input intensity $I_{in}^{(1)} + I_{in}^{(2)} + I_{trig}$, as shown in figure 5*c*. If the input pulses are of intensity slightly less than the critical intensity corresponding to the 'knee' of the curves, the output in the absence of a trigger input will consist solely of $I_{in}^{(1)}$. However, during the time that a small trigger signal, I_{trig}, is present, the output will consist solely of $I_{in}^{(2)}$. Because of the sharp 'knee' in the characteristic curves, only a small I_{trig} is required to accomplish this switching.

In my laboratory we have recently been conducting experiments on the use of m.q.w. material as a saturable absorber for mode-locking semiconductor diode lasers. Stable passively mode-locked pulse trains of 10 ps pulses have already been obtained, and much shorter pulses appear possible.

(*b*) *Parallel processing*

As mentioned earlier in this paper, many optical devices are especially suited for parallel operations such as image processing. As a specific example, consider a 1 cm² Fabry–Perot etalon containing GaAs m.q.w. material. Such an etalon has *ca.* 10^7 resolvable spots, and each spot is a bistable element with a (measured) response time of 30 ns. The throughput of this device is 3×10^{14} bit s^{-1}, which corresponds to simultaneous telephone conversations by the entire population of the world! In principle the total optical power required would be only 0.1 W. Clearly this capability is a powerful one, and it will be a challenge for systems designers to develop optical systems that make use of this potential.

6. Conclusions

As one can see by looking at the papers presented at this meeting, the field of optical switching and logic is still in its infancy, and new mechanisms, materials, devices, and systems are currently being invented and studied. In general we can state the following.

(*a*) The strong points of optical switching devices are:

(1) speed: with an electronic nonlinearity or free-carrier generation in semiconductors, subpicosecond switching times are possible;

(2) capability for parallel processing: with a liquid crystal bistable array, image processing has already been demonstrated;

(3) compatibility with optical fibre systems: i.e. the ability to treat directly signals already in the form of light;

(4) bandwidth: with a non-resonant bistable optical device or a nonlinear interface, a large fraction of the visible light bandwidth can be used.

(*b*) The weak points of optical switching devices are:

(1) high power is required for fast switching: this will tend to create thermal problems unless highly transparent materials are used;

(2) materials do not yet exist that have the ideal combination of properties for these devices;

(3) the minimum size of an optical switching element cannot be reduced below a volume of about λ^3;

(4) theoretical and practical problems involved in waveguide and microresonator formation in λ^3 volumes have yet to be overcome.

In the future, it will be important to design systems that can make full use of the capabilities of optical switching devices. Such systems, which exploit the potential speed and bandwidth capability of optical devices and their capability for parallel processing of information, should find significant applications in communication and computing fields.

References

Bowden, C. M., Gibbs, H. M. & McCall, S. L. (eds) 1983 *Proceedings of the Optical Bistability II Conference.* New York: Plenum.
Keyes, R. W. 1970 *Science, Wash.* **168**, 796–801.
Keyes, R. W. 1975 *Proc. IEEE* **63**, 740–767.
Smith, P. W. 1982 *Bell Syst. tech. J.* **61**, 1975–1993.

Phil. Trans. R. Soc. Lond. A **313**, 357–360 (1984)

Printed in Great Britain

Switching of optically bistable devices by incoherent illumination

By F. A. P. Tooley, A. C. Walker and S. D. Smith, F.R.S.

Department of Physics, Heriot-Watt University, Riccarton, Currie, Edinburgh EH14 4AS, U.K.

We report the first observation of the external switching of an intrinsic optically bistable device by incoherent radiation. The device, consisting of an InSb etalon held at 85 K and optically biased by a continuous wave CO laser operating at 5.5 µm, was switched both from off to on resonance and from on to off resonance by using a photographic flashlamp. Results presented indicate that the required switching energy per unit area can be of the order of 100 fJ µm^{-2}.

The discovery of giant third-order optical susceptibilities combined with the principles of bistable and nonlinear optical devices has opened up many possibilities for all-optical information processing with real-time optical address. However, to make, for example, a real-time optically addressed spatial light modulator based on these effects it is necessary to be able to switch or modulate a two-dimensional all-optical device with an image impressed in incoherent light (Garmire *et al.* 1978). We report here the first steps toward this goal by the demonstration of external switching of a single bistable device by a white-light incoherent pulse.

Crystals of n-type ($n = 1.9 \times 10^{14}$ cm^{-3}) InSb were used to form Fabry–Perot resonators, the natural reflectivity (*ca.* 0.36) of the plane parallel faces providing a measured coefficient of finesse, F, of 0.8–1.5. The sample thickness, L, was in the range 80–500 µm. An Edinburgh Instruments PL3 continuous wave CO laser operating at 1819 cm^{-1} was used to produce bistability with an incident beam diameter, $1/e^2$, of *ca.* 200 µm. The reflected power was monitored by a fast-response InSb detector. The external white-light switching pulses were incident on the back face of the crystal. They were generated by a photographic flash unit (Sunpak 3600) and dynamically monitored by a Si-photodiode (figure 1). A typical characteristic produced with this arrangement is shown in figure 2 for an etalon with $L = 260$ µm. In the input

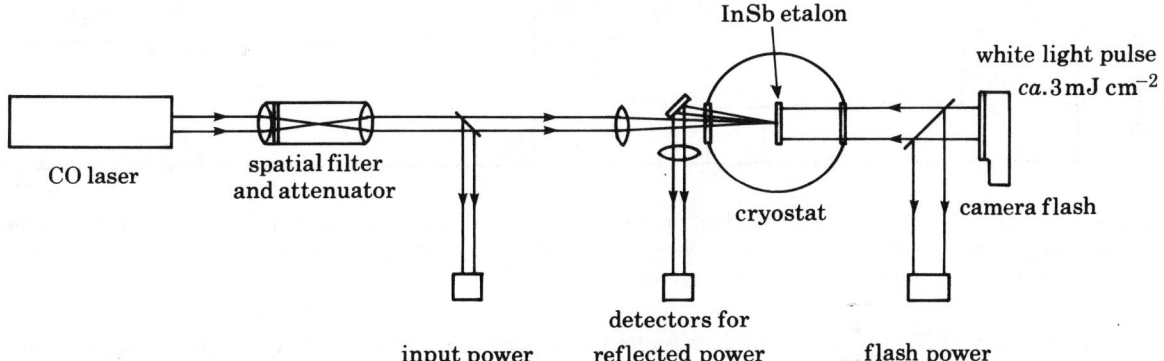

FIGURE 1. Experimental arrangement used to observe optical bistability and allow introduction of external switching pulses.

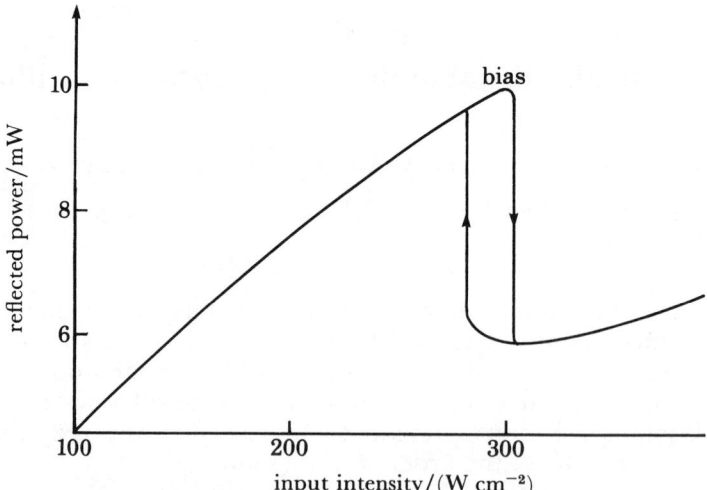

FIGURE 2. Characteristic of a 260 μm thick etalon showing optical bistability in reflection.

power range 47–51 mW the device had two possible output states. Firing the flash unit caused the resonator to switch from off to on resonance.

A critical flash intensity of 39 W cm^{-2} was required when the device was biased with 50 mW (1 mW below switch-on point). Assuming a relaxation time of *ca.* 100 ns for carriers introduced in this manner (Seaton *et al.* 1983), an effective external energy of 1 nJ is calculated as that required to switch the etalon. The total energy involved in switching the etalon should also include that provided by the CO laser during this time. This total energy is comparable to that observed in the only other examples of the external switching of intrinsic bistable systems by Tarng *et al.* (1982) and by Seaton *et al.* (1983). In the latter experiment a similar etalon ($L = 210$ μm) was switched by a 35 ps long pulse of energy 5 nJ, effective over the CO beam area, where the beam derived from a 1.06 μm Nd–YAG mode-locked laser.

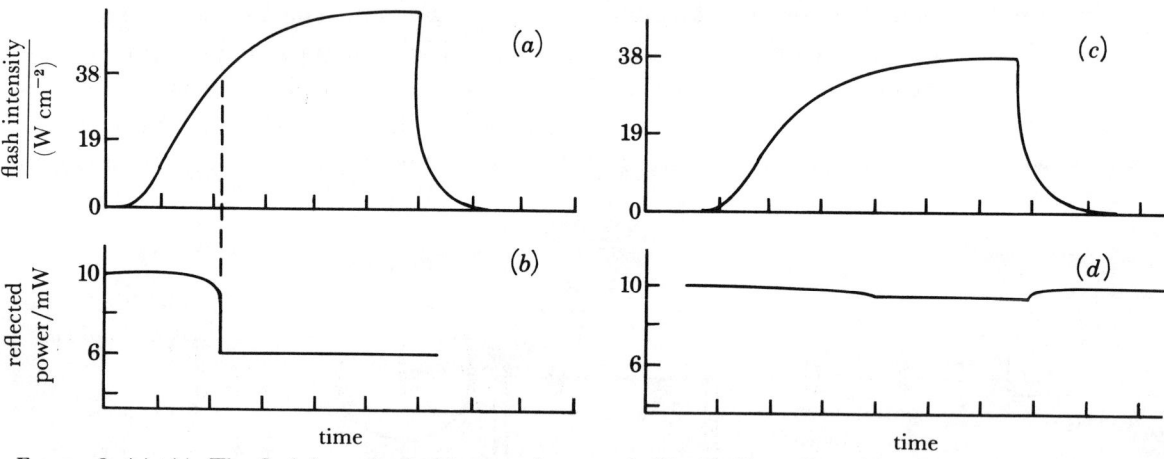

FIGURE 3. (*a*), (*c*). The flash intensity incident on the crystal. (*b*), (*d*) The reflected laser power corresponding to (*a*) and (*c*), respectively. Parts (*c*) and (*d*) show that a high-energy pulse will not switch the etalon if it does not reach a critical intensity. One division represents 50 μs on the time scale.

A thermal effect is concurrent with this electronic one. The absorbed energy heats the sample and this causes a decrease in the energy gap ($d\bar{\nu}_g/dT = 2$ cm^{-1} K^{-1} at 80 K (Camassel & Auvergne 1975)). Consequently, the refractive index increases ($dn/dT = 6 \times 10^{-4}$ K^{-1} (Cardona

1960)) as do the nonlinear refractive index and the linear absorption coefficient (these changes can be estimated from Miller *et al.* 1981). The total effect of all three parameters on the input–output characteristic is shown by figure 4. Clearly the increase in linear refractive index is dominant in a sample of this thickness (260 μm). However, in a thinner sample ($L = 98$ μm) the nonlinear part of the characteristic is shifted toward the origin, which suggests that the increase in nonlinear refractive index is dominant.

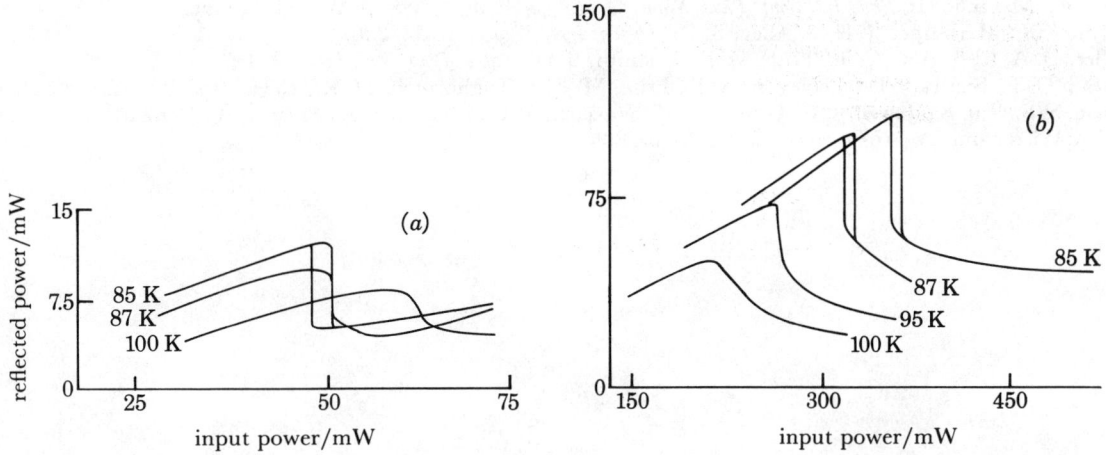

FIGURE 4. The effect of temperature on the characteristics of etalons of thickness (*a*) 260 μm and (*b*) 98 μm.

These shifts can be caused by the heating effect of the flashlamp pulse. For example, figure 5 shows that the absorption of 3 mJ cm⁻² (enough to raise the temperature by about 2 K) causes switching of the 260 μm thick device from on to off resonance. That this is a purely thermal effect is concluded from the dependence of the switching point upon only the energy absorbed, rather than flash intensity. The 98 μm thick cavity was also switched by using this effect. However, for this cavity the thermal effect causes the etalon to switch from off to on resonance.

In conclusion, we have demonstrated for the first time that an intrinsic bistable device can be switched in both directions between its states by low-energy incoherent white-light pulses.

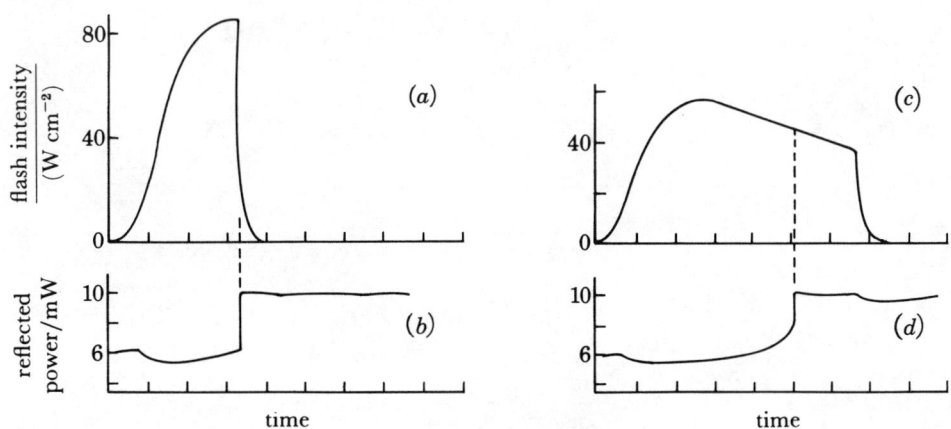

FIGURE 5. As figure 3. The etalon is switched from on to off resonance only when enough energy is absorbed. One division represents 50 μs on the time scale.

The authors would like to acknowledge early contributions by Dr C. T. Seaton to this work and discussions with Dr B. S. Wherrett. One of us (F.T.) acknowledges the support of an S.E.R.C. C.A.S.E. studentship with British Telecom.

REFERENCES

Camassel, J. & Auvergne, D. 1975 *Phys. Rev.* B **12**, 3258.

Cardona, M. 1960 In *Proc. Int. Conf. Phys. Semicond., Prague 1960*, p. 388. Acad. Sci. Prague.

Garmire, E., Marburger, J. H. & Allen, S. D. 1978 *Appl. Phys. Lett.* **32**, 320.

Miller, D. A. B., Seaton, C. T., Prise, M. E. & Smith, S. D. 1981 *Phys. Rev. Lett.* **47**, 197.

Seaton, C. T., Smith, S. D., Tooley, F. A. P., Prise, M. E. & Taghizadeh, M. R., 1983 *Appl. Phys. Lett.* **42**, 131.

Tarng, S. S., Tai, K., Jewell, J. L., Gibbs, H. M., Gossard, A. C., McCall, S. L., Passner, A., Venkatesan, T. N. C. & Wiegmann, W. 1982 *Appl. Phys. Lett.* **40**, 205.

Phil. Trans. R. Soc. Lond. A **313**, 361–362 (1984)

Printed in Great Britain

Bistable behaviour of light waves in a graded-index planar waveguide with nonlinear substrate

By C. Sibilia, M. Bertolotti and D. Sette

Dipartimento di Energertica, Università di Roma, Via Scarpa 14, *Roma* 00161, *Italy*

Bistability of a planar guiding structure with a nonlinear refractive index of the upper substrate and a lower linear substrate with a nonabrupt boundary is studied. The central guiding medium is assumed to be linear.

The condition for the propagation of light along the waveguide (see figure 1) is given in the geometrical optics approximation by the well known dispersion formula

$$\Phi(\beta) = 2Kn_1 \mathrm{d}\psi - 2\phi_{R1} - 2\phi_{R2} - 2\pi N = 0, \tag{1}$$

where N is an integer number selecting the mode, ϕ_{R1} and ϕ_{R2} are the phase shifts due to total reflection at the nonlinear interface '1–2', and the linear interface '1–3' respectively, and the term $2Kn_1 \mathrm{d}\psi$ corresponds to the optical-path phase change in the ray representation and in the small-angle approximation, ψ is the angle between the light ray and $\hat{x}$-axis and d is the waveguide thickness. For a semi-infinite substrate, and in the plane-wave approximation for the guided modes, the nonlinear phase shift ϕ_{R1} can be written (Kaplan 1977) as

$$\phi_{R1} = \mp 2 \arccos |[\Delta\epsilon/\epsilon_1 \pm \{(\Delta\epsilon/\epsilon_1)^2 - 8\psi^2 \epsilon_{n1}|E_0|^2/\epsilon_1\}^{\frac{1}{2}}]^{\frac{1}{2}} (4\epsilon_{n1}|E_0|^2/\epsilon_1)^{-\frac{1}{2}}|, \tag{2}$$

where

$$\epsilon_2 = \epsilon_1 + \Delta\epsilon + \epsilon_{n1}|E_0|^2,$$

and $|E_0|^2$ corresponds to the incident intensity at the interface. In our example $|E_0|^2$ is the intensity of the mode in medium 1.

Instabilities for $\Phi(\beta)$ can be found as a function of the incident intensity, in the small angle approximation, when $\mathrm{d}\Phi(\beta)/\mathrm{d}I_0 = \infty$. After a short calculation it can be shown that instabilities occur at the following two values of incident intensity:

$$I_{0,1} = \frac{\Delta\epsilon}{2\epsilon_{n1}} \left\{ 1 - \left(\frac{\psi}{\psi_c}\right)^2 \right\} \tag{3a}$$

and

$$I_{0,2} = \frac{\Delta\epsilon}{8\epsilon_{n1}} \left(\frac{\psi_c}{\psi}\right)^2, \tag{3b}$$

where $\psi_c^2 \sim \Delta\epsilon/\epsilon_1$ is the critical angle for the total reflection at the boundary 1–2, and we have assumed $\psi < \psi_c$ (guided modes). The instability points correspond to a jump in the phase shift ϕ_{R1}, which makes $\Phi(\beta)$ change from the value $\Phi(\beta) = 0$, corresponding to a guiding structure, to a value $\Phi(\beta) \neq 0$, in which no stable mode is supported. When increasing the input intensity I_0 from a low value for which $\Phi(\beta) = 0$ and a mode is guided in the structure, as soon as $I_0 = I_{0,2}$ we have $\Phi(\beta) \neq 0$ and no mode is propagated any more. When reducing the intensity, at $I_0 = I_{0,1}$ the equality $\Phi(\beta) = 0$ is satisfied again and the structure once more supports a guided mode.

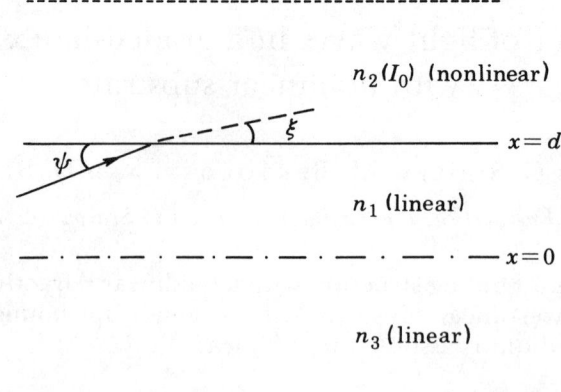

FIGURE 1. Graded-index waveguide with nonlinear substrate, where
$n_2(I_0 = 0) < n_1, n_2(I_0 = 0) \geqslant n_3, n_1 > n_3$.

The intensity values at which instabilities occur (equation (3)) are the same as those of Kaplan for bistability of reflection at the single nonlinear interface; so in the approximation used a wave guiding structure has the same bistable behaviour as a nonlinear interface.

The feedback mechanism needed to create hysteresis is provided here by the evanescent field in the nonlinear medium that reduces the critical angle value and affects the phase shift between incident and reflected beams, so the reduced critical angle further reduces the evanescent field. A more complete analysis with nonlinear Maxwell equations confirms this behaviour.

REFERENCE

Kaplan, A. E. 1977 *Soviet Phys. JETP* **45**, 896.

Phil. Trans. R. Soc. Lond. A **313**, 363–369 (1984)

Printed in Great Britain

Theory of optical hysteresis for TE guided modes

By A. Boardman and P. Egan

Department of Physics, University of Salford, Salford, M5 4WT, U.K.

TE guiding structures consisting of an optically linear dielectric film embedded in dissimilar optically nonlinear unbounded media, are investigated for hysteresis properties. It is shown that many important features can be determined without a knowledge of the electric fields in the nonlinear media and that the eigenvalue equation yields a relation between the guided-wave vector and the field amplitude at the boundary. The power in the system is the important physical quantity and is carried by asymmetric modes that, in the linear limit, are neither odd nor even. This paper explores some of the limitations of such asymmetrically loaded linear dielectric films and shows that they can exhibit optical hysteresis.

Optically nonlinear waveguiding structures

Guiding structures consisting of relatively thin dielectric layers bounded by semi-infinite media have been of interest for some time. Such structures are destined to form part of the integrated optics systems of the future so it is of considerable interest to ask what will happen when one, or all, of the layers becomes optically nonlinear. The latter occurs when the refractive index becomes a function of the intensity of the wave it is carrying. So, if a dielectric layer is optically nonlinear, its dielectric constant will change from its linear value ϵ, say, to $\epsilon + \alpha |\boldsymbol{E}|^2$ because it is supporting an electric field $\boldsymbol{E}$. Optical nonlinearity in symmetrically loaded structures has recently been investigated by Akmediev (1982) in terms of the power flow, and subsequent work by Boardman & Egan (1984) has considerably elucidated this structure and has drawn attention to the fact that the asymmetrically loaded thin film is a non-trivial alternative.

If the guiding structure (see figure 1) consists of media with dielectric constants ϵ_i and nonlinear coefficients α_i then, after some manipulation, it can be shown that the general relation of the electric field amplitude E_0 at $z = 0$ to the electric field amplitude E_b at $z = d$ is the conic section

$$\alpha_3 \left(\frac{\eta_3^2}{\alpha_3} - \frac{\eta_1^2}{\alpha_1}\right)^{-1} \left(E_b^2 - \frac{\eta_3}{\alpha_3}\right)^2 - \alpha_1 \left(\frac{\eta_3^2}{\alpha_3} - \frac{\eta_1^2}{\alpha_1}\right)^{-1} \left(E_0^2 - \frac{\eta_1}{\alpha_1}\right)^2 = 1, \quad \eta_i = \epsilon_2 - \epsilon_i. \tag{1}$$

This equation emerges if it is assumed that E_0 and E_b are real, so that only the part of the conic section appearing in the first quadrant will be significant. Hence, for a particular value of E_0^2, there may exist two, one or no values of E_b^2, a circumstance that is entirely determined by the media parameters.

In general there are four types of conic section that are possible. Type I and type II are hyperbolae. Type I has a horizontal transverse axis and type II has a vertical transverse axis. The other two are ellipses, i.e. type III with a horizontal major axis and type IV with a vertical major axis. Only type II will be considered here, as an illustration, and this is shown in figure 2 for varying degrees of asymmetry between the upper and lower medium.

If the limit of vanishing nonlinearity is taken, then the centre of the conic section, since it

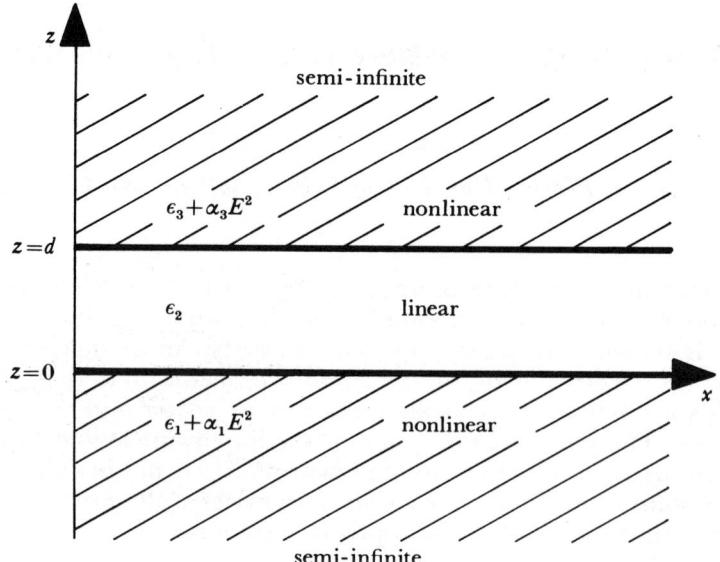

FIGURE 1. Asymmetric linear thin-film optical waveguide with nonlinear bounding dielectrics.

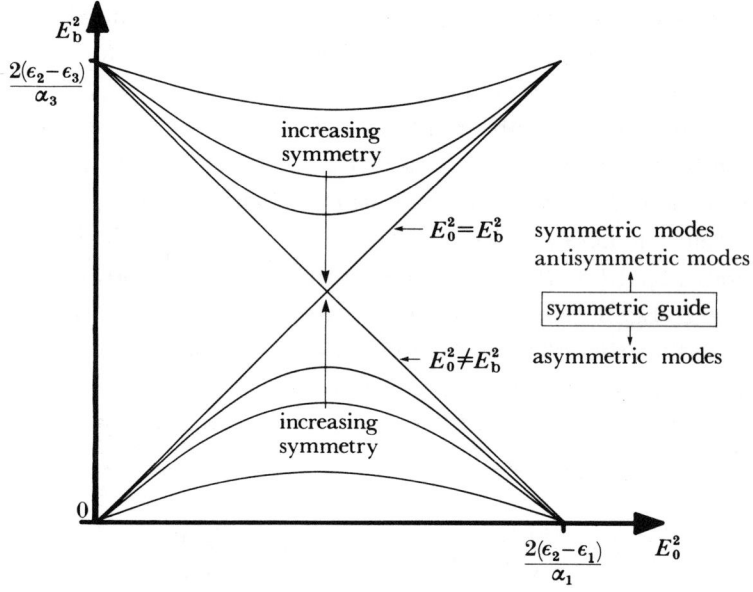

FIGURE 2. Type II boundary-field amplitude relation. In the limit of a symmetrically loaded film the hyperbolae degenerate to straight lines. Note that each set of hyperbolae has a different vertical scale.

has the coordinates $(\eta_1/\alpha_1, \eta_2/\alpha_2)$, moves out diagonally to infinity and the hyperbolae fan out into a set of straight lines of slope $(\epsilon_2 - \epsilon_3)/(\epsilon_2 - \epsilon_1)$. These lines are contained below $E_0^2 = E_b^2$ when $\epsilon_3 > \epsilon_1$. The significant point here is that asymmetric modes for which $E_0^2 \neq E_b^2$ occur both in the all linear structure and the nonlinear structure. This is not true of the symmetrically loaded slab. The limits on the E_b^2, E_0^2 axes show the containment within the guided-wave region.

TE MODE EIGENVALUE EQUATION

For a TE guided mode propagating along the x-axis of figure 1 with angular frequency ω and wavenumber k_x the electric field components in the media are given by:

$$\ddot{E}_2 + k_2^2 E_2 = 0, \quad \ddot{E}_i - (k_i^2 - 2\Lambda_i E_i^2)\, E_i = 0,$$

$$k_2^2 = \frac{\omega^2}{c^2}\epsilon_2 - k_x^2, \quad \Lambda_i = \frac{\omega^2 \alpha_i}{2c^2}, \quad k_i^2 = k_x^2 - \frac{\omega^2}{c^2}\epsilon_i = \frac{\omega^2}{c^2}(\epsilon_2 - \epsilon_i) - k_2^2, \quad i = 1, 3, \tag{2}$$

where a dot over a variable denotes differentiation with respect to z, and c is the velocity of light in vacuo. It is also useful, at this stage, to define the quantities $\gamma_1^2 = k_1^2 - \Lambda_1 E_0^2$, $\gamma_3^2 = k_3^2 - \Lambda_3 E_b^2$ so that the gradient of the fields at $z = 0$ and $z = d$ can be written as $(\dot{E}_1)_{z=0} = \gamma_1 E_0$ and $(\dot{E}_3)_{z=d} = \gamma_3 E_b$, where γ_1 and γ_3 can be positive or negative.

After the application of the boundary conditions the eigenvalue equation reduces to

$$\cos{(k_2 d)} = \pm \left[\frac{\dfrac{c^2 k_2^2}{\omega^2} \pm \left(\epsilon_2 - \epsilon_1 - \tfrac{1}{2}\alpha_1 E_0^2 - \dfrac{c^2 k_2^2}{\omega^2}\right)^{\frac{1}{2}} \left(\epsilon_2 - \epsilon_3 - \tfrac{1}{2}\alpha_3 E_b^2 - \dfrac{c^2 k_2^2}{\omega^2}\right)^{\frac{1}{2}}}{(\epsilon_2 - \epsilon_1 - \tfrac{1}{2}\alpha_1 E_0^2)^{\frac{1}{2}} (\epsilon_2 - \epsilon_3 - \tfrac{1}{2}\alpha_3 E_b^2)^{\frac{1}{2}}} \right]. \tag{3}$$

If the outer sign is taken as $+$ in equation (3) then even parity solutions, for which E_0 and E_b have the same sign, yield asymmetric modes that, in the symmetric structure, give an even mode and an associated asymmetric mode. Now the slopes of the fields at each boundary determine whether the field immediately decays exponentially in the nonlinear medium or first rises to a maximum and then decays exponentially (i.e. whether a bulge appears). It is possible to distinguish various possibilities for E_3 and E_1 without actually needing to know their explicit form. Since the gradients of the fields at the slab boundaries are continuous then use of the field in the linear layer gives

$$\frac{k_2}{\sin{(k_2 d)}} [E_b - E_0 \cos{(k_2 d)}] = \gamma_1 E_0, \quad \frac{k_2}{\sin{(k_2 d)}} [E_b \cos{(k_2 d)} - E_0] = \gamma_3 E_b. \tag{4}$$

Hence if $\sin{(k_2 d)} > 0$, $E_0 > 0$, $E_b > 0$ and $E_b/E_0 > \sec{(k_2 d)}$ then $\gamma_1 > 0$, $\gamma_3 > 0$ so that $E_1 = \gamma_1 E_0$ and $E_3 = \gamma_3 E_b$. This shows, immediately, that the field in the lower medium decays exponentially while the field in the upper nonlinear medium possesses a maximum (i.e. a bulge). On the other hand, if $0 < E_b/E_0 < \cos{(k_2 d)}$ then $\gamma_1 < 0$, $\gamma_3 < 0$ and $E_1 = -|\gamma_1| E_0$, $E_3 = -|\gamma_3| E_b$, which shows that the lower nonlinear medium now possesses a bulge while the field in the upper medium decays exponentially. The intermediate case

$$\cos{(k_2 d)} < E_b/E_0 < \sec{(k_2 d)} \quad \text{gives} \quad \gamma_1 > 0, \gamma_3 < 0$$

and the fields in either nonlinear medium decay exponentially without bulges. All of these arguments refer to the range $0 < k_2 d < \tfrac{1}{2}\pi$. In the other ranges possibilities exist for bulges to appear in both the upper and lower media. It will depend upon the thickness of the layer and the other parameters as to whether these are TE_0, TE_1, ... mode ranges.

Figure 3 shows the dependence of ck_x/ω, for some specimen data, upon $\tfrac{1}{2}\alpha_1 E_0^2$. The upper and lower branches of figure 2 are analysed separately, but each shows a cut-off locus that depends upon E_0.

If $E_0 > 0$, $E_b > 0$, the first allowed range of k_2, for which the outer sign is positive and the inner sign is negative in (3), ends when $\gamma_1 = 0$ at

$$\frac{k_2}{\omega(\epsilon_2 - \epsilon_3 - \frac{1}{2}\alpha_3 E_b^2)^{\frac{1}{2}}/c} = \cos{(k_2 d)} = \pm\frac{E_b}{E_0}. \tag{5}$$

This is a transition from $\gamma_1 > 0$ to $\gamma_1 < 0$ so, since the slope of the field at the boundary in medium 1 is $E_1 = \gamma_1 E_0$, no calculation is needed to understand that this is possible only if a maximum (i.e. a *single bulge*) occurs in the electric field of the nonlinear medium. A further interesting phenomenon occurs, however. This transition point is also where k_2 begins to satisfy (3) with positive inner sign. Thus the transition point is a cusp and allowed eigenvalues do not cease to exist as the single-bulge range is entered.

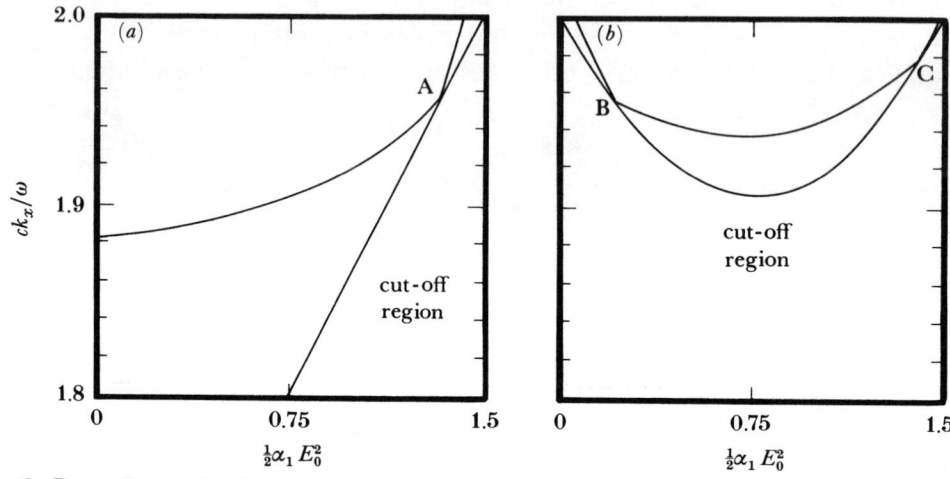

FIGURE 3. Dependence of guided wavenumber on the optical nonlinearity. Data: $\epsilon_1 = 2.5$, $\epsilon_2 = 4.0$, $\epsilon_3 = 2.0$, $d = 0.5\lambda$, $\lambda = 2\pi c/\omega$, $\beta = 1.4$; (a) corresponds to a lower hyperbola of figure 2; (b) corresponds to an upper hyperbola. A indicates the point of the appearance of maximum field in the lower medium and B, C, in the upper medium.

For the upper curve in the (E_0^2, E_b^2) plot, E_b^2 is, initially, at a finite value and is substantially larger than E_0^2 for all E_0. Cut off will, therefore, occur at $c^2 k_2^2/\omega^2 = \epsilon_2 - \epsilon_1 - \frac{1}{2}\alpha_1 E_0^2$ or $\epsilon_2 - \epsilon_3 - \frac{1}{2}\alpha_3 E_b^2$, whichever is the smaller. This leads to a very curved cut-off locus involving both E_0 and E_b.

OPTICAL HYSTERESIS

The possibility that the thin film structure can exhibit hysteresis and possible optical bistability is determined from the behaviour of the power flow down the guide. This power flow can be calculated without knowing the fields in the outside nonlinear media, a feature that was seen above to be also true of the dispersion equation. Indeed, up to now, the fields in the bounding media have not been used at all.

For the outside bounding media

$$(\dot{E}_i/E_i)^2 = k_i^2 - \Lambda_i E_i^2. \tag{6}$$

Differentiation with respect to z gives

$$2\left(\frac{\dot{E}_i}{E_i}\right)\frac{\mathrm{d}}{\mathrm{d}z}\left(\frac{\dot{E}_i}{E_i}\right) = -2\Lambda_i E_i \dot{E}_i, \quad \int_a^b E_i^2 \, \mathrm{d}z = -\frac{1}{\Lambda_i}\left(\frac{\dot{E}_i}{E_i}\right)_a^b. \tag{7}$$

For $i = 1$, $b = 0$, $a = -\infty$ and for $i = 3$, $b = \infty$, $a = d$ where

$$\lim_{z \to -\infty} \left(\frac{\dot{E}_1}{E_1} \right) = k_1, \quad \lim_{z \to \infty} \left(\frac{\dot{E}_3}{E_3} \right) = -k_3$$

and, for any other value of z, $\dot{E}_i/E_i$ is $(k_i^2 - \Lambda_i E_i^2)^{\frac{1}{2}}$. The power flow integrals in media 1 and 3 therefore evaluate to $(k_{1,3} \pm |\gamma_{1,3}|)/\Lambda_{1,3}$ with the restrictions that $\dot{E}_3/E_3 > 0$, $\dot{E}_1/E_1 < 0$ for the positive sign and $\dot{E}_1/E_1 > 0$, $\dot{E}_3/E_3 < 0$ for the negative sign at the boundaries.

In the linear thin film

$$E_2^2 = \frac{1}{2k_2^2} \left(c_2 - \frac{\mathrm{d}}{\mathrm{d}z}(E_2 \dot{E}_2) \right), \tag{8}$$

so that the power integral becomes

$$\int_0^d E_2^2 \, \mathrm{d}z = \frac{1}{2k_2^2} [c_2 d - (E_2 \dot{E}_2)_d + (E_2 \dot{E}_2)_0], \tag{9}$$

but
$$(\dot{E}_2)_d = \gamma_3 E_b, \quad (\dot{E}_1)_0 = \gamma_1 E_0, \quad c_2 = (k_2^2 + \gamma_1^2) E_0^2 = (k_2^2 + \gamma_3^2) E_b^2, \tag{10}$$

so that the total power flow in the thin film and the nonlinear bounding media is

$$P = P_0 K_x \{ \beta(K_1 \pm G_1) + K_3 \mp G_3 + [D(\epsilon_2 - \epsilon_1 - \tfrac{1}{2}\alpha_1 E_0^2) \tfrac{1}{2}\beta \alpha_1 E_0^2 \mp \tfrac{1}{2} G_1 \beta \alpha_1 E_0^2 \pm \tfrac{1}{2} G_3 \alpha_3 E_b^2]/2K_2^2 \}, \tag{11}$$

where $K_x = ck_x/\omega$, $K_{1,3} = ck_{1,3}/\omega$, $G_{1,3} = c\gamma_{1,3}/\omega$, $\beta = \alpha_3/\alpha_1$, $D = cd/\omega$ and $P_0 = 2c^2 \epsilon_0/\alpha_3 \omega$ has dimensions of watts per metre. Since E_b^2 is expressible in terms of E_0^2 and, through the dispersion equation, E_0^2 is a function of K_x, then P/P_0 can be varied with K_x for a particular value of ω.

Figure 4 shows the power flow variation with K_x for dielectric constants that are fairly close together. This is a TE$_0$ mode and shows two distinct power loops. If the system were to become symmetric these loops would merge into a symmetric and an associated asymmetric mode. These curves only seem to be like the symmetric case. In the latter structure it is the upper parts of the upper loop that would be responsible for any bistability. In fact, even a small amount of asymmetry is sufficient to split the curves and thus lift the degeneracy of symmetric and asymmetric modes into two distinct asymmetric modes. If $E_b/E_0 < 0$ a strong peak develops at much higher power and only if $d > 2\lambda$.

The power curves can now be used to predict any optical hysteresis. Suppose that the nonlinear media are lossy (an assumption that the media are infinitely lossy will suffice for the present purposes), then massive energy dumping through the appearance of electric field bulges can be anticipated and the variation of P_{out}, the power out of the guide, with P_{in}, the input power to the guide, can be calculated. In the first place, the power shown in figure 4 is the input power distributed over the whole of a plane perpendicular to the z axis. This energy is, as a consequence, distributed over both the linear layer and the nonlinear media. If at some distance down the guide no absorption has occurred then the power out through a similar plane must be exactly equal to the input power.

As P_{in} advances up the lower loop of figure 4 the bulk of the power flow is in the linear layer so that P_{out} remains essentially the same as P_{in}. As the maximum is approached deviation from linearity occurs as the field strength in the nonlinear media grows. At the maximum in

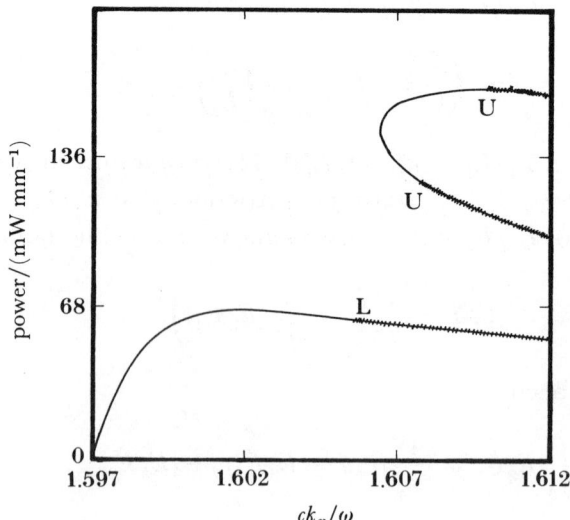

FIGURE 4. Total power flow down the waveguide as a function of ck_x/ω. The hatched part of the curves marks the appearance of a sustained field maximum (bulge) in the upper (U) or lower (L) nonlinear medium. Data: $\epsilon_1 = 2.45$, $\epsilon_2 = 2.6$, $\epsilon_3 = 2.3$, $d = 1.5\lambda$, $\alpha_3 = 6.37 \times 10^{-12}$ m² V⁻², $\beta = 1.05$, $\lambda = 0.515$ μm.

the power curve a further increase in P_{in} causes no further increase in P_{out}, i.e. the power saturates. A much greater increase in P_{in} can cause a transition to the higher power loop.

A reduction of P_{in} from the maximum opens up the option of returning down the first branch, the second branch on the large K_x side of the peak, or both branches. Suppose that to set a limit on the hysteresis curve a descent down the far side of the peak occurs so that more and more energy is dumped into the nonlinear medium as the wavenumber at which a bulge in the field of the lower nonlinear medium is approached. This causes P_{out} to fall below the P_{out} obtained on the outward direction, as P_{in} is reduced. This fall away continues as higher K_x values are accessed until a point is reached where a transition back to a guided wave on the outward branch becomes possible, or a surface mode is generated. It is, of course, possible that a transition to the outward branch could have occurred earlier, but the hysteresis loop shown in figure 5 marks the limit of the possibilities. Note that a hysteresis loop can also be derived for the upper power loop of figure 4 and that the data used for figure 3 would give a highly structured form of hysteresis.

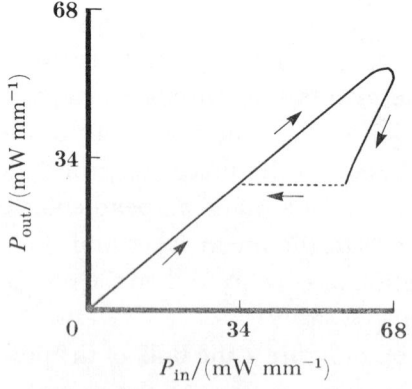

FIGURE 5. Optical hysteresis shown in the output power (P_{out}) against input power (P_{in}) characteristic obtained from figure 4.

[178]

Hysteresis has recently been demonstrated, for TE_1 modes, experimentally by Vach *et al.* (1984), for a nonlinear medium bounded by a linear film on a linear substrate. This paper shows that many other possibilities occur in the type of asymmetric structure discussed above. Guided waves offer the prospect of non-resonant bistable devices, but it should be emphasized that a lot of theoretical work needs to be done before the stability of these transformations and the genuine prospects for bistable operation can be determined.

REFERENCES

Akmediev, N. N. 1982 *Soviet Phys. JETP* **56**, 299–303.
Boardman, A. D. & Egan, P. 1984 *J. Physique Colloq.* (C5) **45**, 291–303.
Vach, H., Seaton, C. T., Stegeman, G. I. & Khoo, I. C. 1984 *Optics Lett.* **9**, 238–240.

Phil. Trans. R. Soc. Lond. A **313**, 371–373 (1984)

Printed in Great Britain

Guided-wave controlled etalons

By D. Sarid

Optical Sciences Center, University of Arizona, Tucson, Arizona 85721, U.S.A.

Optical bistability in a nonlinear Fabry–Perot etalon can be controlled by several optical beams injected through the edges of the etalon. The control beams, which propagate transversely in the etalon, modify the refractive index, which in turn switches the etalon between its two stable states. The theory of operation of the device is presented, possible logical operations are discussed and experimental results from using an InSb etalon and a CO laser are reported.

Probably the most promising application of optical bistability is in the area of optical signal processing (Jewell *et al.* 1984; Seaton *et al.* 1983). Switching light by light requires efficient use of the switching power. Guided waves provide an effective means of applying the intracavity intensity needed to switch bistable etalons between their states, independent of the Fabry–Perot resonances. We have proposed and demonstrated experimentally a bistable InSb etalon at 80 K by using a 5.59 μm CO laser beam in which the switching is controlled by a separate beam endfire coupled through the edge of the etalon (Sarid *et al.* 1984). Here we report on the theory and experimental results of the device and present possible applications for the logic operations basic to signal processing.

When a nonlinear Fabry–Perot etalon is used as a bistable device in reflection together with a control beam, as shown in figure 1, the reflected intensity I_{ref} is given by

$$I_{\text{ref}} = I_{\text{inc}}\left[\!\left[1 - \frac{C_1}{1 + F\sin^2\left\{\gamma\left[(I_{\text{inc}} - I_{\text{ref}})/C_1 C_2 + I_{\text{cont}}\right] + \delta\right\}}\right]\!\right]. \tag{1}$$

Here I_{inc} is the incident intensity, I_{cont} is the intensity of an edge-coupled control beam, C_1, C_2, F and γ are constants depending on the geometry and material properties. The detuning δ can be controlled by angle tuning or by scanning along a slightly wedged etalon.

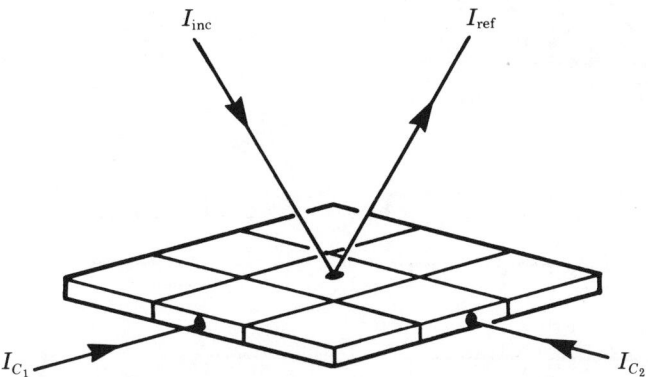

FIGURE 1. Bistable etalon controlled by two injected beams I_{C_1} and I_{C_2} make optical logic operations.

The concept of controlling bistability by using a guided wave was demonstrated with the experimental arrangement shown in figure 2. The InSb etalon is n-doped with Te $(4 \times 10^{14}\ \mathrm{cm}^{-3})$, gold-coated on the back face and cooled to 80 K. A beam from a CO laser tuned to 5.59 μm wavelength is split to form the pump and control beams, which are focused onto the face and edge of the etalon, respectively. The incident, reflected and control beam powers are monitored by pyroelectric detectors and displayed on an oscilloscope.

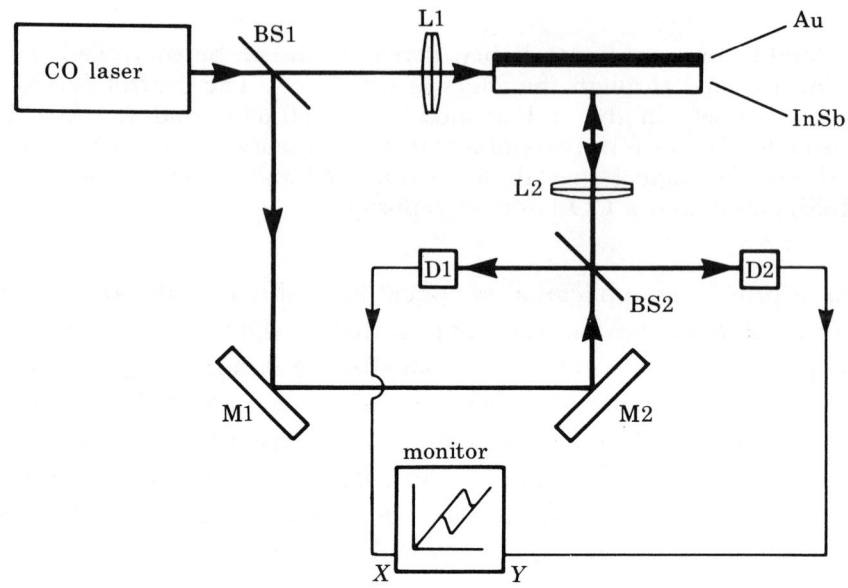

FIGURE 2. The experimental arrangement: BS, beam splitter; L, lens; M, mirror; D, detector.

A typical curve showing the reflected power as a function of control beam power from (1) is given in figure 3. Experimentally, the incident power and detuning can be adjusted to hold the etalon at the 'on' state (point A); by applying the control beam power, the etalon can be switched to the 'off' state (point B).

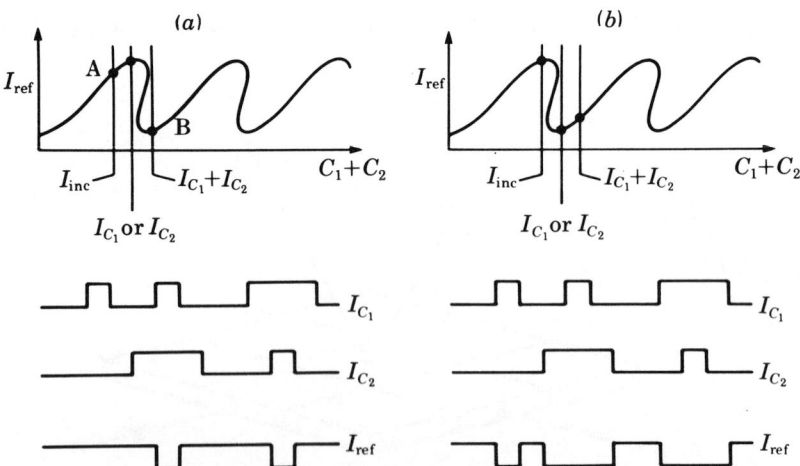

FIGURE 3. Theoretical curves of the reflected intensity as a function of the two control beam intensities, showing operation in reflection as an (a) 'AND' or (b) 'OR' gate.

The advantage of using a beam injected through the edge of the device to control the switching action is that its effect is decoupled from that of the incident beam. As can be seen from (1), the control intensity I_{cont} serves only to change the detuning of the cavity and does not influence the reflected beam directly, as does the incident intensity I_{inc}. Also, the guided control beam is not affected by the resonances of the etalon and therefore is used efficiently.

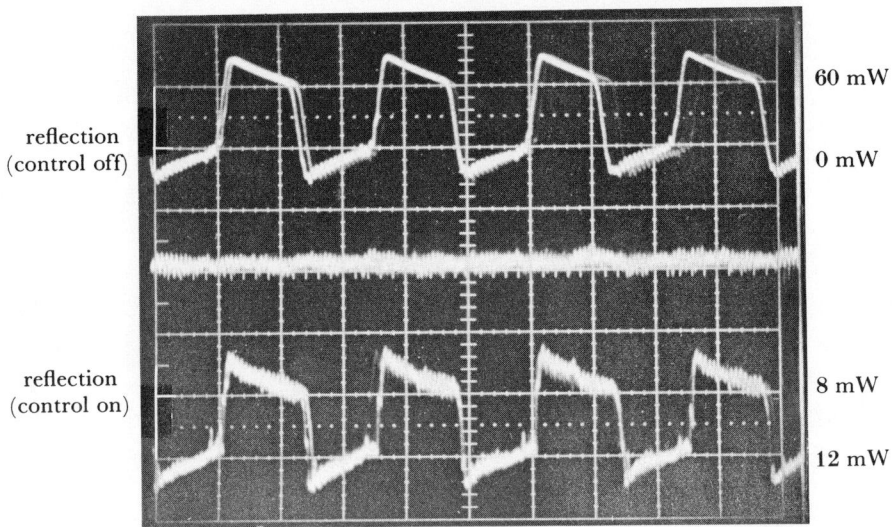

FIGURE 4. Experimental results showing the chopped control beam (upper trace), response of reflected beam to control beam (lower trace, shown inverted) and constant reflected beam (12 mW) in absence of control beam (centre trace). The horizontal scale is 10 ms per division.

The etalon displayed a typical bistable response when operated in reflection. The switching action of the control beam is shown in figure 4, where the upper trace is the modulated control beam power, the centre trace is the constant reflected power in the absence of a control beam, and the lower trace (shown inverted) is the reflected power switched between the 'on' and 'off' states in the presence of a control beam.

By injecting two control beams into a bistable etalon, such as in figure 1, it is possible to make optical logic operations. With the device used in reflection and biased to the appropriate operating point by the incident beam and detuning, 'AND', 'OR' and 'XOR' operations are possible, as illustrated in figure 3. In transmission, 'NAND', 'NOR' and 'NXOR' operations can be made. By fabricating arrays of such devices, more elaborate optical signal processing can be achieved.

This research is supported by the National Science Foundation grant ECS-8117311 and Nato Travel Grant RG.143.81. Helpful discussions with S. D. Smith, N. Peyghambarian and J. Moloney are acknowledged.

REFERENCES

Jewell, J. L., Rushford, M. C. & Gibbs, H. M. 1984 *Appl. Phys. Lett.* **44**, 172–174.
Sarid, D., Jameson, R. S. & Hickernell, R. K. 1984 *Optics Lett.* **9**, 159.
Seaton, C. T., Smith, S. D., Tooley, F. A. P., Prise, M. E. & Taghizadeh, M. R. 1983 *Appl. Phys. Lett.* **42**, 131–133.

Phil. Trans. R. Soc. Lond. A **313**, 375–380 (1984)

Printed in Great Britain

Optical logic on a single etalon

By J. L. Jewell[1]†, M. C. Rushford[1], H. M. Gibbs[1], M. Warren[1],
N. Peyghambarian[1], A. C. Gossard[2] and W. Wiegmann[2]

[1]*Optical Sciences Center, University of Arizona, Tucson, Arizona 85721, U.S.A.*

[2]*AT&T Bell Laboratories, Murray Hill, New Jersey 07947, U.S.A.*

A simple technique for performing all-optical logic (NOR etc.) on a single etalon is described. Computer simulation and experiments with dye-filled and GaAs etalons verify its validity.

Operation

We report a technique for operating a single nonlinear Fabry–Perot etalon that yields the decisions NOR, NAND, XOR, OR and AND, simultaneously if desired, and with minimum time and energy per cycle. The transmission–time characteristics obtained from experiments with dye-filled (Jewell *et al.* 1984) and GaAs etalons qualitatively verify a computer simulation.

Consider two inputs and a probe to be pulses of short duration in comparison to the medium relaxation time τ_R. The nonlinear medium must be such that absorption of one input pulse changes the refractive index enough to shift the Fabry–Perot transmission peak by about one f.w.h.m. To obtain all possible gates, two inputs should produce about twice the phase shift for one input. The peak will, of course, return to its original wavelength in a few τ_R, but if the probe pulse is incident much less than τ_R after the input(s) only this instantaneous transmission determines the output. With appropriate initial detunings of the probe, the various gates are obtained. This is shown graphically in figure 1, from which the approximate transmissions are found for five 'standard' probe detunings with 0, 1, or 2 inputs. Fine adjustment of the detuning and input strength can improve contrast, reliability, etc.

The pulsed operation minimizes both the time and energy per cycle because it *does not require the medium to be kept excited*. So, no extra energy is spent holding the device on and it *relaxes in the dark*, resulting in minimum relaxation time. The logical decision can be made in less than a picosecond for some materials (for example GaAs) (Shank *et al.* 1982; Shank *et al.* 1979), and the gate *self-resets* in a few τ_R.

The transmission against time was calculated and plotted (figure 2) by using the standard Fabry–Perot formula and a refractive index that suddenly changes (as from an imput pulse), then exponentially relaxes to its original value. Two inputs produce twice the phase shift as one in this simulation.

Logic gates

In the dye-etalon experiment two inputs were formed by splitting a continuous wave (c.w.) argon laser beam, passing them through a mechanical chopper and focusing onto the etalon (figure 3). A c.w. probe from a dye laser at lower intensity was focused onto the same spot

† Present address: AT&T Bell Labs, Holmdel, New Jersey 07733, U.S.A.

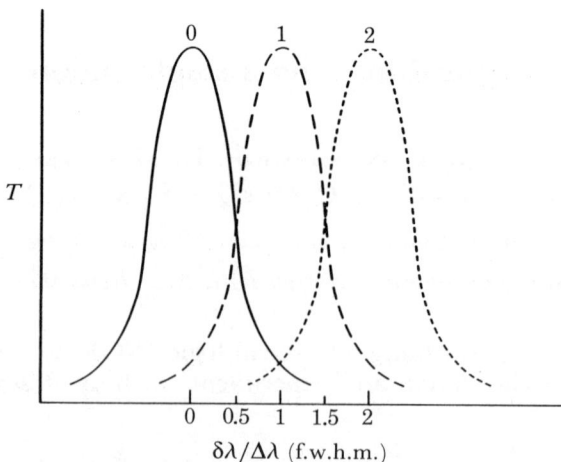

FIGURE 1. Position of the transmission peak with 0, 1, or 2 inputs. With the probe wavelength at one of the five labelled values (expressed by the initial detuning in f.w.h.m.s of the transmission peak) the gates in the table are obtained. The fractional values in the columns below 0, 1, and 2 (number of inputs) are the approximate transmissions when each input shifts the peak by one f.w.h.m. In reflection AND and NOR have poor contrast.

initial detuning	0	1	2	T	R
0	1	0.2	0	NOR	OR
0.5	0.5	0.5	0	NAND	(AND)
1	0.2	1	0.2	XOR	XOR
1.5	0.1	0.5	0.5	OR	(NOR)
2	0	0.2	1	AND	NAND

and transmission against time was recorded. The dye in the etalon absorbed much more strongly at the input than the probe wavelength. Waveforms resembling those in figure 2 were easily obtained (figure 4) for all but the OR gate. An OR waveform was achieved that did not resemble the simulation. Although the relaxation time was milliseconds this technique is applicable for any dispersive nonlinearity. Recently we have observed NOR and OR gating in a GaAs etalon at room temperature (figures 5 and 6) by using mode-locked argon laser pulses as inputs. The 2.05 μm bulk GaAs layer clad by *ca.* 0.6 μm thick AlGaAs windows was sandwiched between *ca.* 90% reflecting mirrors. This construction is far from optimized. The phase shift was not linear with respect to input energy, so this other gates were not observed. Most of the energy in the *ca.* 1 nJ mode-locked argon laser pulses was absorbed by the AlGaAs window, thus increasing the energy requirements as well as accumulative heating. The high optical density of GaAs (and AlGaAs) at 514 nm prevented the input pulses from penetrating very far into any of our 3–5 μm thick samples. We believe the reason our bulk GaAs sample showed superior performance over our multiple quantum well structure (m.q.w.s.) devices is that carrier diffusion spreads the index change across the entire GaAs layer in the former, whereas in the latter, the AlGaAs barriers prevent diffusion between the GaAs layers. Despite its inefficiency, the device operated at 82 MHz with 5–10 ns recovery times. The *ca.* 150 mW probe beam (detuned sufficiently far from the band edge to have low absorption, but close enough to see a change in refractive index from the argon pulses) was c.w. to show the relaxation characteristics. Simple theoretical calculations indicate that an optimized GaAs gate may be able to work with input energies below picojoule levels. Previous work (Gibbs *et al.* 1979) with bulk GaAs at 80 K showed saturation energy densities of 10 μJ cm^{-2} (0.1 pJ μm^{-2}).

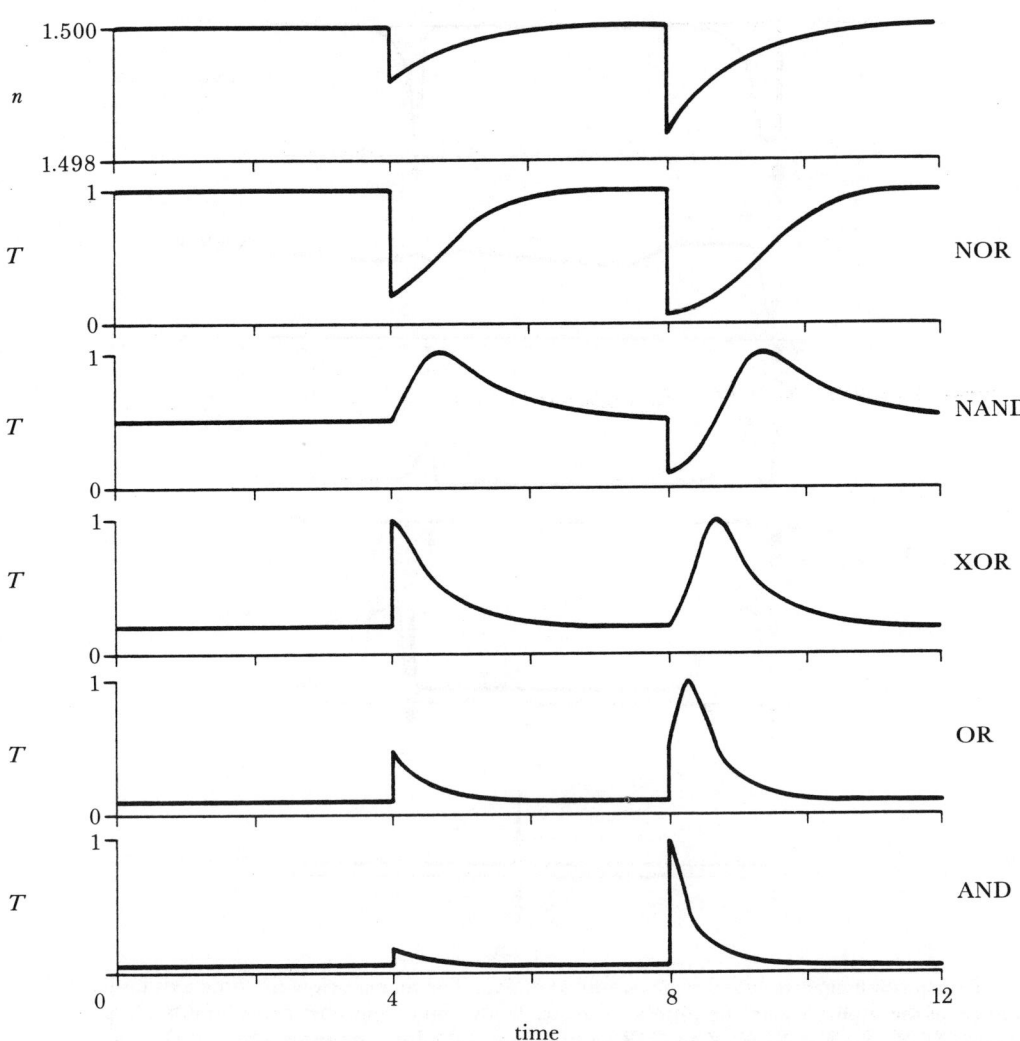

FIGURE 2. Computer-calculated transmission against time for five initial detunings with 0, 1, and 2 input pulses incident at times 0, 4, and 8. Time units are $(1/e)$ recovery times for the refractive index. The top trace is refractive index against time. Mirror reflectivities are 90 % with 10 μm spacing, wavelength 500 nm, and losses are zero.

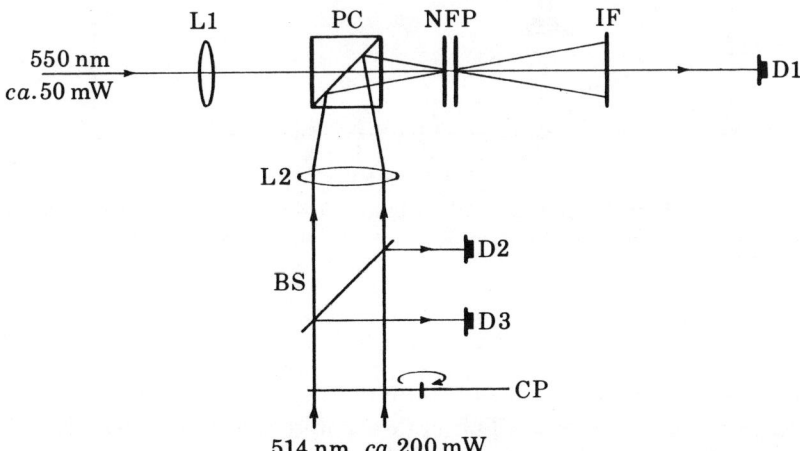

FIGURE 3. Experimental layout for monitoring dye-etalon gate transmission. Components: L1, L2, lenses; D1, D2, D3, detectors; PC, polarizing cube; NFP, nonlinear Fabry–Perot; IF, interference filter; CP, chopper; BS, beamsplitter.

[187]

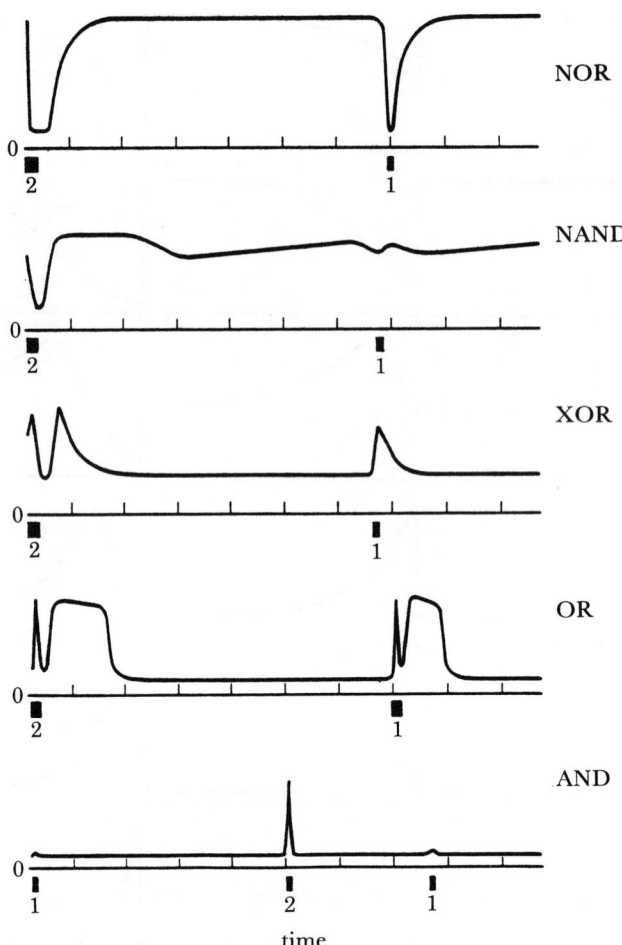

FIGURE 4. Transmission against time for the dye-filled etalon. The marks below the time axis indicate the time and duration of the input(s) with the number directly below indicating how many inputs are present. Time per division: NOR, NAND, XOR, 2 ms; OR, AND, 5 ms. OR does not agree with predictions.

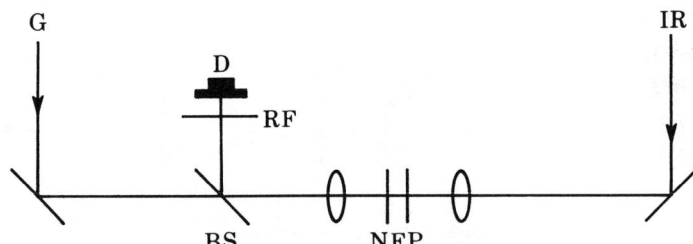

FIGURE 5. Layout for GaAs gatings: G, mode-locked Ar pulses, 514 nm; IR, c.w. infrared dye-laser beam; NFP, GaAs nonlinear Fabry–Perot; BS, beamsplitter; RF, deep-red filter blocks for any Ar light; D, photodiode.

FLIP-FLOP

These gates can also work in a c.w. or mixed mode (c.w. inputs, pulse train probe), although accumulative heating would increase. The etalon could then be considered to be a kind of flip-flop. Another flip-flop that was conceived and tested at Arizona (Jewell *et al.* 1983; Rushford *et al.* 1983) employs two etalons in series in a c.w. beam (figure 7a). The first etalon

[188]

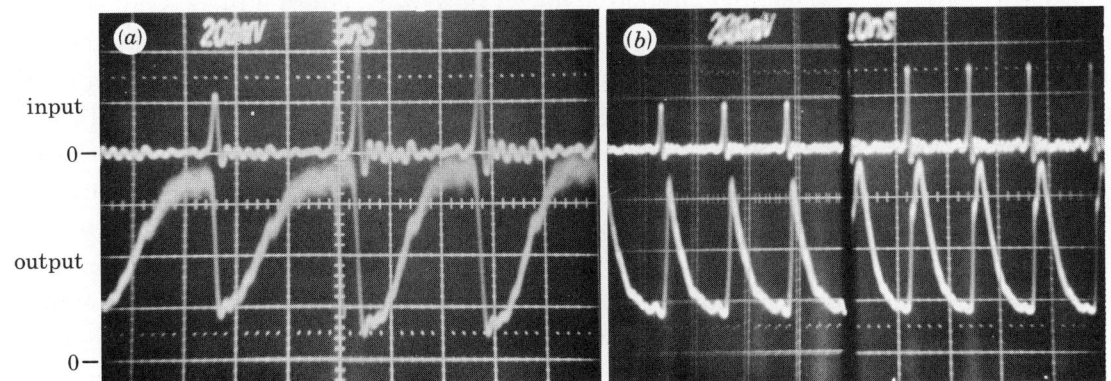

FIGURE 6. Split oscilloscope traces with one input on the left side and two inputs on the right show (a) NOR, and (b) OR capability. Top traces are the inputs; bottom traces are etalon transmissions.

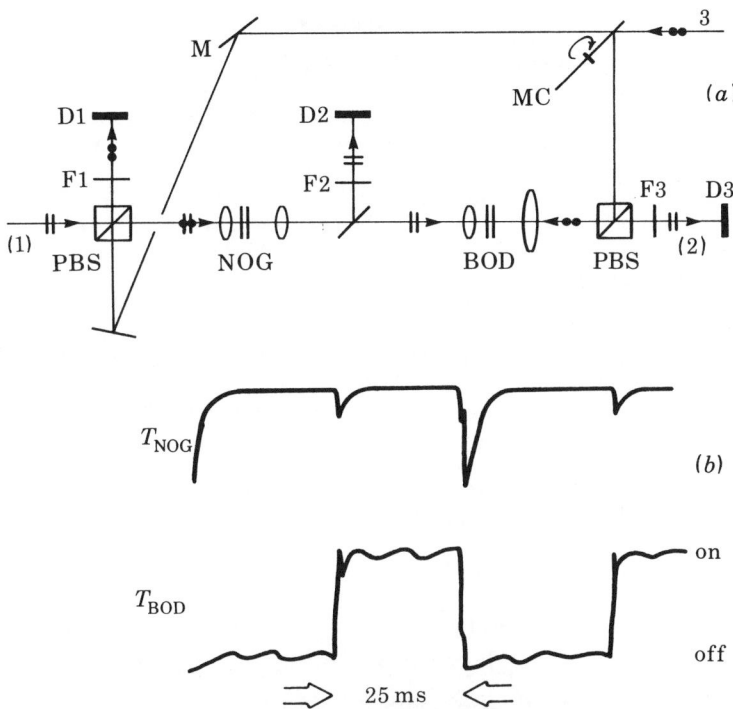

FIGURE 7. (a) Optical flip-flop. (1) Continuous-wave dye-laser bias beam (ca. 550 nm). (2) Transmission of bias beam through bistable optical device (BOD) switched on or off by 514 nm control beam directed by rotating mirror–chopper MC. D1 uses 'off' pulses to trigger the oscilloscope; D2 monitors negative optical gate (NOG) transmissions; D3 shows total flip-flop output. Filters F1, F2, F3 pass the appropriate wavelengths. Beams are combined by polarization beamsplitters (PBS). (b) Response of optical flip-flop; top trace from D2, bottom trace from D3. Large (negative) pulse switches the BOD off; the small pulse is a timing reference from mirror–chopper transmission and BOD transmission affecting the NOG. Bottom trace is on–off intensity monitored by D3. The on–off intensity ratio is ca. 10. The ripple is due to room lights.

is tuned for maximum transmission and the second uses the throughput to operate as a bistable optical device (Gibbs *et al.* 1980; Miller 1982) in the usual sense. The second etalon alone can be switched on by an optical pulse (Tarng *et al.* 1982), but switch-off requires a second nonlinearity of a drop in the incident intensity. The first etalon accomplishes such a drop if it is hit by a switch-off pulse that detunes its transmission peak away from the incident wavelength. This process is identical to that of the NOR gate except that there is only one input pulse and the probe is c.w. This flip-flop was tested with dye-filled etalons yielding light-by-light control (figure 7b). An optical memory or pulse–c.w. conversion are two possible applications for the flip-flop. Another application is to generate short rectangular-shaped optical pulses of precisely controllable length. Since switching on and off are both accomplished by *excitation* of the nonlinear media, the rise and fall times can be extremely short. The relative time separation between the 'on' and 'off' pulses gives the length of the transmitted pulse.

A technique for making logic operations on a single etalon has been presented. Some other optical logic gates have been proposed (Fork 1982) or tested (Seaton *et al.* 1983; Collins *et al.* 1980; Soffer *et al.* 1980), but those presented here are simple, use the high-speed capability of optics with low power requirements, and represent a new mode of operation for existing devices.

REFERENCES

Collins, S. A. Jr, Fatehi, M. T. & Wasmundt, K. C. 1980 *SPIE* **232**, 168–173.
Fork, R. L. 1982 *Phys. Rev.* A**26**, 2049–2064.
Gibbs, H. M., McCall, S. L. & Venkatesan, T. M. C. 1980 *Opt. Engng* **19**, 463–468.
Gibbs, H. M., Venkatesan, T. N. C., McCall, S. L., Passner, A., Gossard, A. C. & Wiegmann, W. 1979 *Appl. Phys. Lett.* **34**, 511–513.
Jewell, J. L., Rushford, M. C. & Gibbs, H. M. 1984 *Appl. Phys. Lett.* **44**, 172–174.
Jewell, J. L., Tarng, S. S., Gibbs, H. M., Tai, K., Weinberger, D. A., Ovadia, S., Gossard, A. C., McCall, S. L., Passner, A., Venkatesan, T. & Weigmann, W. 1983 *Topical Meeting on Opitcal Bistability, Rochester, New York.*
Miller, D. A. B. 1982 *Laser Focus* **18**, 79–84.
Rushford, M. C., Gibbs, H. M., Jewell, J. L., Peyghambarian, N., Wienberger, D. A. & Li, C. F. 1983 *Topical Meeting on Optical Bistability, Rochester, New York.*
Seaton, C. T., Smith, S. D., Tooley, F. A. P., Prise, M. E. & Taghizadeh, M. R. 1983 *Appl. Phys. Lett.* **42**, 131–133.
Shank, C. V., Fork, R. L., Greene, B. I., Weisbuch, C. & Gossard, A. C. 1982 *Surf. Sci.* **13**, 108–111.
Shank, C. V., Fork, R. L., Leheney, R. F. & Shah, J. 1979 *Phys. Rev. Lett.* **42**, 112–114.
Soffer, B. H., Boswell, D., Lackner, A. M., Chavel, P., Sawchuck, A. A., Strand, T. C. & Tanguay, A. R. Jr 1980 *SPIE* **232**, 128–136.
Tarng, S. S., Tai, K., Jewell, J. L., Gibbs, H. M., Gossard, A. C., McCall, S. L., Passner, A., Venkatesan, T. N. C. & Wiegmann, W. 1982 *Appl. Phys. Lett.* **40**, 205–207.

Note added in proof (1 August 1984). Very recently we have observed, in a m.q.w.s. device, all five gates of figure 2 with only 8 pJ per input pulse (0.1 pJ μm^{-2}) incident on the device producing 6:1 contrast in the NOR gate. This low energy was made possible by a high-finesse etalon whose mirrors were designed to transmit at the input wavelength.

Phil. Trans. R. Soc. Lond. A **313**, 381–384 (1984)

Printed in Great Britain

381

Photonic logic based on molecular reorientation of nematic liquid crystals

By J. A. Martin-Pereda, F. J. López, M. A. Muriel and J. M. Otón

Departamento Electronica Cuantica, E.T.S. Ing. Telecommunicacion U.P.M.,
Ciudad Universitaria, 28040-Madrid, Spain

A new type of photonic logic, based on the use of nematic liquid crystals is proposed. The system takes advantage of the refractive-index changes induced by laser beams. Examples of AND, OR and NOR functions are presented.

It is nowadays a well-known fact that nematic liquid crystals offer a very interesting set of large optical nonlinearities. As has been shown, both theoretically (Khoo 1981) and experimentally (Khoo 1980, 1982), these nonlinearities come from the optical field-induced, collective reorientation of the nematic molecules. A continuous wave (c.w.) laser of moderate power can be intense enough to yield a significant refractive index change in the medium. Moreover, this change has an appreciable effect on laser beam propagation and, hence, output laser beam characteristics can be very different from the input ones. One of the more promising effects to be applied in optically bistable devices is the one reported by Zolot'ko *et al.* (1980) and by Durbin *et al.* (1981). They have reported the observation of a multiple-ring pattern of laser diffraction from a nematic liquid crystal film. This phenomenon appears for a certain laser intensity level and its time responses depend on both cell thickness and molecular orientation, the configuration employed most often being the homeotropic one. If the laser power is below the threshold, no diffraction phenomenon is obtained and both laser beams, namely input and output beams, have similar characteristics. Moreover, as we have shown Martín-Pereda (1982, 1983), if two laser beams cross the cell at the same point, one beam being more intense than the other, the output characteristics of the smaller one can be affected by the larger one. This effect has been reported as opto-optical modulation, with possible applications in all-optical devices.

The above facts have been employed to develop optical logic gates; AND, OR, NOR and NAND functions have been achieved. The basis of the operation is the ability of a nematic homeotropic cell to alter its molecular orientation according to the crossing total laser intensity regardless of whether this intensity comes just from one laser beam or from more than one. A cooperative effect is the result.

Planar 120 µm thick structures were used. The homeotropic alignment was obtained by surface treatment with hexadecyltrimethyl ammonium bromide (HTAB); p-methoxybenzylidene-p′-n-bulylaniline (MBBA) was employed as nematic liquid crystal, and the experiments were made at room temperature (about 25 °C), i.e. within the nematic range of MBBA.

The experiments were made with c.w. Ar⁺ ion lasers working at their green line (514.5 nm). The optical propagation was, in every case, at an angle of 0° with the liquid crystal molecular axis.

The system shown in figure 1 gives, as will be shown, AND and OR logical functions. LC_1

25-2

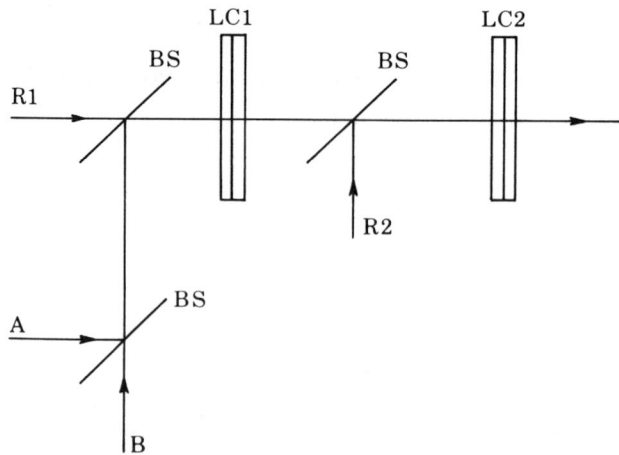

FIGURE 1. A system for AND and OR logical functions.

and LC_2 are two homeotropic liquid crystal cells. BS are beam splitters and R_1 and R_2 two read-out laser beams. A and B are the two possible inputs. The behaviour of the system is now described.

AND FUNCTION

If A and B inputs are non-zero, the light crossing LC_1 has a level corresponding to the addition of A, B and R_1. Under these circumstances, the system is adjusted in such a way that the total power impinging on the nematic cell is above the threshold. Hence, the output from LC_1 is diffraction rings. The light crossing LC_2 is just R_2 because no appreciable contribution comes from the exit of LC_1. If the light intensity of R_2 is below the LC_2 threshold, the output of the system will be R_2. So, the system output corresponds to one.

If either A or B are zero, the light crossing LC_1 is $R_1 + A$ (or B). If it is below the LC_1 threshold, the light going onto LC_2 will be $R_2 + R_1 + A$ (or $R_2 + R_1 + B$). Because this power level is now over the LC_2 threshold, the system output is zero.

If both A and B are zero, the light crossing LC_2 is $R_1 + R_2$. And because the system is adjusted in such a way that this intensity is over the threshold level, the system output will be zero.

OR FUNCTION

The system works in a similar way to that described for the AND function. The only difference is the threshold level of the LC_1 cell. In this case, just when the crossing light is R_1, the power is below the threshold. When A or B are different from zero, LC_1 will form diffraction rings and the system output will be 1.

NOR FUNCTION

A system like that shown in figure 2 is able to work as a NOR function. The behaviour is similar to that described for AND and OR functions. The difference is that just one read-out laser beam is needed. When the light crossing LC is just R_1, no diffraction rings appear

[192]

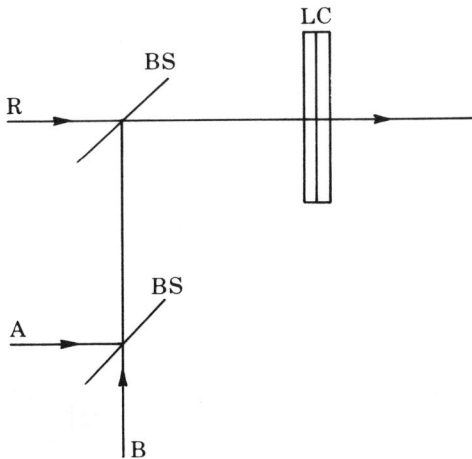

FIGURE 2. A system for a NOR function.

because it is below the threshold. Hence the system output will be 1. But when any one of A or B (or A and B) are present the light intensity level is over the threshold so that a zero is obtained. The advantage of this system is that, as is well known from logic, with this function any other function can be achieved.

To get some feeling of the intensity levels required from each of the three working laser beams some calculations need to be done. If we arbitrarily assign 100 to the power level needed for diffraction rings to arise the following conditions are obtained:

$$\tfrac{1}{4}A + \tfrac{1}{2}R > 100;$$
$$\tfrac{1}{2}R < 100; \quad \tfrac{1}{4}A < 100;$$
$$\tfrac{1}{2}R > A.$$

The first condition comes from the fact that A + B must give 0. The second one is imposed by the need to have no diffraction rings when any one of them is working alone. The last one is obtained from the fundamental requirement for any computer that the output of a gate be strong enough to drive other gates. As can be seen, the major drawback of the system comes from the beam splitters, because a lot of optical power is lost at them.

With these conditions a diagram, like the one shown in figure 3, can be drawn. As can be seen, there is just one region where the system can work. The numbers express the power needed by each beam, 100 being the threshold level for the appearance of diffraction rings at the liquid crystal cell output. The beam splitters have been designed in such a way that 50% of the input power is reflected and 50% transmitted. Hence, the laser intensity crossing the cell is given by

$$I_{LC} = \tfrac{1}{2}R + \tfrac{1}{4}A + \tfrac{1}{4}B.$$

If some other beam-splitter characteristics are assumed, this relation must be modified.

According to our previous work, Ar^+ intensities of about 3 kW cm^{-2} are needed for the rings to be formed. The ranges of intensities of laser beams A, B and R are

$$4.5 \text{ kW cm}^{-2} < R < 6 \text{ kW cm}^{-2},$$
$$A, B < 3 \text{ kW cm}^{-2}.$$

These values are in very good agreement with our experimental values.

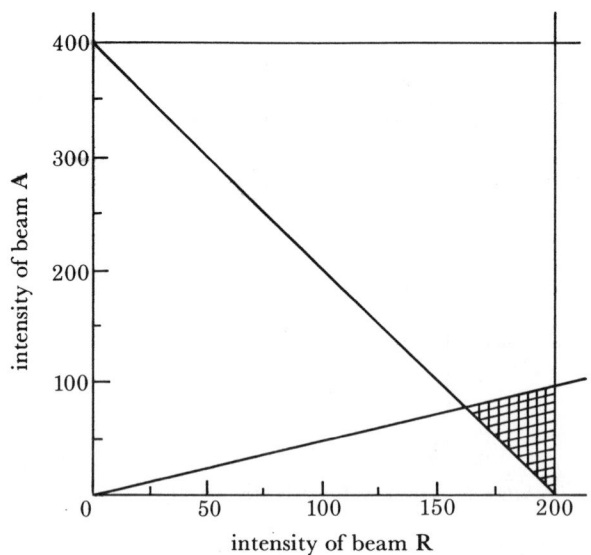

FIGURE 3. Diagram to indicate the working zone (hatched area).

An advantage of this type of logic gate is that its working conditions are almost absolutely independent of the acting wavelengths. In fact, we have also made our experiments with two types of lasers, namely an Ar^+ laser as a read-out laser and a He–Ne laser for data entry. This allows the possibility of a clear separation between both types of beams.

This work was supported by the Spanish Comision Asesora de Investigación Científica y Técnica Grant no. 1564-82.

REFERENCES

Durbin, S. D., Arakelian, S. M. & Shen, Y. R. 1981 *Optics Lett.* **6**, 411.
Khoo, I. C. 1982 *Phys. Rev.* A**25**, 1040.
Khoo, I. C. & Zhuang, S. L. 1980 *Appl. Phys. Lett.* **37**, 3.
Khoo, I. C., Zhuang, S. L. & Shepard, S. 1981 *Appl. Phys. Lett.* **39**, 937.
Martín-Pereda, J. A. & López, F. J. 1982 *Optics Lett.* **7**, 590.
Martín-Pereda, J. A., Lopez, F. J. & Muriel, M. A. 1983 *Molec. Cryst. liq. Cryst.* **99**, 1.
Zolot'ko, A. S., Kitaeva, V. F., Sobolov, N. N. & Csillag, L. 1980 *Soviet Phys. JETP Lett.* **32**, 158.

Phil. Trans. R. Soc. Lond. A **313**, 385–387 (1984)

Printed in Great Britain

Optical phase conjugation via degenerate four-wave mixing in copper chloride

By M. L. Claude, L. L. Chase†, D. Hulin and A. Mysyrowicz

Groupe de Physique des Solides de l'E.N.S., Tour 25, 2 place Jussieu, 75251 Paris, France

We present a study of optical phase conjugation through degenerate four-wave mixing in CuCl, as a function of light frequency, intensity, polarization and sample temperature.

The phase-conjugate signal exhibits two maxima at the frequencies of the Γ_5 exciton (one-photon resonance) and of the biexciton (two-photon resonance). By appropriate combinations of the input beam polarizations, it is possible to single out the various terms contributing to the signal (small-spaced, large-spaced population grating and two-photon coherent excitation).

The absolute value of $\chi^{(3)}$ responsible for the two-photon resonance has been determined ($\chi^{(3)}(-\omega, \omega, \omega, -\omega) = 3.10^{-7}$ e.s.u.‡). It leads to 'mirror' efficiencies of the order of 10 % for input pump intensities $I_0 \approx 10^5$ W cm^{-2} and sample thickness $d \approx 10^{-4}$ cm. At higher I_0, a saturation of the reflection takes place. The variation of the conjugate beam with temperature has been studied in the range $15\ \mathrm{K} < T < 70\ \mathrm{K}$. Finally, the phase-conjugate nature of the signal has been verified by inserting an aberrator in the path of the input probe beam, and by checking the reconstruction of the reflected beam.

The mechanism of conjugate wavefront generation has stirred a considerable interest, in view of its possible applications in optical imaging, owing to the fact that the generated wave has the property of time reversal with respect to the incident wave (Fisher 1983). This allows, in particular, the reconstruction of aberrated wavefronts. A practical method of creating a conjugate mirror is via degenerate four-wave mixing in a nonlinear medium. A current thrust of research is to discover materials with high efficiency. This can be achieved by making use of the resonant enhancement of the third-order nonlinear susceptibility $\chi^{(3)}(-\omega, \omega, \omega, -\omega)$ describing the four-wave mixing process. Among the resonant terms, those based on a two-photon coherence (t.p.c.) are particularly attractive, since they have potentially very fast response times.

In a recent letter (Chase *et al.* 1983), we have reported an experimental study of phase conjugation in CuCl. A t.p.c. signal has been observed in samples as thin as 1 μm, with efficiencies exceeding 5 %. In this paper, we present additional results concerning the polarization and temperature variation of the signal. The beams configuration is shown in the inset of figure 1. The two counter-propagating pump beams 2 and 3, of equal intensity I_0 (maximum intensity 10^7 W cm^{-2}), as well as the probe beam 4, of intensity $5 \times I_0/100$, are derived from the same nitrogen-pumped dye laser (pulse duration, 5 ns). Propagation directions of the beams are shown in figure 1, where the external angle α is either 3° or 90° with respect to pump beam 2. The generated beam 1 (the signal beam) is directed, through

† Present address: Physics Department, Indiana University, Bloomington, Indiana 47401, U.S.A.

‡ 1 e.s.u. = 1 cm³ erg⁻¹ ≡ 1.4×10^{-8} m² V⁻².

a small aperture, to a photomultiplier tube. To discriminate the signal from scattered pump light, mechanical chopping of the probe beam is used together with phase-sensitive detection.

Figure 1 shows the spectral dependence of the signal intensity, recorded in the vicinity of the exciton region of CuCl, for various configurations of polarization of the beams. In case $V_1 V_2 V_3 V_4$ (where V_i((H_i) stands for vertical (horizontal) polarization direction of beam i) two

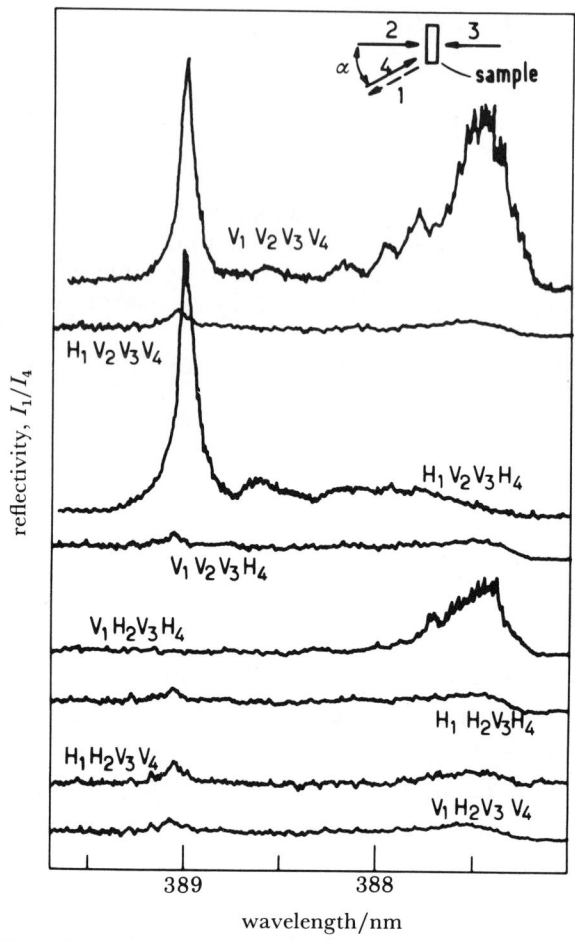

FIGURE 1. Wavelength dependence of the phase-conjugate signal at $T = 2$ K, for various polarization configurations of the optical beams. V_i(H_i) stands for vertical (horizontal) polarization of beam i; $I_p = 1$ MW cm^{-2}, $t = 1.25$ μm, $\alpha = 3°$.

maxima are apparent. From their position and polarization characteristics, they may be readily interpreted. The peak at $\lambda \approx 387.5$ nm falls near the $n = 1$ exciton absorption line, formed with a hole of the upper Γ_7 valence band and an electron of the lowest Γ_6 conduction band. This peak disappears if the probe polarization is orthogonal to both pumps. Such behaviour is expected from a one-photon resonance term, resulting in the formation of a real exciton population volume grating. Of more direct concern to us is the peak located at 389 nm, the position of which corresponds exactly to half the biexciton energy $\hbar\omega_B$. As is well known (Hanamura 1973; Gale *et al.* 1974), the biexciton state leads to a giant two-photon transition cross-section, and therefore should also be effective in t.p.c. phase conjugation. Given the biexciton symmetry Γ_1, one can show that the signal from the biexciton two-photon resonance should be proportional to the scalar product of the fields $E_2 \cdot E_3$, independent of the orientation of E_4. Furthermore, E_1 should be parallel to E_4. These predictions are well obeyed and give support to the assignment of the peak.

In figure 2, we have plotted the intensity dependence of the signal as a function of the lattice temperature. The data were recorded on a 1 μm thin sample, with a pump intensity

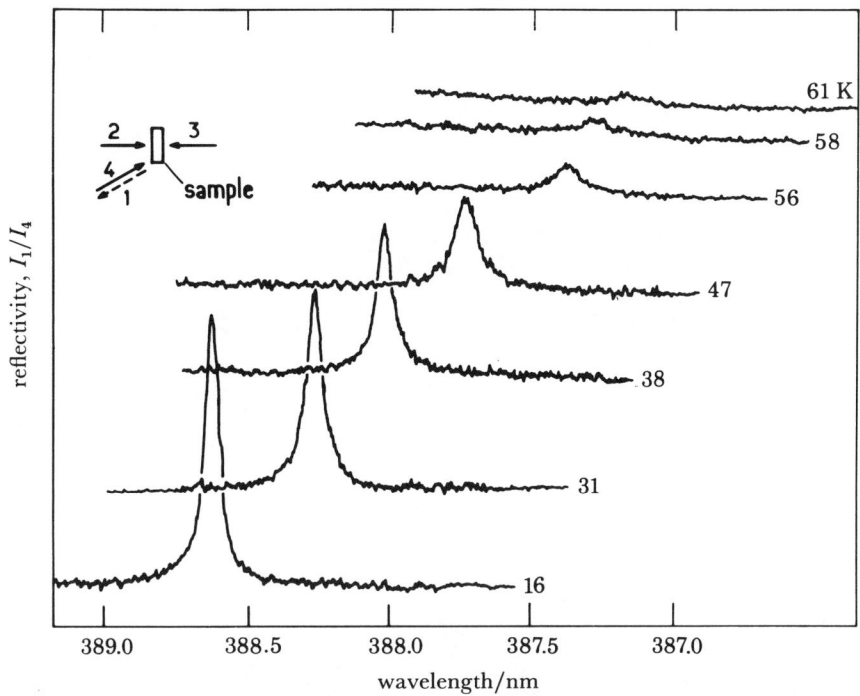

FIGURE 2. Temperature dependence of the t.p.c. conjugate signal in a 1 μm thick CuCl sample for $H_1 V_2 V_3 H_4$.

$I_0 \leqslant 10^6$ W cm^{-2}. As can be seen, the signal is clearly observed up to temperatures of the order of 60 K, with a noticeable broadening above $T = 40$ K. A similar behaviour in function of temperature has been reported and discussed by Itoh *et al.* (1980) for the biexciton two-photon absorption line of CuCl.

Finally, as mentioned by Chase *et al.* (1983), we have verified the phase-conjugate nature of the reflected wave by inserting an aberrator in the path of the probe. Also, we have determined the absolute value of $\chi^{(3)}(-\omega, \omega = \frac{1}{2}\omega_B, \omega, -\omega) = 3 \times 10^{-7}$ e.s.u. This very large nonlinearity suggests the possibility of achieving mirror efficiencies exceeding 100% with samples of thickness 10^{-3} cm and pump beam intensities $I_0 \approx 10^6$ W cm^{-2}. So far, we have obtained efficiencies of the order of 20% without taking any special precaution to ensure good quality and spatial overlap of the optical beams.

REFERENCES

Chase, L. L., Claude, M. L., Hulin, D. & Mysyrowicz, A. 1983 *Phys. Rev.* A **28**, 3696.
Fisher, R. A. 1983 (ed.) *Optical phase conjugation.* New York: Academic Press.
Gale, G. M. & Mysyrowicz, A. 1974 *Phys. Rev. Lett.* **32**, 727.
Hanamura, E. 1973 *Solid St. Commun.* **12**, 951.
Itoh, T., Watanabe, S. & Ueta, M. 1980 *J. phys. Soc. Japan* **48**, 542.

[197]

Phil. Trans. R. Soc. Lond. A **313**, 389–394 (1984)

Printed in Great Britain

Stationary properties of dispersive optical bistability in a dense exciton–biexciton system: copper chloride

By C. M. Bowden[1], J. W. Haus[1], C. C. Sung[2] and W. K. Chiu[3]

[1] *Research Directorate, U.S. Army Missile Laboratory, U.S. Army Missile Command, Redstone Arsenal, Alabama 35898, U.S.A.*

[2] *Department of Physics, The University of Alabama in Huntsville, Huntsville, Alabama 35899, U.S.A.*

[3] *Department of Physics, Indiana Central University, Indianapolis, Indiana 46227, U.S.A.*

A semiclassical exciton–biexciton model for CuCl is used to obtain the stationary solutions for optical bistability (o.b.) by numerical integration of the second-order Maxwell equation in the limit of large Fresnel numbers. The results obtained from the calculation, which we call exact (ex.), are compared with corresponding numerical results by using the slowly varying envelope approximation (s.v.e.a.) as well as corresponding analytical results obtained by using the mean-field approximation (m.f.a.). The results obtained with the s.v.e.a. are shown to be in close quantitative (within 5%) agreement with the ex. results for each point in the parameter space considered, whereas the m.f.a. gives reasonably good qualitative results only (more than 20% quantitative discrepancy with ex.). Furthermore, o.b. is predicted from these calculations for incident laser field detunings on either side of the two-photon biexciton resonance as well as in the neighbourhood of the resonance. The nonlinearity in the dielectric function that causes o.b. is seen to be very nearly of the Kerr medium type, and consequently, the two-photon biexciton resonance contributes only weakly to the o.b. characteristics.

Introduction

Previous calculations for optical bistability (o.b.) in CuCl assumed the dielectric function to be constant throughout the cavity; this version of the so-called 'mean-field approximation' (m.f.a.) has the advantage of yielding analytical results. Results from the m.f.a. were obtained by several authors (Koch & Haug 1981; Sarid *et al.* 1983; Sung & Bowden 1984). Other examples can be found in Bowden *et al.* (1984). Since the m.f.a. limit is idealized, there are always questions concerning the accuracy and reliability of the results. Therefore, we solve here, in addition, the second-order Maxwell equation, which requires numerical integration (ex. results). A large Fresnel number is assumed so that transverse contributions can be neglected. Bearing in mind that numerical integration of the second-order equation is expensive, we also calculate results from the so-called slowly varying envelope approximation (s.v.e.a.) (Icsevgi & Lamb 1969; Fleck 1970). It is found that, with the exception of rather esoteric conditions, the s.v.e.a. can be used to produce results from our model that are within 5% agreement with the ex. results, whereas the m.f.a. yields results that differ quantitatively by more than 20%. Thus, cost-effectively, the s.v.e.a. will be very useful for our future work; for example, the calculation of the dynamics and switching without recourse to adiabatic elimination (Geigenmüller *et al.* 1983).

1. Model Hamiltonian

The model Hamiltonian for CuCl, which has been widely used to calculate the nonlinear interaction with light for an incident laser field tuned near the exciton or two-photon biexciton resonance, is given in the rotating wave and electric dipole approximations, by

$$H = H_0 + H', \tag{1.1}$$

where
$$H_0 = \omega_x b^+ b + \omega_m B^+ B, \tag{1.2a}$$

and
$$H' = ig_1 E^+ b^+ + ig_2 E^+ B^+ b + \text{h.c.} \tag{1.2b}$$

Here, $B^+(B)$ and $b^+(b)$ are the creation (annihilation) operators for the biexciton and excitons, respectively, and ω_m and ω_x are their respective transition energies (units such that $\hbar = 1$ are used); g_1 and g_2 are the coupling constants, whose numerical values are inferred from experiment. The electric field E with positive and negative frequency components, E^+ and E^-, respectively, is assumed to be the superposition of rightward and leftward propagating monochromatic plane waves in the dielectric medium and is understood to have both spacial and temporal dependence in general. The exciton and biexciton operators are correspondingly understood to have k-vector dependence consistent with the electric field. We take the electric field E to be classical and proceed with the semiclassical model calculation where the field and dielectric are coupled by the Maxwell equation.

2. Numerical integration of the second-order Maxwell equation at steady state: ex. results

The field-dependent nonlinear dielectric function obtained by solving the equations of motion derived from (1.2) at steady state, and to lowest order in the field–exciton–biexciton interaction is (Koch & Haug 1981; Sung & Bowden 1984a)

$$\epsilon^+(E^+) = \epsilon_\infty + 4\pi g_1^2/(\delta' - g_2^2 |E|^2/\Delta'), \tag{2.1}$$

where unity has been replaced in the first term in (2.1) on the right side by the high-frequency dielectric constant ϵ_∞. Here, $\delta' = \omega_x - \omega - i\gamma_x$, $\Delta' = \omega_m - 2\omega - i\gamma_m$ and $|E|^2 = E^+ E^-$. In these relations ω is the frequency of the incident electric field, where we have used

$$E^+(x, t) = e^{-i\omega t} E^+(x) \tag{2.2}$$

and x is the longitudinal direction of propagation in the dielectric medium of length L. Also γ_x and γ_m are the exciton and the biexciton relaxation rates, respectively, which have been added phenomenologically.

The stationary solution of the Maxwell equation based upon (2.1) and (2.2) is obtained by solving (we omit the superscripts on ϵ and E from here on)

$$\partial^2 E(x)/\partial x^2 + \epsilon(E) k_0^2 E(x) = 0, \tag{2.3}$$

where $k_0 = \omega/c$ is the wavenumber in vacuum. If the incident laser beam of amplitude E_I is given by

$$E_{\text{INC}} = E_I \exp(-i\omega t + ik_0 x) \tag{2.4}$$

for $x \leqslant 0$, then the boundary conditions are represented by the following set of equations at $x = 0$ for a given surface reflective dielectric film mirror of reflectivity r_m,

$$(1-r_m)^{\frac{1}{2}} E_{INC} + (1+r_m^{\frac{1}{2}}) E_1 = E, \qquad (2.5a)$$

$$ik_0[(1-r_m)^{\frac{1}{2}} E_{INC} - (1-r_m^{\frac{1}{2}}) E_1] = \partial E/\partial x, \qquad (2.5b)$$

where E_1 is the amplitude of the reflected light at $x = 0$, i.e. between the material (CuCl) and the cavity mirror. Similar boundary conditions prevail at $x = L$,

$$(1+r_m^{\frac{1}{2}}) E_3 = E, \qquad (2.6a)$$

$$ik_0(1-r_m^{\frac{1}{2}}) E_3 = \partial E/\partial x, \qquad (2.6b)$$

where E_3 is the amplitude of the forward-propagating wave at $x = L$ between the material (CuCl) and the mirror. The transmitted wave intensity I_T is given by

$$I_T = |E_T|^2 = (1-r_m) |E_3|^2. \qquad (2.7)$$

These boundary conditions can be combined to solve (2.3) by numerical integration. The results we call ex. results.

3. MEAN-FIELD APPROXIMATION

The m.f.a. is often used in the literature to describe o.b. in a Fabry–Perot cavity filled with a Kerr medium, and avoids the necessity for numerical integration by imposing the ansatz

$$E(x) = E_R^{(0)} e^{ikx} + E_L^{(0)} e^{-ikx}, \qquad (3.1)$$

where $E_R^{(0)}$ and $E_L^{(0)}$ are rightward and leftward propagating components of the electric field E and are taken to be independent of x. Here k is the complex wavevector defined by

$$k = k_0 \epsilon^{\frac{1}{2}}, \qquad (3.2)$$

where ϵ is given by (2.1). If (3.1) is used in conjunction with the boundary conditions (2.5) and (2.6), the result is (Sung & Bowden 1984b)

$$\tau \equiv |E_T|^2/|E_I|^2 = |4(1-r_m) \epsilon^{\frac{1}{2}}\{[(1+r_m^{\frac{1}{2}}) \epsilon^{\frac{1}{2}} + 1 - r_m^{\frac{1}{2}}]^2 e^{\alpha} - [(1+r_m^{\frac{1}{2}}) \epsilon^{\frac{1}{2}} - (1-r_m^{\frac{1}{2}})]^2 e^{-\alpha}\}^{-1}|^2,$$
$$(3.3)$$

where $\alpha = k_0 L\epsilon^{\frac{1}{2}}$ is the complex phase.

4. SLOWLY VARYING ENVELOPE APPROXIMATION SOLUTION

The s.v.e.a. has been extensively used in the literature (Icsevgi & Lamb 1969; Fleck 1970) and we will not elaborate on its construction here. It essentially results in a linearization of the Maxwell second-order partial differential equation by removal of the rapidly varying temporal and spacial components of the field. The novel aspect of our calculation is that we remove the spacially varying part according to the approximate wavevector in the material

$$\hat{k} = k_0 \text{ Re } (\epsilon(0))^{\frac{1}{2}}, \qquad (4.1)$$

rather than k_0 in free space, as is usually the procedure. In (4.1) $\epsilon(0) \equiv \epsilon(E = 0)$. The details of our s.v.e.a. calculation will be presented elsewhere (Sung et al. 1984).

[201]

From the s.v.e.a. and the solution of the equations of motion from (1.2) in steady state, we have

$$\frac{\partial \epsilon_F^+}{\partial z} = \frac{2\pi i \omega \Delta' g_1^2}{cS} (\delta' \Delta' - g_2^2 |\epsilon_F^+|^2) \epsilon_F^+ - \frac{i\omega}{2} \frac{(\epsilon_R(0) - \epsilon_\infty)}{\epsilon_R(0)} \epsilon_F^+, \tag{4.2a}$$

$$\frac{\partial \epsilon_B^+}{z} = -\frac{2\pi i \omega \Delta' g_1^2}{cS} (\delta' \Delta' - g_2^2 |\epsilon_B^+|^2) \epsilon_B^+ + \frac{i\omega}{2} \frac{(\epsilon_R(0) - \epsilon_\infty)}{\epsilon_R(0)} \epsilon_B^+. \tag{4.2b}$$

In these equations ϵ_F^+ and ϵ_B^+ are the slowly varying forward and backward propagating field amplitudes, respectively, and

$$S = (\delta' \Delta')^2 - 2g_2^2 \delta' \Delta' (|\epsilon_F^+|^2 + |\epsilon_B^+|^2) + g_2^4 (|\epsilon_F^+|^4 + |\epsilon_B^+|^4 + |\epsilon_F^+|^2 |\epsilon_B^+|^2). \tag{4.3}$$

It is to be noted that if the last term in (4.3) on the right side were a perfect square, equations (4.2) could be integrated immediately, by using the boundary conditions (2.5)–(2.7). However, this is not the case, so we proceed with numerical solution of (4.2) with the boundary conditions already mentioned.

5. Results and discussion

The ex., s.v.e.a. and m.f.a. results for a sample length $L \approx 10$ μm are presented in figure 1. We have fixed the frequency $\omega = 3.177$ eV and varied L to obtain a well-defined o.b. condition in m.f.a. This is why $L = 9.98165$ μm in these results. We have chosen a value for ω far from resonance, as was done by Sarid et al. (1983). Since γ_x is small compared to γ_m, we neglect it altogether to proceed more easily. The value for γ_m varies considerably in the literature and may well vary from sample to sample. It is also claimed to be field-dependent (Sarid et al. 1983). It has been pointed out by Abram (1983), that at least part of the apparent field dependence of the shape of the absorption spectrum for CuCl can arise from spacial 'chirp' due to the field-dependent shift of the peak of the absorption spectrum dictated by (2.1). Since this condition is obviously present in the ex. and s.v.e.a. calculations from our model, we do not impose further conditions extraneous to our model.

The remarkable agreement between the s.v.e.a. and ex. results is due largely to the choice of the wavevector $\hat{k}$, (4.1) in the s.v.e.a. The m.f.a. results give only qualitatively reliable results. The comparison between the s.v.e.a., ex. and m.f.a. depicted in figure 1 is consistent throughout the parameter space tested, including the cases at $L = 1$ μm and $L \approx 30$ μm.

Displayed in figure 2 are results for the incident field frequency tuned to a value approximately as far above the two-photon resonance as the value chosen in figure 1 was below resonance. This shows at least as strong a hysteresis condition as for the previous case below resonance. The s.v.e.a. and m.f.a. obviously compare in the same way as before and the ex. result is not plotted since the agreement is not different from that depicted in figure 1. Also shown in the box on the right in figure 2 is the s.v.e.a. at $L = 30$ μm. The change in length produces a change in intensity threshold by two orders of magnitude. Hence it is possible to reduce the threshold by searching for optimal values of the length and detuning.

A major conclusion from these results is that the nonlinearity that causes o.b. in excitonic CuCl is largely of the Kerr medium type. We have tested the o.b. conditions near the two-photon

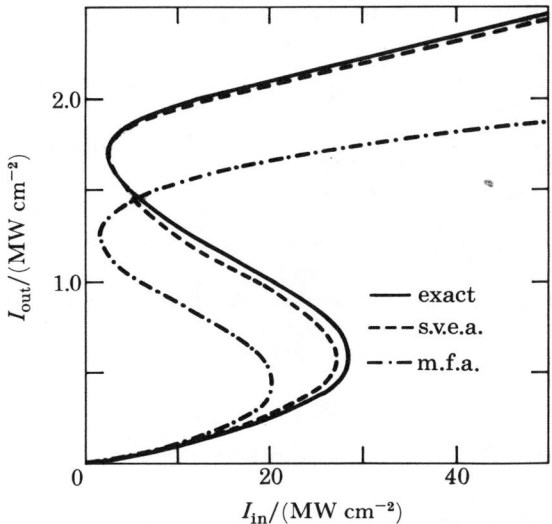

FIGURE 1. Input against output intensities for the laser frequency $\omega = 3.177$ eV and sample length $L = 9.98165$ μm. Other relevant parameters are: $r_m = 0.9$, $\gamma_m = 0.3$ meV, $\gamma_x = 0$, $4\pi g_1^2 = 27.5$ meV^2, $|M|^2 = 1.57 \times 10^{-16}$ meV^2 cm^3 and $g_2^2 = |\epsilon| |M|^2/\omega$, $\omega_x = 3.2027$ eV, $\omega_m = 6.3725$ eV and $\epsilon_\infty = 5.0$.

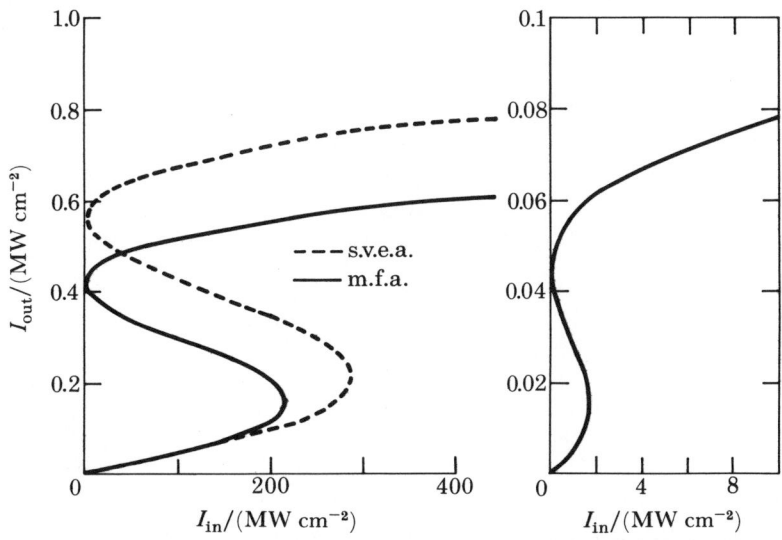

FIGURE 2. Input against output intensity for the laser frequency $\omega = 3.1956$ eV and $L = 10$ μm. The box on the right is the s.v.e.a. for a sample length $L = 30$ μm; note that the output intensity scale is reduced by a factor of 10 and the input intensity scale has been changed. All other parameters are the same as for figure 1 except that $\gamma_m = 0.1$ meV.

resonance and the o.b. is relatively weak. The fact of the Kerr medium behaviour and the choice of (4.1), explains the close agreement between our s.v.e.a. and ex. results.

Based upon these results, we are now in the process of calculating the dynamics and switching times for this system by using the s.v.e.a. and without using adiabatic elimination, which is not valid for CuCl.

J.W.H. is a National Research Council research associate.

REFERENCES

Abram, I. 1983 *Phys. Rev.* B **28**, 4433.
Bowden, C. M., Gibbs, H. M. & McCall, S. L. (eds) 1984 *Optical Bistability* 2. New York: Plenum Press.
Fleck, J. A. 1970 *Phys. Rev.* B **1**, 84.
Geigenmüller, U., Titulaer, U. M. & Felderhof, B. U. 1983 *Physica* A **119**, 411.
Icsevgi, A. & Lamb, W. E. 1969 *Phys. Rev.* **185**, 517.
Koch, S. W. & Haug, H. 1981 *Phys. Rev. Lett.* **46**, 450.
Sarid, D., Peyghambarian, N. & Gibbs, H. M. 1983 *Phys. Rev.* B **28**, 1184.
Sung, C. C. & Bowden, C. M. 1984*a* *Phys. Rev.* A **29**, 1957.
Sung, C. C. & Bowden, C. M. 1984*b* *J. opt. Soc. Am.* B **1**, 395.
Sung, C. C., Bowden, C. M., Haus, J. W. & Chiu, W. K. 1984 *Phys. Rev* A **30**. (In the press.)

Phil. Trans. R. Soc. Lond. A **313**, 395–399 (1984)

Printed in Great Britain

Intrinsic polarization bistability in nonlinear media

BY J. A. GOLDSTONE AND E. GARMIRE

*Center for Laser Studies, University of Southern California, Los Angeles,
California* 90089-1112, *U.S.A.*

This paper calculates the effect of a bistable polarization on the optical fields in a nonlinear medium. Bistabilities in effective refractive index and absorptivity are calculated. Bistability on reflection from a single surface is also described.

INTRODUCTION

In the talk presented at this symposium by the same authors, and published earlier in these proceedings, the way in which a nonlinear constitutive relation can lead to optical bistability was outlined. The constitutive relation for polarization in a medium was described by using the model of a Duffing oscillator, and a bistability in the polarization as a function of local electric field was derived. This paper describes the macroscopic manifestations of this microscopic bistability by calculating the effect of the bistable polarization on the optical fields. The calculation proceeds by using the nonlinear polarization derived in the earlier paper as the independent variable in the nonlinear wave equation. The optical fields are then determined from the constitutive relation. To describe observable effects, the incident wave is traced from the moment it enters a nonlinear medium. The bistable polarization creates a bistable phase relation of the light and polarization within the medium. This bistable phase relation results in bistable refractive index and absorptivity. A particularly simple method of observing these effects is in reflection from a single interface, by applying the usual boundary conditions.

CONSTITUTIVE RELATION

The nonlinear constitutive relation that exhibits bistability was derived in the previous paper in this symposium and can be written in terms of the microscopic vibration parameter x. For simplicity we ignore local field corrections and define the macroscopic polarization P through $P = Nex$. We have

$$\{m(\omega_0^2 - \omega^2) - i\gamma\omega + \tfrac{3}{4}\beta|x|^2\} x = eE, \tag{1}$$

where ω_0 is the resonant frequency, γ is the damping and β is the coefficient of the cubic component of the nonlinear restoring force, $F = -Kx + \beta x^3$.

NONLINEAR POLARIZATION

The propagation of light within the nonlinear medium is characterized by the wave equation, with the constitutive relation providing the driving term. As discussed previously, the field will be considered a function of the polarization. The wave equation therefore becomes an equation

[205]

for the polarization. Under the assumption of plane-wave propagation in the z direction, we define the spatial and time dependence of the polarization by

$$P(z,t) = Ne\Gamma(z) \exp\left[i\{\phi(z)+kz-\omega t\}\right].\tag{2}$$

By using the slowly varying envelope approximation, the wave equation may be solved for $P(z,t)$, which yields

$$\ln\{\Gamma(z)/\Gamma_0\} + \left[\tfrac{3}{2}\beta\Delta\{\Gamma^2(z)-\Gamma_0^2\} + \tfrac{27}{64}\beta^2\{\Gamma^4(z)-\Gamma_0^4\}\right](\Delta^2+\gamma^2\omega^2)^{-1} = -\alpha z\tag{3}$$

and

$$\phi(z) = \phi_0 + \frac{1}{2k}\left(\frac{\omega^2}{c^2}-k^2\right)z - \frac{1}{\gamma\omega}\left[\Delta\ln\{\Gamma(z)/\Gamma_0\} + \tfrac{9}{8}\beta\{\Gamma^2(z)-\Gamma_0^2\}\right],\tag{4}$$

where, for small loss, k and α satisfy the usual (linear oscillator) conditions:

$$k = \frac{\omega}{c}\left[1+\frac{\omega_p^2\Delta}{\Delta^2+\gamma^2\omega^2}\right]^{\frac{1}{2}} \equiv \frac{n_0\omega}{c}\tag{5}$$

and

$$\alpha = \gamma\omega^3\omega_p^2/\{2kc^2(\Delta^2+\gamma^2\omega^2)\}.\tag{6}$$

Here Γ_0 and ϕ_0 are the amplitude and phase of $P(z)$ at $z=0$ and $\omega_p = (4\pi Ne^2/m)^{\frac{1}{2}}$ is the plasma frequency of the material.

Transmitted and incident flux

The equations given express the spatial and temporal dependence of the phase and amplitude of the polarization in terms of the initial conditions Γ_0 and ϕ_0, which express the amplitude and phase of the polarization at $z=0$. Because the polarization may have two stable states, this initial condition is not uniquely defined in terms of the incident field. Use of this polarization as the independent variable in the constitutive relation (1) allows us to calculate the energy flux in the nonlinear medium:

$$\bar{S} = \frac{c_m^2\Gamma^2}{8\pi e^2}\left[\left(n_0+\frac{c\phi'}{\omega}\right)\{(\Delta+\tfrac{3}{4}\beta\Gamma^2)^2+\gamma^2\omega^2\} + \tfrac{3}{2}\beta\gamma c\Gamma\Gamma'\right],\tag{7}$$

where $\Gamma' = d\Gamma/dz$ and $\phi' = d\phi/dz$ are obtained from differentiation of (3) and (4).

The incident flux (which is defined for $z \leqslant 0$) that produces the flux $\bar{S}$ in the nonlinear medium is defined through boundary conditions at the incident plane ($z=0$), the boundary between a linear and a nonlinear medium. Expressing the incident flux in terms of the flux in the nonlinear medium, and then representing the latter in terms of the nonlinear polarization through (7) gives, for normal incidence and for a linear medium of refractive index n_L,

$$\bar{S}_I = \left(\frac{cn_L}{8\pi}\right)\left(\frac{m^2\Gamma_0^2}{16n_0^2n_L^2e^2}\right)\left[\{\gamma\omega(n_0^2+2n_Ln_0+1)\}^2 + \{(n_0^2+2n_Ln_0+1)(\Delta+\tfrac{3}{4}\beta\Gamma_0^2)+\omega_p^2\}^2\right].\tag{8}$$

The reflected flux is obtained in the same manner.

The independent parameter in these expressions, which describes their nonlinear behaviour, is the magnitude of the microscopic oscillation at $z=0$; i.e. Γ_0. Nonlinear transfer curves of transmitted against incident, and reflected against incident, flux are obtained by varying Γ_0.

These are shown in figure 1. The loops in the transfer curve are characteristic of the hysteresis seen in the field variables. The lower branch in transmission and the upper branch in reflection are unstable. Note that the polarization bistability, shown in the inset, does not have these loops. These curves are valid except close to the transition points, where the fluxes jump to the other branch. While the curves may not be entirely valid at these turning points, the analysis is qualitatively correct.

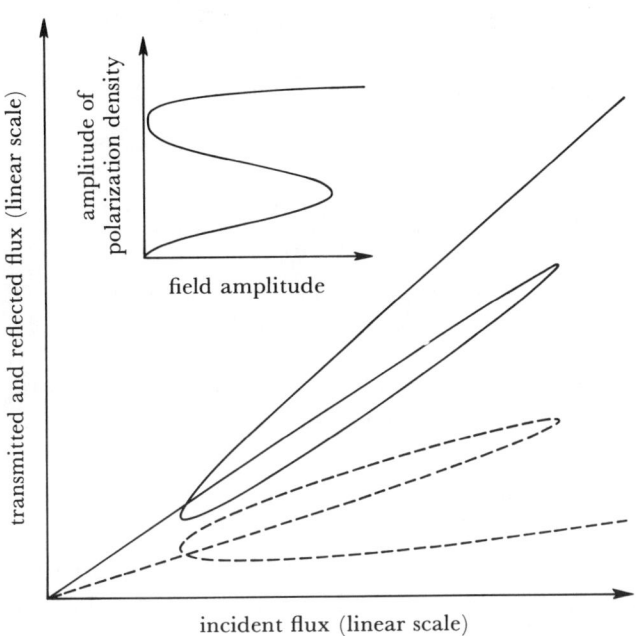

FIGURE 1. Transmitted and reflected flux (solid and broken lines, respectively), as a function of incident flux, showing bistability and hysteresis loops. Inset shows the polarization as a function of field amplitude in the nonlinear medium. This bistable curve does not show loops.

EFFECTIVE REFRACTIVE INDEX AND PHASE

The effective refractive index in the nonlinear medium may be determined in terms of the effective wavelength or by the relation between the field and the polarization. It is most convenient, however, to define the effective refractive index at $z = 0$ in terms of the ratio of the transmitted and incident flux

$$\frac{\bar{S}(0)}{\bar{S}_{\mathrm{I}}(0)} = \frac{4n_{\mathrm{eff}}n_2}{(n_{\mathrm{L}}+n_{\mathrm{eff}})}. \tag{9}$$

The effective index is bistable with respect to the input flux in the interesting way shown in figure 2. The arrows show the direction in which the refractive index follows the curve. The unstable branch is below either stable branch, in this case. An unusual crossing of two stable branches can be seen. In this analysis, we restrict the change in the value of the nonlinear index over which the hysteresis exists to 10%. However, by using the phenomenon of interference, this small percentage difference can be turned into a large intensity difference. This is most easily seen by studying the phase of the transmitted and reflected waves.

26-2

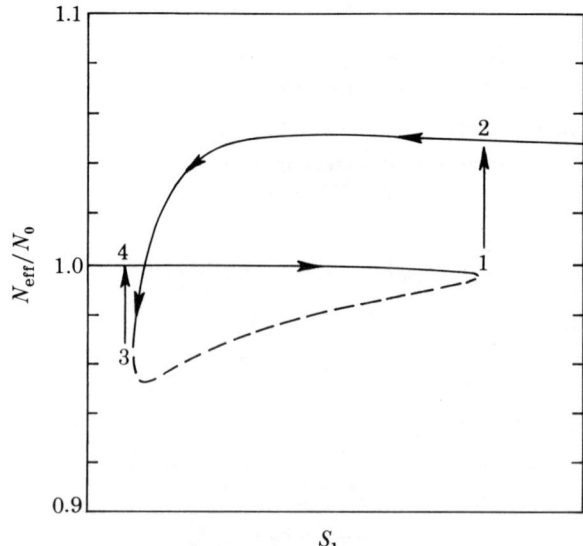

FIGURE 2. Bistable effective index of refraction n_{eff} as a function of incident flux, $\bar{S}_I$. The broken line represents the unstable portion of the curve.

The phase difference between the incident field and the polarization at $z = 0$ is also expressed in terms of the amplitude of the microscopic vibration, Γ_0:

$$\tan \phi_0 = \frac{(2n_0 n_L + n_0^2 + 1)\,\gamma\omega}{(2n_0 n_L + n_0^2 + 1)\,(\Delta + \frac{3}{4}\beta\Gamma_0^2) + \omega_p^2}. \tag{10}$$

Figure 3 shows this phase difference against incident flux, again obtained by varying Γ_0 simultaneously in ϕ_0 and $\bar{S}_I$. It can be seen that the change in the phase at switch-up and

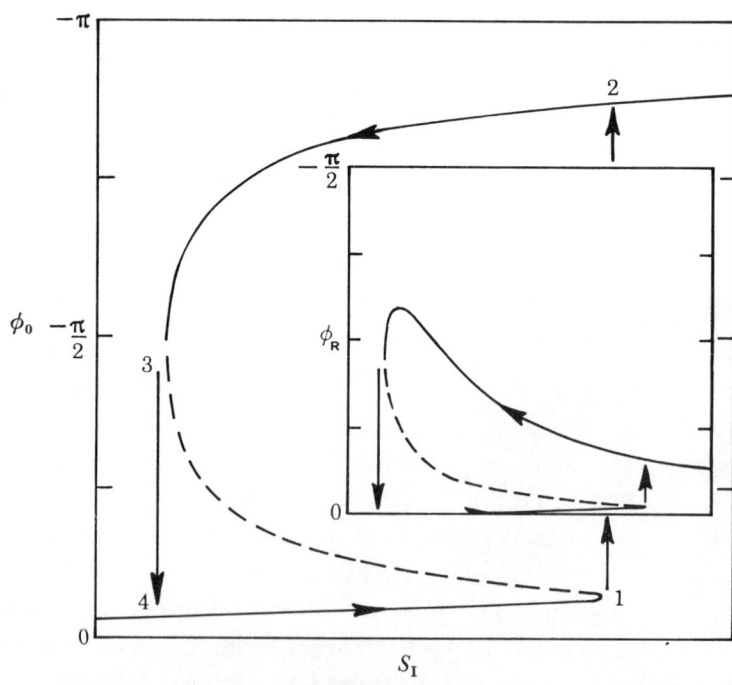

FIGURE 3. Bistable phase between the incident and transmitted field, ϕ_0 plotted as a function of the incident flux. The reflected phase ϕ_R is shown in the inset. The broken lines represent unstable solutions.

switch-down is near π (for small loss). The inset in figure 3 shows the phase of the reflected field, also against incident flux.

This phase jump in the reflected wave can be observed as an intensity jump by arranging an interferometric experiment. This is shown in figure 4, in which a bistable reflected flux was obtained through interference with a linear reference beam. This shows that even though the nonlinearities in the amplitude variations should be small, the phase variations may be large enough to be observed.

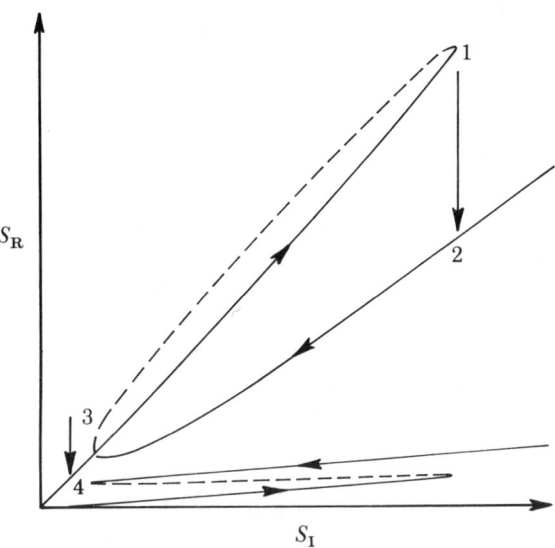

FIGURE 4. Bistable reflected flux obtained by interfering the reflected beam with a reference beam, plotted as a function of the incident flux.

CONCLUSIONS

We have described some of the macroscopic manifestations of microscopic bistability due to one form of the nonlinear constitutive relation. These include bistability in the refractive index and absorptivity, as well as in the intensity and phase of transmitted and reflected waves at a linear–nonlinear interface. It should be stressed that the interface is only an artifice to keep track of incident and transmitted beams and is not fundamental to the observation of bistability. Bistability can be observed at all angles of incidence to the interface; in fact, the angle of refraction itself is bistable. These characteristics set this effect apart from previous nonlinear interface effects (Smith *et al.* 1979).

As discussed by Garmire, Poole & Goldstone (this symposium), observation of bistability of this nature will require large nonlinearities and narrow lines (small damping constants). Perhaps the first place to look for such bistability is in the microwave region.

This work was supported by the National Science Foundation.

REFERENCE

Smith, P. W., Hermann, J.-P., Tomlinson, W. J. & Moloney, P. J. 1979 *Appl. Phys. Lett.* **35**, 846–848.

Phil. Trans. R. Soc. Lond. A **313**, 401–403 (1984)

Printed in Great Britain

Resonant modulation

By W. J. Stewart, I. Bennion and M. J. Goodwin

Plessey Research (Caswell) Limited, Allen Clark Research Centre, Caswell, Towcester, Northants. NN12 8EQ, U.K.

Devices that offer fast modulation and switching of light waves currently attract widespread research efforts for potential applications in high-speed signal processing. The guided-wave techniques of integrated optics have demonstrated the capability to achieve drive power/bandwidth levels of a few milliwatts per gigahertz modulation bandwidth. Most recently, efforts have been concentrated on electro-optic travelling-wave configurations (Alferness *et al.* 1983; Gee & Thurmond 1983; Cross *et al.* 1984) to achieve operation at higher frequencies with higher bandwidths than are usually obtainable from lumped-electrode devices, wherein the dominant restriction of the speed of operation is the RC constant associated with electrode charging and discharging. Here, we discuss the potential merits of an alternative approach, which employs resonant optical waveguide cavities.

Electrical power supplied to an electro-optic waveguide modulator is generally lost, although it is not necessarily dissipated in the modulation process itself. A reduction in the power that must be supplied is clearly attractive in itself and, through thermal considerations, becomes almost essential when closely packed devices are considered. In practice, the reduction amounts to an improvement in the modulator response. It can be achieved by reducing (i) the interaction volume, (ii) the electrical bandwidth or (iii) the optical bandwidth. In adopting the integrated optical approach, the first step is essentially taken in the direction indicated by (i). However, there remains scope for further improvement by several orders of magnitude, theoretically, over the performance demonstrated by currently favoured device geometries. In figure 1 we indicate possible directions for improvement achievable in a simple electro-optic phase modulator by following (ii) and (iii).

Reduction of the electrical bandwidth can be accomplished by introducing an electrical resonator into the structure (Molter-Orr *et al.* 1983), assuming, of course, that the device is not required to operate at d.c., which is the case in many of the applications envisaged. The electrical, modulating field for a given input power is thereby increased in proportion to the Q-factor of the resonant cavity. Alternatively, we may choose the approach indicated by (iii) above and reduce the *optical* bandwidth of the device, and here the gains are potentially very much greater.

The waveguide phase modulator can be placed within a Fabry–Perot cavity, as shown in figure 1, with reduction in the optical bandwidth according to the finesse of the cavity. Such an optical waveguide resonator containing an electro-optic element has been used to form a hybrid bistable optical device wherein feedback is derived from an optical detector monitoring the cavity throughput (Smith *et al.* 1978). If, however, the feedback loop is disconnected and the resonator and active element is considered independently as a modulator, the cavity can be seen to perform two functions. First, the applied phase (delay) modulation is converted to an intensity modulation: this is well known and is readily performed in other, arguably superior ways. Secondly, the cavity enhances the modulation for a given drive power by a factor of the order of the Q-factor of the resonator. But it is significant to note that this applies to phase modulation as well as amplitude modulation: indeed, in this device phase modulation is generally more convenient, since the accompanying gain is maximal on the resonance peak and is in the same sense throughout.

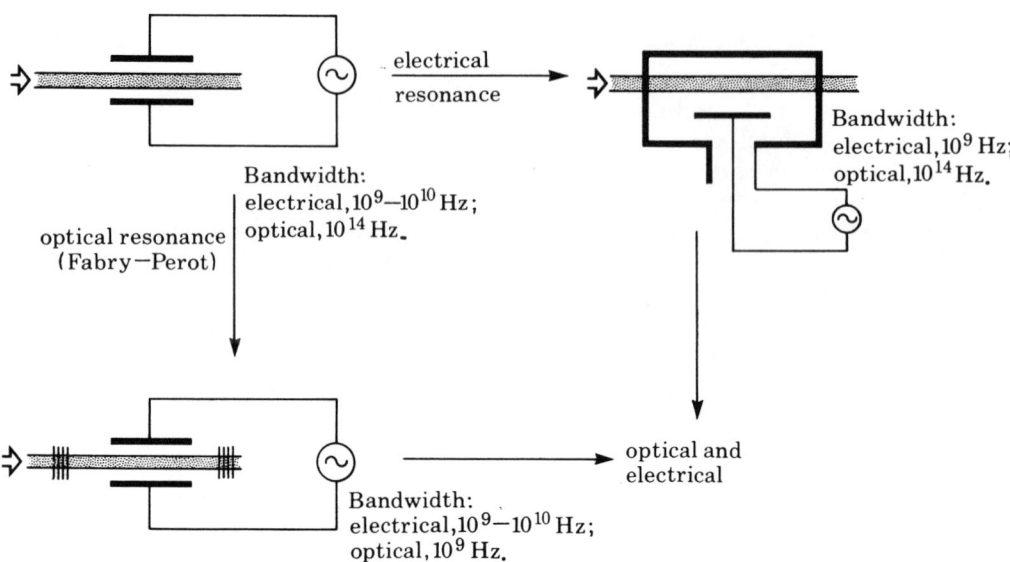

FIGURE 1. High-speed electro-optic phase modulation.

The foregoing arguments are most significant for high-speed modulation of an optical wave, permitting considerable gain enhancement to be achieved without reducing the overall modulation bandwidth to r.f. levels. This gain would generally be accessed by reducing the size of the device. Higher modulation frequencies can be achieved by driving the device between successive cavity resonances, as indicated by figure 2, achieving an unrestricted frequency range

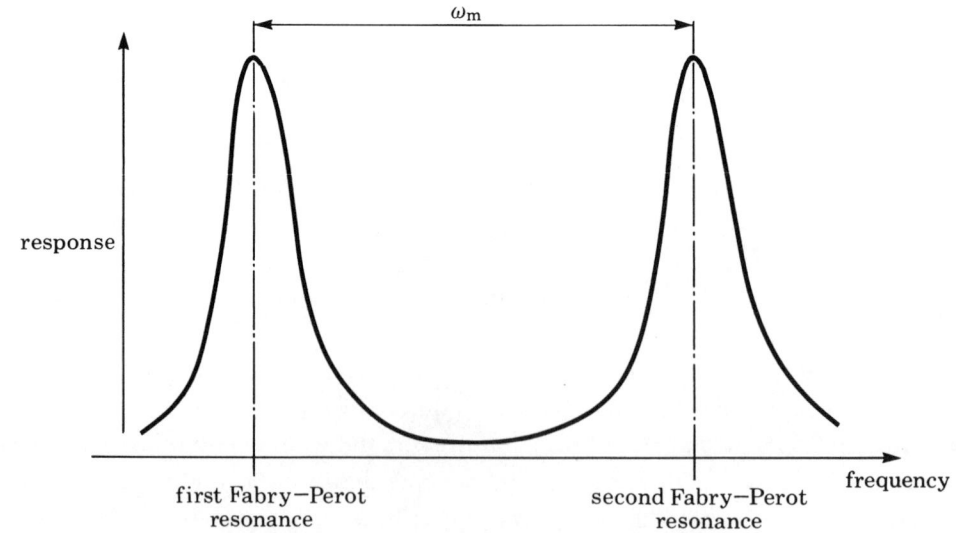

FIGURE 2. Phase modulation at frequency ω_m by sweeping through successive Fabry–Perot resonances.

with limited, but still large, modulation bandwidth. By reducing the optical bandwidth from 10^{14}–10^{15} Hz to that actually required, there is an obvious potential gain amounting to several orders of magnitude.

Several configurations might be adopted for the resonant waveguide phase modulator including Fabry–Perot and distributed feedback resonators, and the ring resonator. It is worthy of note that in those resonators exhibiting multiple resonance peaks the gain in sensitivity is reduced . In this context, the use of a distributed feedback resonator demonstrating a simple sharp resonance may be preferable, although clearly over-restriction of the bandwidth will impose severe tolerance requirements on the stability and control of the optical source linewidth.

We have achieved qualitative confirmation of the principle described by employing electro-optic waveguides formed by thermal diffusion of titanium into $LiNbO_3$. By using the Z-cut crystal orientation, waveguides supporting a single mode at a (free-space) wavelength of 1.15 µm were prepared. The ends of the crystal were carefully cut and polished normal to the waveguide axes and the resonator was formed by cementing multilayer dielectric mirrors to these ends. The drive voltage was applied to electrodes deposited on the crystal, separated from the waveguides by an isolating layer of SiO_2. Figure 3 shows a test device mounted on

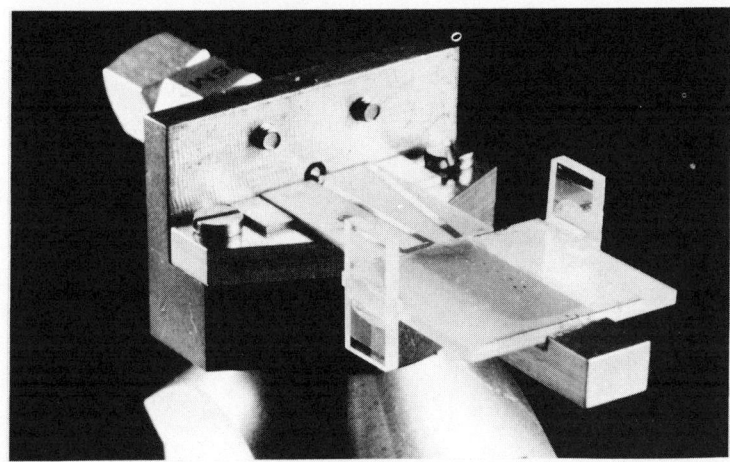

FIGURE 3. Experimental waveguide electro-optic resonant modulator in $Ti:LiNbO_3$.

a carrier, which supplies the drive signal to the electrodes via a terminated microstrip line. The test devices were less than ideal, exhibiting finesse of only 4–5, ascribable largely to optical losses at the cemented mirrors. However, application of an alternating voltage of increasing frequency appears to confirm the principle described, and new higher finesse devices, which we are now preparing, are expected to demonstrate the effect fully.

We gratefully acknowledge the support of this work by the U.K. Ministry of Defence (Procurement Executive).

REFERENCES

Alferness, R. C., Joyner, C. H., Buhl, L. L. & Korotky, S. K. 1983 *IEEE Jl Quant. Electronics* **19**, 1339.
Cross, P. S., Baumgartner, R. A. & Kolner, B. H. 1984 *Appl. Phys. Lett.* **44**, 486.
Gee, C. M. & Thurmond, G. D. 1983 In *Proceedings 2nd European Conference on Integrated Optics, Firenze, Italy, 17–18 October 1983*, p. 118.
Molter-Orr, L. A., Haus, H. A. & Leonberger, F. J. 1983 *IEEE Jl Quant. Electronics* **19**, 1877.
Smith, P. W., Kaminow, I. P., Maloney, P. J. & Stulz, L. W. 1978 *Appl. Phys, Lett.* **33**, 24.

Phil. Trans. R. Soc. Lond. A **313**, 405–409 (1984)

Printed in Great Britain

Observation of bifurcation to chaos in a passive all-optical Fabry–Perot resonator: a high-frequency optical modulator

By R. G. Harrison, I. A. Al-Saidi and W. J. Firth

Department of Physics, Heriot-Watt University, Riccarton, Edinburgh EH14 4AS, U.K.

We report the first observations of bifurcation routes to chaos in an all-optical resonator. Generation of associated deep and sustained Ikeda oscillation of the smooth CO_2 laser input pulses at twice the round-trip time of the Fabry–Perot resonator provides a high-frequency (*ca.* 0.1 GHz) passive optical modulator device. Results are in excellent agreement with our adaption of optical bistability theory to the time-dependent régime.

The transition to chaos is of current interest throughout physics. Among systems exhibiting chaos, 'optically bistable' devices merit special attention as basically simple systems (Ikeda 1979; Nakatsuka *et al.* 1983). Of these a two level medium within an optical resonator is the most fundamental and furthermore is fully quantizable (Harrison *et al.* 1983). We report here first experimental evidence for chaotic behaviour in such a system. NH_3 gas is used as the nonlinear medium contained in a Fabry–Perot resonator optically pumped by the smooth pulses from a CO_2 laser. Deep Ikeda oscillation at twice the 'round-trip' time ($2t_r$) is obtained and, by varying the parameters, we observe $\frac{2}{3}t_r$ modulation, and period-doubling to $4t_r$ and $\frac{4}{3}t_r$. These phenomena, coupled with theory, suggest that the aperiodic pulses that we also observe represent the first evidence for all-optical chaos in a quantum system.

We also report first observations of optical hysteresis in this system by using SF_6, a precursor to possible generation of nonlinear instabilities under (continuous wave) (c.w.) conditions in this and similar molecules with low saturation intensities.

The Fabry–Perot cavities containing ammonia gas at *ca.* 5–40 Torr were pumped by smooth TEA CO_2 laser pulses of *ca.* 100 ns duration and peak intensity *ca.* 10–20 MW cm^{-2}. Oscillation was observed by using a number of laser transitions near-coincident with NH_3 absorption lines. Here we concentrate on the 10R(14) CO_2 line, pumping the aR(1,1) NH_3 transition 1.23 GHz above line centre (Garing *et al.* 1959). This transition is pressure broadened above *ca.* 5 Torr (27 MHz Torr^{-1}† at f.w.h.m.), where it acts as a homogeneously broadened two-level system pumped off resonance (Harrison *et al.* 1984*a*).

The resonators of length 40–100 cm comprised a single-surface Ge flat input coupler of reflectivity $R_0 = 36$–85 % and a single surface Ge output coupler, of 2 m radius of curvature and reflectivity $R_L = 76$ %. The input and output signals were monitored by photon-drag detectors and a Tektronix model 7104 oscilloscope; total resolution $\lesssim 1$ ns. The output coupler was equipped with PZT tuning encompassing one free spectral range of the Fabry–Perot resonator. The transverse intensity profile of the input signal was Gaussian with a 1/e spot diameter of *ca.* 3 mm. An optical delay line prevented feedback of the smooth 100 ns (f.w.h.m.) CO_2 laser pulses from the nonlinear resonator to the laser.

---

† 1 Torr $\approx$ 133.322 Pa.

Representative examples of the modulated output for PZT tuning are shown in figure 1*a* for cavity length 86 cm, cell length 70 cm and pressure 10 Torr, with input coupler reflectivity 67 %. Strong Ikeda oscillation (period *ca.* 13 ns, close to $2t_r = 11.5$ ns), persistent throughout the pulse, is evident in the neighbourhood of minimum transmission, consistent with the four-wave mixing interpretation of this instability (Firth 1981; Firth *et al.* 1982). (Note that in contrast to the ring resonator the Fabry–Perot geometry does not prescribe $2t_r$ as the basic period for Ikeda oscillation.) PZT tuning of the cavity leads progressively to 'switching' behaviour with high peak transmission followed by damped oscillation of longer period.

At lower pressures (4–8 Torr), where inhomogeneous broadening may be important, we also obtained strong and sustained $4t_r$ oscillation. At higher pressures (20–30 Torr) much more complex pulse shapes were obtained. These features were enhanced for reduced input coupler reflectivity ($R_0 = 36 \%$), since large input coupling is needed to bleach the high absorption

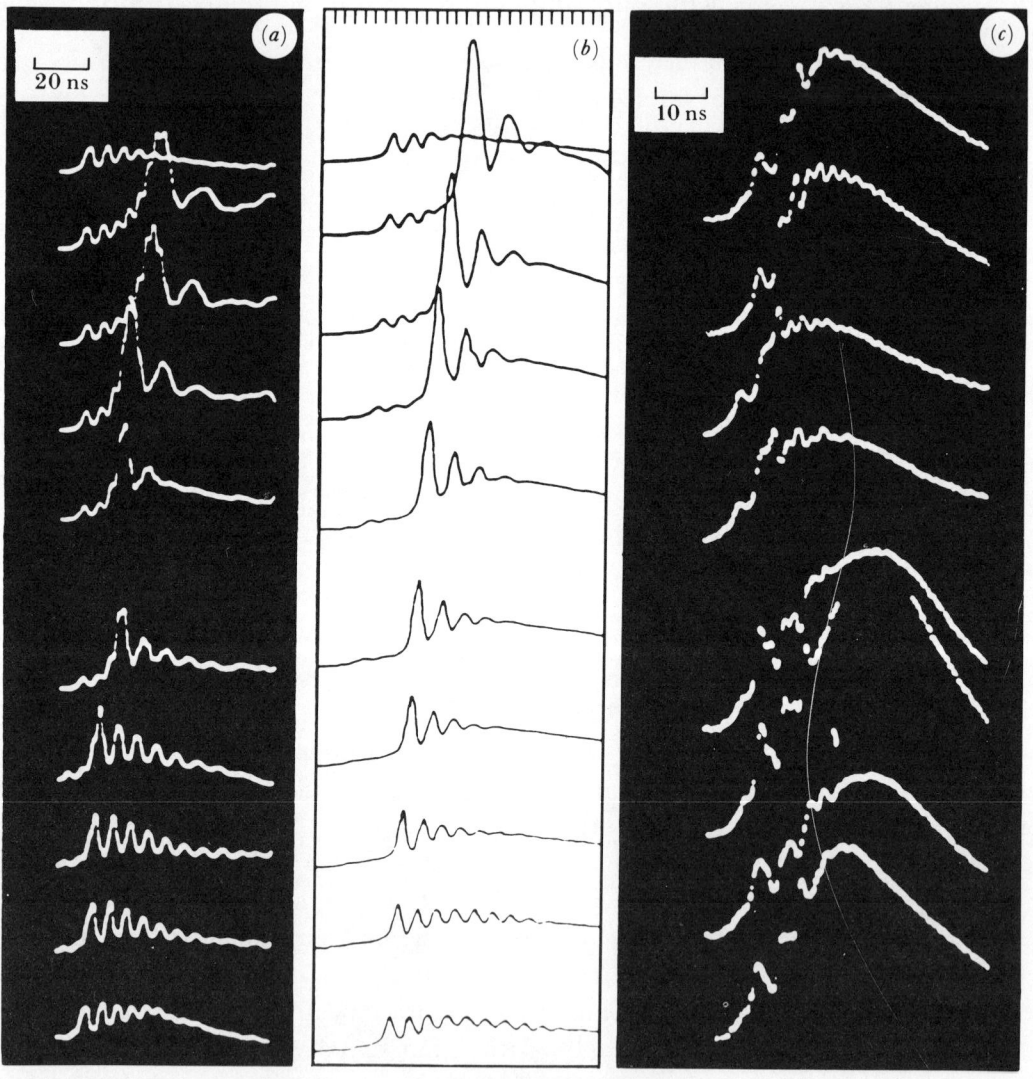

FIGURE 1. (*a*) PZT scan showing Ikeda oscillation at 10 Torr; cavity length = 86 cm; $R_0 = 67 \%$; $R_L = 76 \%$. NH₃ transition aR(1, 1). (*b*) Computed traces for parameters given for (*a*); $\alpha l = 1.75$, $\Delta = 10$; one scale division is equivalent to t_r (cavity round-trip time). (*c*) PZT scan showing $2t_r$, $\frac{2}{3}t_r$, $\frac{4}{3}t_r$ modulation and aperiodic pulses at 19 Torr. Cavity parameters as given for (*a*) with $R_0 = 36 \%$.

($\alpha L = 6$) to achieve adequate cavity feedback. The PZT sequence shown in figure 1c for a pressure of 19 Torr shows $2t_r$ oscillation (top trace), developing to $\frac{2}{3}t_r$ oscillation on the higher branch, bifurcating to $\frac{4}{3}t_r$ before again evolving to lower-branch $2t_r$ modulation (bottom trace). Aperiodic pulse shapes, characteristic of chaos, are also evident here and in other data taken at similar pressures.

Observation of these effects under c.w. conditions requires molecules with low saturation intensities, such as SF_6 (ca. 6 W cm^{-2} Torr^{-1}). As a precursor to this we report observation of optical hysteresis effects in this gas over a wide range of parameter conditions. We note that operation here is in the 'bad cavity' limit and so precludes generation of Ikeda instabilities. Fabry–Perot resonators, ranging in length from 2 to 15 cm, with intra-cavity gas cells as short as 1 mm were operated at SF_6 gas pressures ranging from ca. 1 to ca. 200 Torr. Various lines in the 10P band of the TEA CO_2 laser were used to pump the dense and broad spectral features of the ν_3 vibrational mode of SF_6 near resonance (Harrison & Butcher 1980). Effects of switching, power limiting, and overshoot with nanosecond response times, limited only by cavity decay time, were routinely obtained at gas pressures commensurate with self focusing. Sample traces of the transmitted signal for PZT tuning the cavity through half a spectral range are shown in figure 2b, cavity length 3 cm, cell length 1 cm with SF_6 at a pressure of ca. 10 Torr pumped at 947.7 cm^{-1} (10P16 CO_2 line). Corresponding recordings of instantaneous input (x-axis) and output (y-axis) intensity (figure 2c) show pronounced optical hysteresis effects. Equivalent recordings for the empty cavity are shown in figure 2a for reference.

Stability analysis of nonlinear Fabry–Perot resonators has apparently only been examined in the dispersive (Kerr) limit (Firth 1981). Our experiments, however, necessarily involve the bleaching of a rather substantial absorption. We have therefore adapted the model of Carmichael & Hermann (1980), which treats steady-state optical bistability in a Fabry–Perot resonator, handling the time dependence by replacing the gas cell by N thin slices symmetrically placed within the cavity. Assuming the adiabatic limit, we apply steady-state theory to find the forward and backward transmissions of each slice (unequal due to the phase–population grating). We can then follow the field around the cavity, where it interacts in the slices with N earlier and N later fields, and keep track of its attenuation at each stage. At the output and input couplers we apply the usual Fabry–Perot boundary conditions, and thus have, in effect, a $2(N+1)$ parameter mapping problem. Full details of this analysis will be reported elsewhere (Harrison et al. 1984b).

Application of this procedure to NH_3 and use of an input pulse of the form shown in figure 2a yields the transmitted pulse shapes in figure 1b, in pleasing agreement with the observed pulse shapes, especially since only measured parameters are used: α is 0.025 cm^{-1} at 10 Torr and scales as p^2 (this, plus the pressure broadening rate and constants gives a saturation intensity $I_s \approx 2.3$ MW cm^{-2} at 10 Torr (2.5 MW cm^{-2} at 30 Torr)). The value $I/I_s = 7$ is thus in line with the measured input intensities in the range 10–20 MW cm^{-2}.

This good agreement is rather surprising in view of the omission of reservoir (Harrison et al. 1983, 1984a,b) and transverse effects in the analysis. Inclusion of such effects, in particular transverse effects (Moloney et al. 1982), will, however, be necessary in predicting the more complex pulse shapes of figure 1c, though qualitative agreement still exists from the plane-wave approach. It is emphasized that for c.w. inputs these models give steady-state oscillation and chaos within our experimental parameter range and we have identified period doubling to $16t_r$ en route to chaos in one case.

[217]

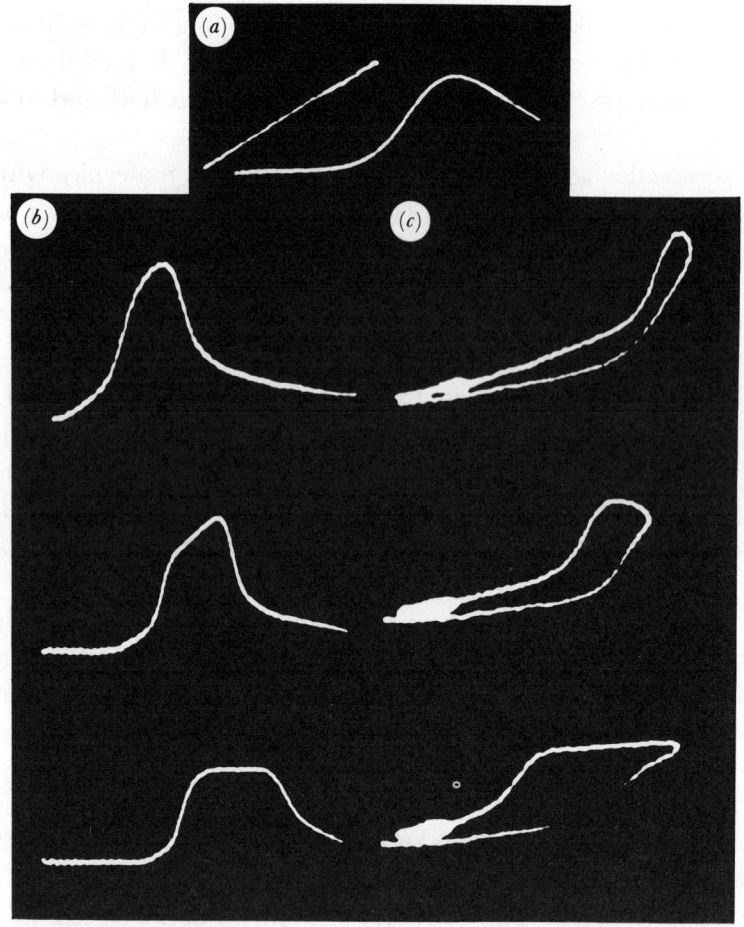

FIGURE 2. (*a*) Typical single-mode input pulse together with *x–y* trace of instantaneous input and output intensity for empty Fabry–Perot resonator. (*b*) PZT scan (over half of the free spectral range) of transmitted signal, SF_6 pressure cavity length 3 cm, cell length 1 cm. (*c*) Corresponding *x–y* traces of input and transmitted signal intensity.

In conclusion, our observations of $2t_r$ oscillation period doubling to $4t_r$ and, over a more limited parameter range, $\frac{2}{3}t_r$, $\frac{4}{3}t_r$ modulation and aperiodic pulse shapes in NH_3 gas confirm the predicted behaviour of this system and support the conclusion that we have driven an all-optical two-level system through oscillation to chaos. Already of use as a passive optical modulator ($2t_r \approx 100$ MHz) of pulsed signals, prospects for modulation in the gigahertz range are promising. The possibility of generating these effects under c.w. conditions is supported by our observations of strong optical hysteresis in this system by using SF_6 which, along with many other molecular gases, exhibits low saturation intensity.

REFERENCES

Firth, W. J. 1981 *Optics Commun.* **39**, 343–346.
Firth, W. J., Abraham, E. & Wright, E. M. 1982 *Appl. Phys.* B **28**, 170.
Carmichael, H. J. & Herman, J. A. 1980 *Z. Phys.* B **38**, 365–380.
Garing, J. S., Nielsen, H. H. & Rao, K. N. 1959 *J. molec. Spectrosc.* **3**, 496–527.
Harrison, R. G. & Butcher, S. R. 1980 *Contemp. Phys.* **21**, 19–41.
Harrison, R. G., Firth, W. J. & Al-Saidi, I. A. 1984*a* *Appl. Phys. Lett.* **44**, 716–718.

Harrison, R. G., Firth, W. J. & Al-Saidi, I. A. 1984*b* *Phys. Rev. Lett.* **53**, 258–261.
Harrison, R. G., Firth, W. J., Emshary, C. & Al-Saidi, I. A. 1983 *Phys. Rev. Lett.* **51**, 562–565.
Ikeda, K. 1979 *Optics Commun.* **30**, 257–261.
Moloney, J. V., Hopf, F. A. & Gibbs, H. M. 1982 *Phys. Rev.* A **25**, 3442–3445.
Nakatsuka, H., Asaka, S., Itoh, M., Ikeda, K. & Matsuoka, M. 1983 *Phys. Rev. Lett.* **50**, 109–112.

Phil. Trans. R. Soc. Lond. A **313**, 411–415 (1984)

Printed in Great Britain

Stable oscillations in a bistable, bidirectional CO_2 ring laser

By J. R. Tredicce, G. L. Lippi, F. T. Arecchi and N. B. Abraham†

Istituto Nazionale di Ottica, Largo E. Fermi 6, 50125 Firenze, Italy

We show experimentally and theoretically the existence of spontaneous self-modulation and mode-switching in a bidirectional ring laser. The process involves at least two different characteristic frequencies. Experimental records of the temporal behaviour and associated power spectra for various operating conditions are shown.

Introduction

The dynamical behaviour of bidirectional ring dye lasers has recently been studied (Roy & Mandel 1980 a, b; Mandel *et al.* 1981). In these cases the intensity output was stable (as one mode quenched the other). Stochastic switching between the dominance of the two counter-propagating modes was observed and explained by additive and multiplicative noise (Lett *et al.* 1981).

In the presence of an external back-reflecting mirror, regular pulsations have also been observed in such a laser (Kühlke 1982). However, in a dye laser only the field variables play a relevant role in the dynamics.

In contrast, we have shown in previous work that the population inversion must be taken into account to describe the dynamical behaviour of a CO_2 laser (Arecchi *et al.* 1982, 1984). The interaction between the two fields and the population inversion leads to spontaneous pulsations in a bidirectional CO_2 ring laser.

Experimental results

The experimental arrangement is shown in figure 1. The parameters to be changed are pressure, current, cavity tuning, and tuning of the external mirror. The results reported here are for the arrangement without the external mirror. For a wide range of pressure and current, we have the previously observed bistable behaviour between two stable states, in which one mode has a constant, non-zero intensity while the other is completely extinguished. For lower pumping rates (by changing current or pressure) an unstable state appears in which a regular spiking in both modes simultaneously is possible (figure 2a).

For lower pressures the giant spikes are followed by damped relaxation oscillations, as shown in figure 2b. The spikes occur at regular intervals with a repetition rate of the order of the slow population relaxation rate. The damped oscillations can also be observed during the transients in single-mode lasers and their frequency is of the order of the geometric mean of the population and field relaxation rates. The repetition rate of the giant pulses seems to be related to the population relaxation rate. The power spectra in these two cases show a sharp peak at a low

† Present address: Department of Physics, Bryn Mawr College, Bryn Mawr, Pennsylvania 19010, U.S.A.

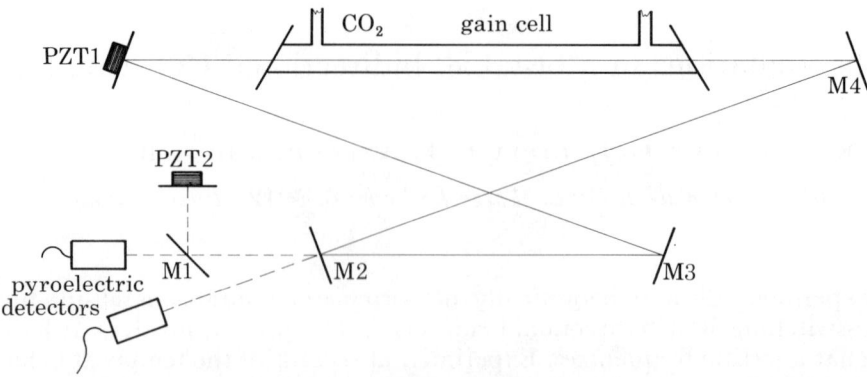

FIGURE 1. Experimental arrangement showing bidirectional CO_2 ring laser (10.6 μm wavelength) with optional external back-reflecting mirror. The total 'round-trip' cavity length is 4.2 m. The reflectivity and R for each is: M1, 80%, ∞; M2, 80%, 5 m; M3, 100%, 5 m; M4, 100%, 2 m; PZT1, 100%, 2 m; PZT2, 100%, 5 m.

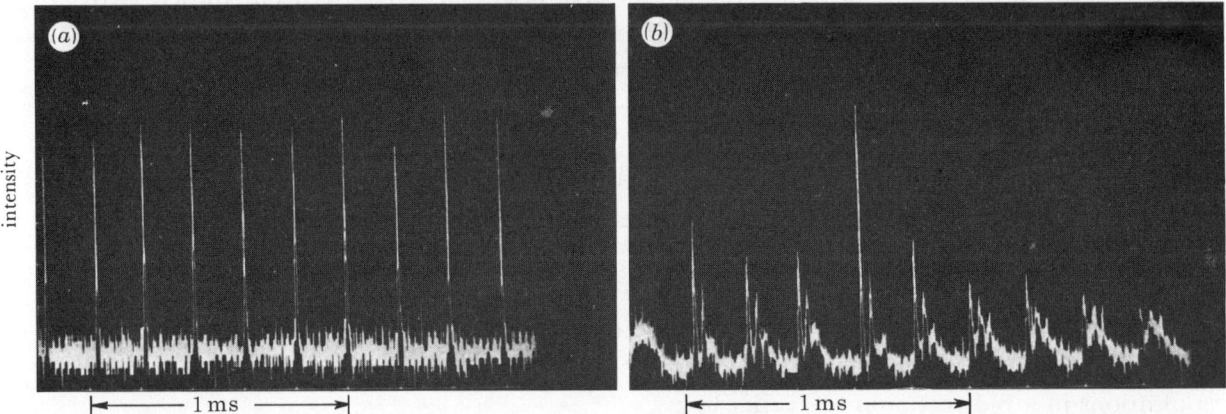

FIGURE 2. Intensity against time for one of the two modes: (a) current $I = 4$ mA per division, pressure $P = 1330$ Pa; (b) current $I = 4$ mA per division, pressure $P = 933$ Pa.

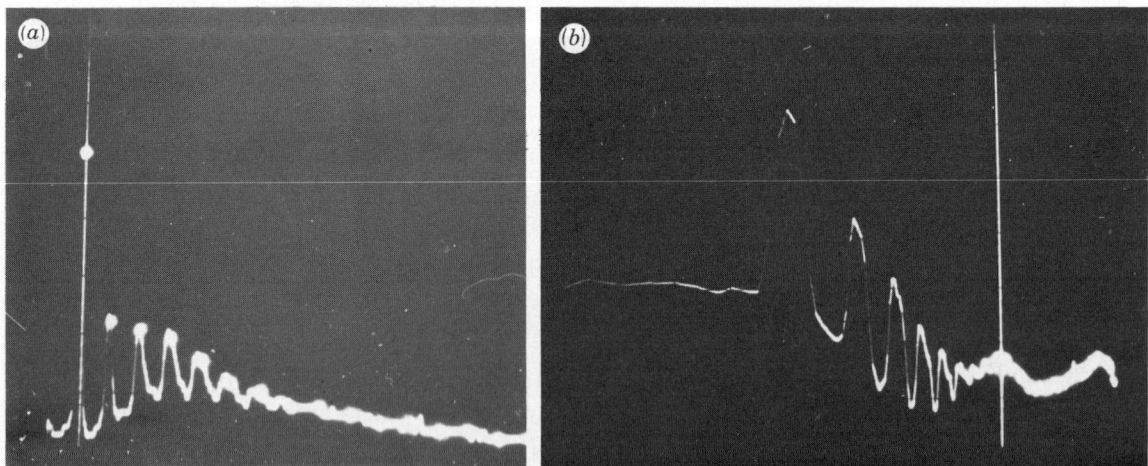

FIGURE 3. Intensity–power spectra for the current and pressure conditions of figure 2, where a and b correspond in the two cases. The frequency scale in a is plotted linearly, while in b it is logarithmic, as can be seen from the spacing of successive harmonics.

frequency (of order 2 to 7 kHz) and peaks at its harmonics, and a second broad peak at the higher frequencies corresponding to the relaxation oscillations (figure 3a, b).

In figure 4 we are able to compare the behaviour of the two modes. In figure 4a the modes pulse synchronously, but their relaxation oscillations are 180° out of phase, which indicates a common origin of the principal pulsation and some type of competition in the damped ringing. In figure 4b we observe the switch-on of one mode (upper trace) and the switch-off the second mode (lower trace) again with out-of-phase higher frequency oscillation in the modes.

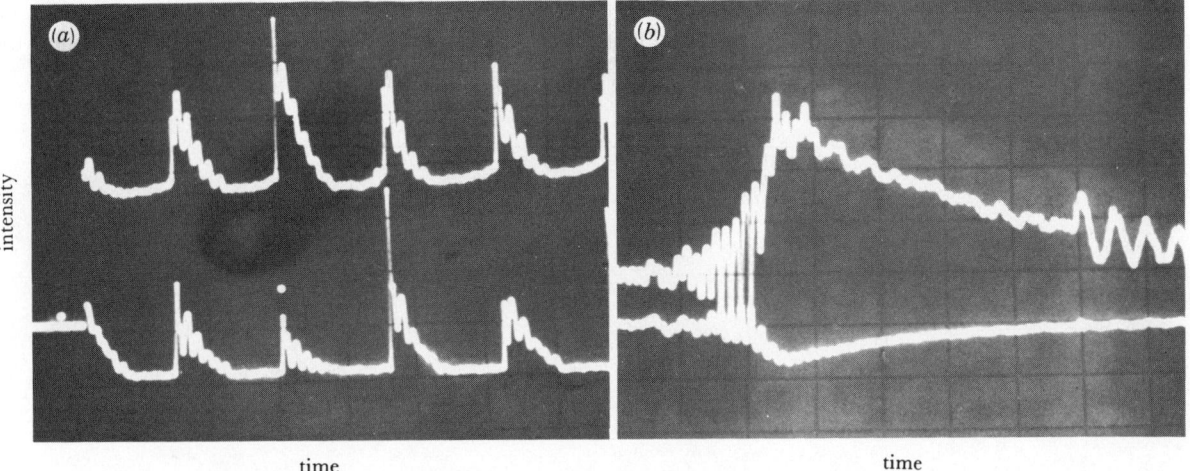

FIGURE 4. Intensity against time to show the two-mode behaviour for two different values of the cavity tuning: (a) synchronous pulsing with out-of-phase relaxation oscillations; (b) the mode switching with one on and one off.

THEORETICAL RESULTS

Theoretically, the system can be described by the Maxwell–Bloch type of equations for a two-level system interacting with two counter-propagating waves having the same spatial mode structure. By assuming that the spatial harmonic patterns formed by the two interfering modes are relatively weak, we can truncate the equations to the normal single-mode set with only lowest order corrections. Adiabatically eliminating the polarization, we arrive at a set of seven equations similar to those derived elsewhere when back-scattering was included (Perevedentseva *et al.* 1980). Here, in contrast, we retain the effects of detuning. (Previous analyses by Zhelnov *et al.* (1970) and by Mandel & Agrawal (1982) have shown that without detuning or other coupling mechanisms the asymmetric stationary states are stable.) Taking into account the complex electric fields of the two modes, the population inversion, and the complex amplitude of the spatial population grating formed in the medium by the interference of the counter-propagating fields, the equation set may be written as

$$\dot{x} = (1+i\delta)^{-1}(zx+w^*y)-x,$$

$$\dot{y} = (1+i\delta)^{-1}(zy+wx)(k_2/k_1)y+xR\exp(i\phi),$$

$$(k_1/\gamma_\|)\dot{z} = -(z-z_0)-(1+\delta^2)^{-1}\{z(|x|^2+|y|^2)+w^*x^*y+wxy^*\},$$

$$(k_1/\gamma_\|)\dot{w} = -w-(1+\delta^2)^{-1}[zx^*y+\tfrac{1}{2}w\{|x|^2+|y|^2+i\delta(|y|^2-|x|^2)\}],$$

where x and y are the complex amplitudes of the two modes, respectively, z is the population inversion and w is the complex amplitude of the population grating; δ is the detuning between

27-2

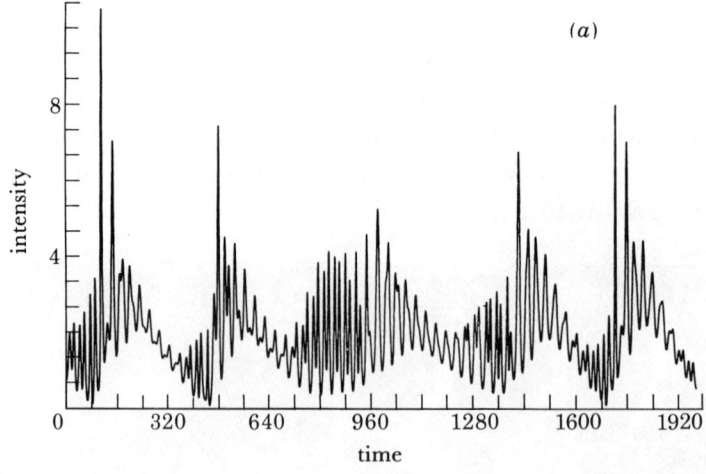

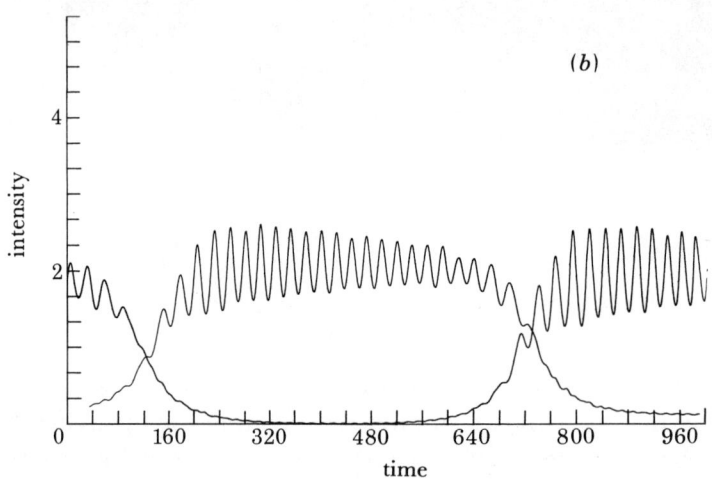

FIGURE 5. Results from numerical integration of the seven-equation model. Normalized intensity against dimensionless time (units of k_1^{-1}) for typical CO_2 laser parameters. Characteristic features of the experimental results shown in figure 4a and 4b are reproduced in (a) and (b), respectively.

the cavity and atomic frequencies, normalized to the polarization relaxation rate. R and ϕ are the amplitude and phase of the back reflection; k_1 and k_2 are the loss rates for the two modes and $\gamma_\parallel$ is the loss rate for the population inversion. Time has been renormalized to $\tau = k_1 t$ and the dot signifies differentiation with respect to τ. The parameters x, y, z and w are the slowly varying amplitudes of the rapidly varying carrier wave at a frequency equal to the atomic resonance frequency. Finally, z_0 is the equilibrium population inversion in the absence of the two fields.

Principal theoretical results include the coexistence of the two stable stationary states for zero detuning, which are destabilized with detuning in favour of spontaneous switching between them. The dynamical behaviour, when the system is unstable, is in good agreement with the experimental results, showing both simultaneous pulsing and alternate switching of the two modes (figure 5) for no back-reflection ($R = 0$), and suitable detuning.

Since completing this work we have observed similar solutions in a recent paper by Polushkin *et al.* (1983), which uses rather more complicated models for the material variables and the field coupling.

Quasiperiodic and chaotic behaviour, observed in our experimental and theoretical results, will be described elsewhere.

We wish to acknowledge the technical assistance of G. P. Puccioni, L. Albavetti and C. Castellini.

REFERENCES

Arecchi, F. T., Lippi, G. L., Puccioni, G. L. & Tredicce, J. R. 1984 In *Coherence and quantum optics* V (ed. L. Mandel & E. Wolf). (In the press.)

Arecchi, F. T., Meucci, R., Puccioni, G. & Tredicce, J. R. 1982 *Phys. Rev. Lett.* **49**, 1927.

Kühlke, D. 1982 *Acta phys. pol.* A **61**, 547.

Lett, P., Christian, W., Singh, S. & Mandel, L. 1981 *Phys. Rev. Lett.* **47**, 1892.

Mandel, P. & Agrawal, G. P. 1982 *Optics Commun.* **42**, 269.

Mandel, L., Roy, R. & Singh, S. 1981 In *Optical bistability* (ed. C. M. Bowden, M. Ciftan & H. R. Robl), pp. 127–150. New York: Plenum.

Perevedentseva, G. V., Khandokhin, P. A. & Khanin, Ya. I. 1980 *Kvantovaya Elektron (Moscow)* **7**, 128 (*Sov. J. quant. Electron.* **10**, 71 (1980)).

Polushkin, N. I., Khandokhin, P. A. & Khanin, Ya. I. 1983 *Kvantovaya Elektron (Moscow)* **10**, 1461 (*Sov. J. quant. Electron.* **13**, 950 (1983)).

Roy, R. & Mandel, L. 1980a *Optics Commun.* **34**, 133.

Roy, R. & Mandel, L. 1980b *Optics Commun.* **35**, 247.

Zhelnov, B. L., Smirnov, V. S. & Fadeev, A. P. 1970 *Optika Spectrosk.* **28**, 744 (*Optics Spectrosc.* **28**, 400 (1970)).

Phil Trans. R. Soc. Lond. A **313**, 417–420 (1984)

Printed in Great Britain

Experimental study of the response of a bad cavity bistable system to fast light switch-on

By W. E. Schulz, W. R. MacGillivray and M. C. Standage

School of Science, Griffith University, Nathan 4111, Australia

Preliminary results are reported on the response of an intrinsic bistable system to the rapid switch-on of an injected light field. The bistable system consists of a linear array of 25 sodium atomic beams contained in a low finesse Fabry–Perot etalon. The beams are aligned perpendicular to the direction of propagation of the cavity field and are collimated sufficiently to resolve the D-line ground state $3^2S_{\frac{1}{2}}$ hyperfine structure.

The etalon conditions are such that the cavity lifetime is an estimated 6 ns compared with the atomic lifetime of 16 ns; hence the system approaches the bad cavity limit in which the atom dynamics should be observable.

The radiation from a single-mode ring dye laser is switched on to the bistable system in approximately 1 ns. Although preliminary results show low branch to high branch switching there is no evidence of the expected nutational oscillations.

1. Introduction

One aspect of optical bistability that has received scant experimental attention is that of the spectrum. An interesting feature of an intrinsically bistable system consisting of an optical cavity with an atomic medium is the prediction that the fluorescence from the low branch would be single-peaked while that from the upper branch would exhibit the three peaks of the dynamic Stark effect (Bonifacio & Lugiato 1976; Carmichael & Walls 1977). To do an experiment where the fluorescent spectrum was analysed would be technically difficult, owing to the high degree of frequency stability required of the laser and optical cavity over a long period of time. However, the three-peaked spectrum has a corresponding feature in the time domain. If radiation with intensity above the threshold for a three-peaked spectrum is suddenly switched on, the response of an atomic medium on resonance is to produce a coherent optical transient oscillating at the Rabi frequency, i.e. the frequency of separation of the central fluorescent peak from its two sidebands. This has been demonstrated for sodium atoms in a vapour by MacGillivray *et al.* (1978).

Numerical simulation by Bonifacio & Meystre (1978) and Abraham & Hassan (1980) of the response of an optically bistable system that is stepwise excited to the upper branch indeed verified the existence of the Rabi oscillations in the Maxwell–Bloch model. The conditions of the calculation required the cavity lifetime, t_c, to be very much smaller than T_1 and T_2, the atomic population and dipole relaxation times, respectively. This is the so-called 'bad cavity' limit. The first attempts to observe this transient phenomenon are reported here.

Recently, the 'good cavity' limit case, where the cavity lifetime is dominant, has been studied experimentally by Grant & Kimble (1983).

2. The experiment

An abridged schematic diagram of the apparatus is shown in figure 1. The Fabry–Perot etalon mirrors are flat and have a reflectivity of approximately 93%. The cavity tuning is controlled by a set of driven piezo-ceramics mounted on one mirror. A ring cavity is constructed by the addition of one 100% reflecting flat mirror. The empty cavities have a free spectral range of approximately 750 MHz and a finesse of about 10. So t_c is estimated to be 6 ns.

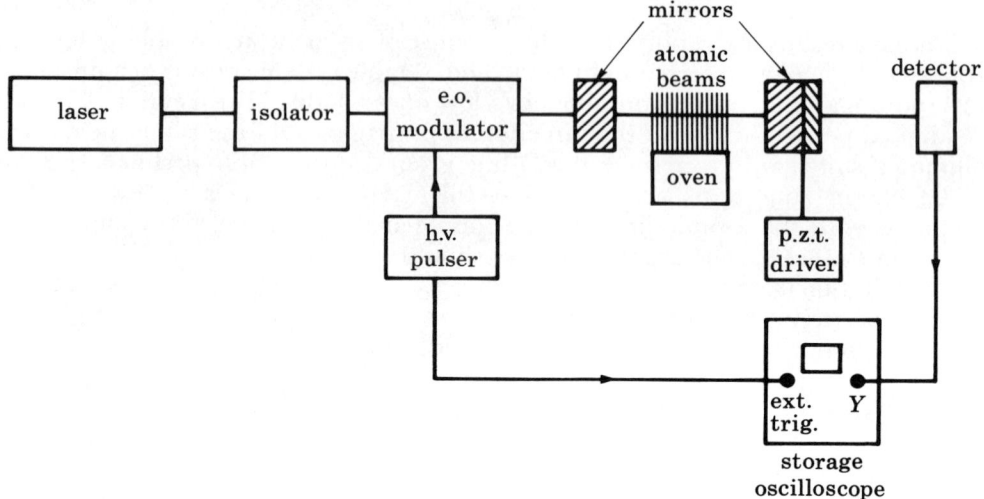

FIGURE 1. Experimental configuration: e.o., electro-optic; h.v., high voltage; p.z.t., piezoelectric.

The cavity medium consists of a linear array of 25 atomic beams of sodium. Unlike the experimental arrangements of Weyer *et al.* (1981) and Grant & Kimble (1983), high collimation of the atomic beams is not possible since greater absorption in the medium is required than in their experiments. This is to maintain the system in a bistable condition as described by the bistability coefficient

$$C = \alpha l F / 2\pi.$$

That is, a decrease in the finesse of the cavity must be compensated by an increase in the absorption in the medium. The D lines are being investigated and the atomic beam collimation is just sufficient to resolve the transitions associated with the two hyperfine levels of the $3^2S_{\frac{1}{2}}$ ground state. This resolution is necessary to remove the condition for creating a non-absorption resonance in the Fabry–Perot system as observed by Schulz *et al.* (1983), when sodium vapour was used as the cavity medium.

The radiation source is a Spectra Physics 380D actively stabilized ring laser pumped by a 171 Ar$^+$ laser. Up to 200 mW of power is available at the input port in a collimated beam of just greater than 1 mm diameter.

The electro-optic device acts as a switch capable of making square pulses of light variable between 5 ns and 500 ns long with rise-times of about 2 ns. The leading edge of the driving pulse triggers the oscilloscope for a transmission against time plot.

The light transmitted by the cavity is detected by a PIN photodiode. The data is recorded photographically from the storage oscilloscope.

[228]

3. Results and discussion

Figure 2a is the transient response of the atomic medium to the rapid switch on of the radiation. There is no cavity in this case. Fifty milliwatts of linearly polarized light was injected, tuned to the $F = 2 \to F'$ hyperfine transition of the D_2 line. An almost identical result was obtained for the $F = 1 \to F'$ transition. The Rabi oscillations are clearly evident with a frequency of approximately 200 MHz.

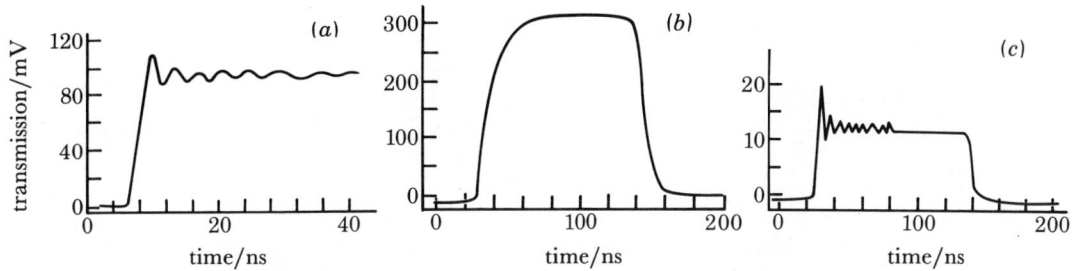

FIGURE 2. The transient response of (a) the atomic medium (b), (c) the empty cavity. The transmission is in units of detector volts. For (c) the cavity frequency has been detuned from that of the injected light.

The response of the resonant, empty Fabry–Perot etalon to a similar stepwise change in the injected intensity is shown in figure 2b. The cavity filling time of approximately 20 ns (0–90 %) is consistent with the lifetime $t_c = 6$ ns (Grant & Kimble (1983), equation 3). If the cavity is detuned, a transient oscillation occurs due to the interference of non-phase-matched cavity fields, which have completed different numbers of round trips (figure 2c). The frequency of this oscillation increases with detuning. Similar results to these were obtained with the ring cavity.

Figure 3 shows the response of the atom-filled cavities to the rapid switch on in intensity. The laser is tuned to the D_2 line and to the frequency that gives the optimum hysteresis for a slow intensity cycle. The varied parameter in figure 3 is the cavity frequency. The data is interpreted in the following way. In figure 3a the Fabry–Perot is detuned from the bistable region. The oscillations are those observed previously in the empty cavity since they increase in frequency with further cavity detuning, whereas a Rabi oscillation would decrease in frequency due to the lowering of the effective cavity field. Figure 3b illustrates switching from the low branch to the high branch as the cavity is 'pulled' into resonance by the changing refractive index of the atomic medium. Figure 3c depicts the same phenomenon for the cavity with greater initial detuning. Essentially the same behaviour is evident in the ring cavity (figures 3d,e,f).

There is no indication of Rabi oscillations. The cavity intensity is estimated to be 150 mW so that the effective Rabi frequency should be less than 400 MHz, which is within the bandwidth of the detector equipment.

The most plausible reason for the non-appearance of the Rabi oscillations is that the light field inside the cavity is not switching on fast enough to obtain a coherent transient. From figure 2a it can be seen that the nutations from the atomic beams alone have decayed away in approximately 15 ns.

There are two options for the pursuit of these oscillations. Either a lower finesse cavity will

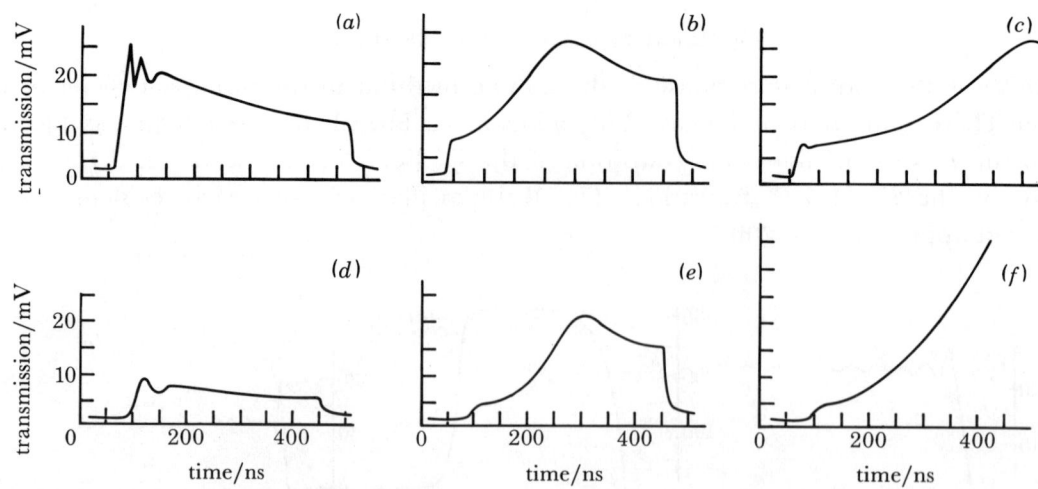

FIGURE 3. The transient responses of the bistable system: (*a*)–(*c*) for the Fabry–Perot etalon; (*d*)–(*f*) for the ring cavity. The transmission is in units of detector volts. In each sequence the cavity frequency is the varied parameter.

have to be manufactured with, subsequently, a faster cavity filling time, or an atomic system with a longer lifetime will have to be used. The problem associated with the first alternative is that if the finesse is too low a bistable condition may become impossible to attain.

Financial support for this project has come principally from the Australian Research Grants Scheme. One of us (W.E.S.) is supported by a Griffith University Postgraduate Scholarship.

REFERENCES

Abraham, E. & Hassan, S. S. 1980 *Optics Commun.* **35**, 291.
Bonifacio, R. & Lugiato, L. A. 1976 *Optics Commun.* **19**, 172.
Bonifacio, R. & Meystre, P. 1978 *Optics Commun.* **27**, 147.
Carmichael, H. J. & Walls, D. F. 1977 *J. Phys.* B **10**, L685.
Grant, D. E. & Kimble, H. J. 1983 *Optics Commun.* **44**, 415.
MacGillivray, W. R., Pegg, D. T. & Standage, M. C. 1978 *Optics Commun.* **25**, 355.
Schulz, W. E., MacGillivray, W. R. & Standage, M. C. 1983 *Optics Commun.* **45**, 67.
Weyer, K. G., Wiedenmann, H., Rateike, M., MacGillivray, W. R., Meystre, P. & Walther, H. 1981 *Optics Commun.* **37**, 426.

Phil. Trans. R. Soc. Lond. A **313**, 421–423 (1984)

Printed in Great Britain

Sodium atoms in Fabry–Perot resonators: studies of static and dynamic behaviour in transverse magnetic fields

By W. Lange, F. Mitschke, R. Deserno and J. Mlynek

Institut für Quantenoptik der Universität Hannover, Welfengarten 1, D-3000 Hannover, F.R.G.

In experiments on a sodium-filled Fabry–Perot resonator optical bistability, tristability and critical slowing-down were observed, as well as magnetically induced self-pulsing (*Phys. Rev. Lett.* **50**, 1660 (1983)). Our experimental results can be explained by the model of Kitano, Yabuzaki & Ogawa (*Phys. Rev.* A **24**, 3156 (1981)) in a qualitative way. In a more detailed treatment we obtain modified equations of motion that make phase space three-dimensional. The results of a stability analysis and of numerical solutions of the equations of motion are compared to experimental observations. We discuss the limitations of the adiabatic elimination of the cavity and demonstrate that not only tristable but also bistable systems can evolve into oscillatory states under the action of a static magnetic field; though the range of allowed experimental parameters is more restricted.

Many experiments on bistability in all-optical systems have been made in Fabry–Perot resonators containing sodium atoms. These experiments are facilitated by the fact that sodium atoms are not at all ideal two-level atoms, but have hyperfine structure and level degeneracy; this introduces hyperfine pumping and Zeeman pumping as very efficient nonlinear mechanisms. We have studied the static and dynamic behaviour of this type of system under conditions of transverse optical pumping.

Under static conditions the non-absorbing resonance related to ground state coherence allows the observation of optical bistability at very low power levels of the laser light source. Both absorptive and dispersive bistability can be obtained. The bistability can be controlled by a static magnetic field (Mlynek *et al.* 1982). Our experimental data can be described in a semiquantitative way by a simple model based on the assumption of three-level atoms (Λ-type scheme) (Mlynek *et al.* 1984).

It should be noted that the nonlinearity produced by Zeeman coherence can be used in related nonlinear optical experiments to advantage. Very recently, for example, multistabilities in intracavity phase conjugation through degenerate four-wave mixing have been observed (Mlynek *et al.* 1983).

We also studied the transient response of the device to step-inputs of light. Our measurements demonstrate the crucial role of critical slowing-down for switching time. We find that time-dependent diffusion processes can strongly affect the dynamics of the bistable device. Numerical calculations yield results in satisfactory agreement with experimental data (Mitschke *et al.* 1983 *a*).

While in the experiments mentioned so far the light was always circularly polarized, optical tristability connected with polarization switching is obtained with linearly polarized light. It has been pointed out by Kitano *et al.* (1981) that under the action of the external magnetic field magnetically induced optical self-pulsing should occur, and the phenomenon has been observed recently by Mitschke *et al.* (1983*b*), when the sodium density was sufficiently high. While the basic features of the observations are well explained by the theory of Kitano *et al.* (1981), the following modifications have to be applied to obtain a more quantitative description of the experiment.

(i) In Kitano *et al.*(1981) absorptive losses in the nonlinear Fabry–Perot resonator are not taken into account. On the other hand, the experiment has to be made not too far from resonance to facilitate optical pumping. We find a strong influence of absorption on the pulse shapes. For example, the smoothness and symmetry of the pulses displayed in figure 1 are a consequence of absorption, while the pulse shape displayed in figure 6 of Kitano *et al.* (1981) is typical for the purely dispersive case.

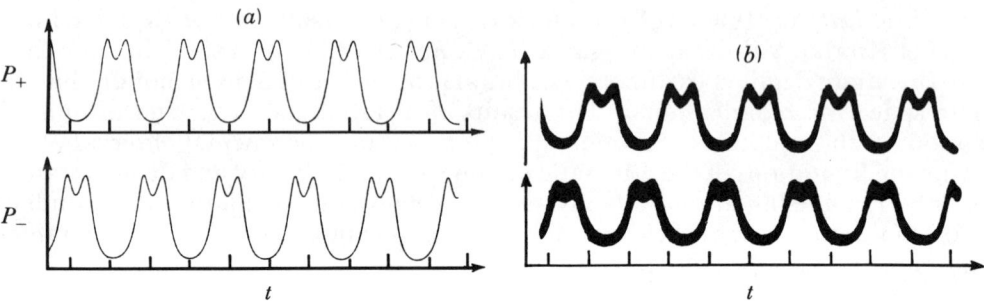

FIGURE 1. Typical time dependence of right (upper trace) and left (lower trace) circularly polarized component of output intensity: (*a*) numerical solution; (*b*) experimental. The scale of the abscissas is 200 ns per division.

(ii) Obviously, adiabatic elimination of the cavity fails when the repetition rate of the pulses is increased by applying a strong magnetic field. We find that the oscillation is quenched by increasing the magnetic field above a limiting value.

(iii) Using the model and abbreviations of Mitschke *et al.* (1983*b*), we find the following equations of motion for the normalized components of magnetization of the sample, which play the crucial role in the explanation of the experiment:

$$\mathrm{d}m_x/\mathrm{d}t = -\Omega_0 m_z - (\Gamma + P_+ + P_-)\, m_x - \Delta(P_+ - P_-)\, m_y, \tag{1a}$$

$$\mathrm{d}m_y/\mathrm{d}t = -(\Gamma + P_+ + P_-)\, m_y + \Delta(P_+ - P_-)\, m_x, \tag{1b}$$

$$\mathrm{d}m_z/\mathrm{d}t = \Omega_0 m_x - (\Gamma + P_+ + P_-)\, m_z + (P_+ - P_-). \tag{1c}$$

Here Ω_0 is the Larmor frequency, P_+ and P_- represent the pumping rates induced by right or left circularly polarized light, respectively, Γ is the relaxation constant of the orientation and Δ is the detuning of the laser with respect to the atomic resonance normalized to the spectral width of the absorption line. The 'dispersive' terms containing Δ are not present in (4*a*) to (4*c*) of Kitano *et al.* (1981), since these are based on an analysis assuming 'broad band' excitation.

The dispersive terms reflect the fact that in a strong off-resonant light field the spins do not simply precess about the magnetic field, but a nutation is superimposed. They strongly influence the shape of the signals, especially in the starting period of the oscillation. We found, for example, that signals like the one in figure 2, where switching occurs via a pronounced oscillatory process, can be described reasonably only if the dispersive terms are taken into account.

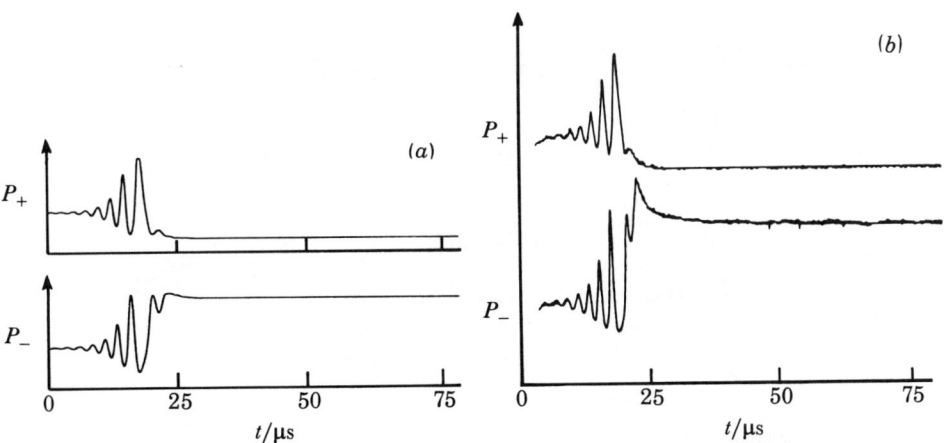

FIGURE 2. Time dependence of output intensity under conditions where the system leaves an unstable focal point and runs into the domain of attraction of a stable fixed point: (a) numerical solution; (b) experimental.

It should be noted that the dispersive terms make phase space three-dimensional; this might give rise to a much more complicated and possibly chaotic behaviour of the system. There are, however, no indications of chaos in the range of parameters explored in the experiment, neither experimentally nor in numerical solutions of the equations of motion.

Though the concept of magnetically induced optical self-pulsing has been developed in the context of optical tristability, the phenomenon can also occur in bistable systems with circularly polarized light input. In both cases an analytic treatment allowing linear stability analysis can be given and the topology of phase space can be studied in some detail.

REFERENCES

Kitano, M., Yabuzaki, T. & Ogawa, T. 1981 *Phys. Rev.* A **24**, 3156.
Mitschke, F., Deserno, R., Mlynek, J. & Lange, W. 1983 a *Optics Commun.* **46**, 135.
Mitschke, F., Mlynek, J. & Lange, W. 1983 b *Phys. Rev. Lett.* **50**, 1660.
Mlynek, J., Mitschke, F., Deserno, R. & Lange, W. 1982 *Appl. Phys.* B **28**, 135.
Mlynek, J., Mitschke, F., Köster, E. & Lange, W. 1983 In *Coherence and quantum optics* V (ed. L. Mandel & E. Wolf), pp. 1179–1186. New York: Plenum Press.
Mlynek, J., Mitschke, F., Deserno, R. & Lange, W. 1984 *Phys. Rev.* A **29**, 1297.

Phil. Trans. R. Soc. Lond. A **313**, 425–428 (1984)

Printed in Great Britain

Transient noise-induced optical bistability

By G. Broggi and L. A. Lugiato

Dipartimento di Fisica dell'Università di Milano, Via Celoria 16, 20133 Milano, Italy

We show that, as a consequence of critical slowing-down and noise, the intensity probability distribution becomes double-peaked during its approach to the single-peak steady-state distribution. Even for very low noise levels, the switching time undergoes remarkable fluctuations. On average it is shorter than predicted by the deterministic theory, hence noise counteracts, in part, critical slowing-down.

The phenomenon of critical slowing down in optical bistability (o.b.), theoretically predicted by Bonifacio & Lugiato (1976), was extensively studied theoretically (Bonifacio & Meystre 1978; Benza & Lugiato 1979) and observed experimentally (Garmire *et al.* 1979; Barbarino *et al.* 1982; Grant & Kimble 1983; Mitschke *et al.* 1983). However, no analysis of the effects of noise on critical slowing-down has been made so far, even though it is quite important for the switching behaviour of the system. So we consider the simplest model that describes amplitude fluctuations in absorptive o.b.:

$$\frac{\partial P(x, \tau)}{\partial \tau} = \left\{ \frac{\partial}{\partial x}\left(x - y + \frac{2Cx}{1 + x^2}\right) + q\frac{\partial}{\partial x^2} \right\} P(x, \tau). \tag{1}$$

In this Fokker–Planck equation $x(y)$ is the normalized amplitude of the transmitted (incident) field, $P(x, \tau)$ is the probability distribution of the variable x at time τ, where τ is normalized to the cavity build-up time, $C = \alpha L/(2T)$ is the bistability parameter, where α is the absorption coefficient per unit length, L the length of the atomic sample and T the mirror transmissivity coefficient. The form of the diffusion term corresponds to Gaussian white noise and the diffusion coefficient q measures the noise level.

We solved (1) numerically with the conditions $P(x, 0) = \delta(x)$, $C = 20$, $y = 21.04$ and several values of q. The operating value $y = 21.04$ is slightly larger than the switching-up threshold $y_M = 21.0264$ (figure 1). The three-dimensional figure 2 shows the time evolution of $P(x, \tau)$ for $q = 0.1$. Initially the distribution is single-peaked, but soon it develops a long tail and subsequently becomes double-peaked. Finally the left peak disappears and the distribution approaches the steady-state single-peaked configuration. Hence there is an observable time interval during which the probability distribution becomes double-peaked. We call this phenomenon *transient noise-induced optical bistability*.

It is well known that the steady-state intensity probability distribution is double-peaked when y is in the bistable region $y_m < y < y_M$ (Bonifacio & Lugiato 1978; Schenzle & Brand 1978). However, the observation of the steady-state bimodality for $y_m < y < y_M$ is hard because the lifetime of the two metastable states is tremendously long. Here, instead, the observation is accessible because the phenomenon arises in the transient.

We stress that the transient bimodality, which arises exclusively from noise, is a phenomenon of general type. Nicolis and colleagues (Baras *et al.* 1983; Frankowitz & Nicolis 1984) first

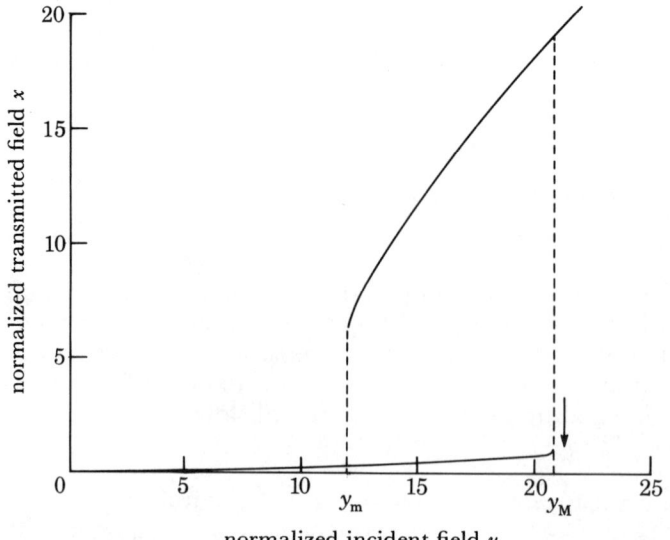

FIGURE 1. Hysteresis cycle of normalized transmitted field x as a function of normalized incident field y for $C = 20$. The arrow indicates a value of the incident field slightly larger than the switching-up threshold y_M.

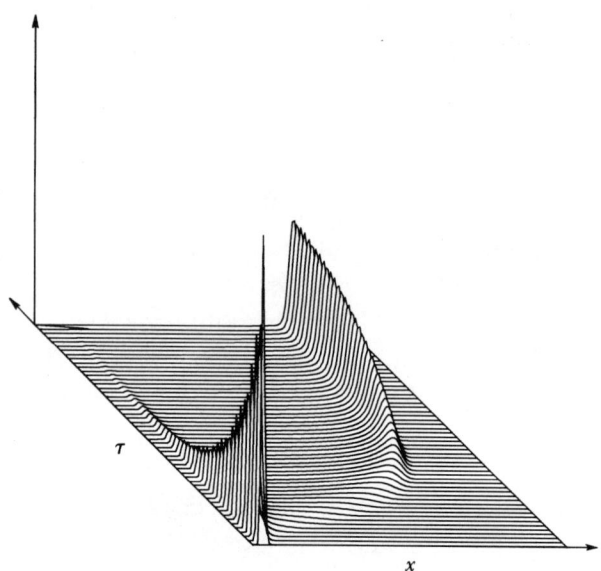

FIGURE 2. Time evolution of the probability distribution $P(x, \tau)$ for $q = 0.1$. Arbitrary units. In this, and in figures 3 and 4, $C = 20$, $y = 21.04$.

predicted it for combustion, and suggested that the same phenomenon arises whenever the solution involves a slowing-down stage followed by a rapid switching to a final single stable attractor. Hence a large variety of physical, chemical, biological, etc., systems can exhibit this phenomenon. The specific interest of o.b. in this context is that it is a promising candidate for an observation of this effect.

Another main result of our analysis is that the switching time undergoes remarkable fluctuations. As shown in figure 3, even when q is as small as 10^{-3} the switching time distribution is very broad, and the average switching time is sensibly smaller than the one predicted by the deterministic theory. This is evident from figure 4, on comparing the deterministic time

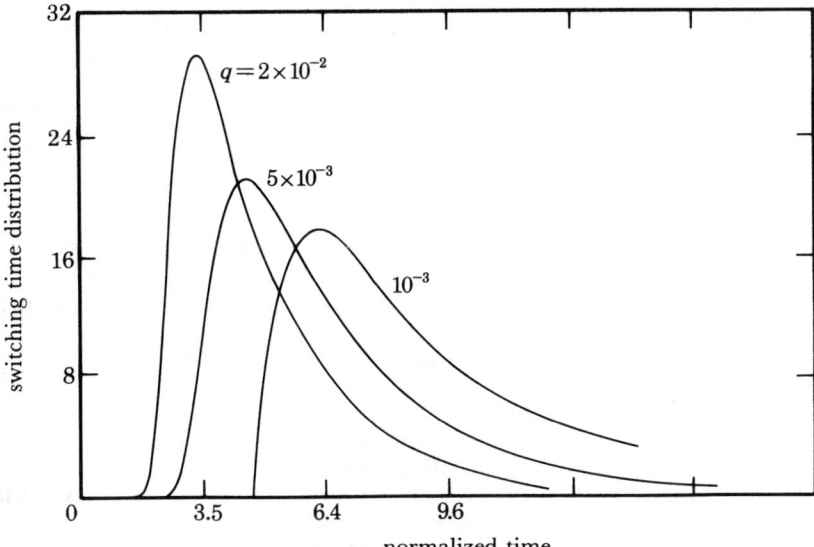

FIGURE 3. Switching time distribution for different values of q.

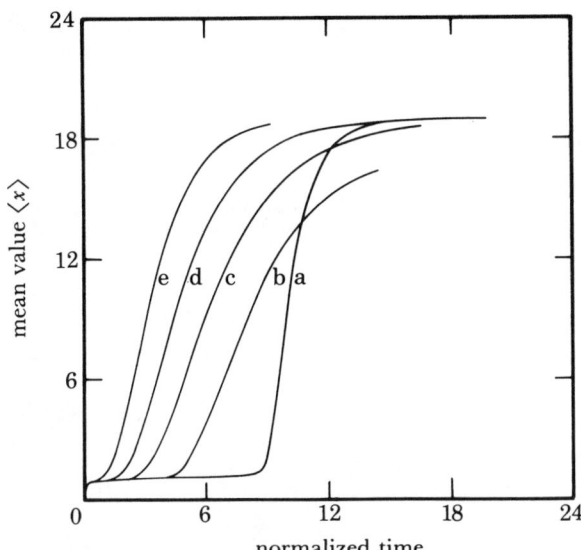

FIGURE 4. The evolution of the mean value $\langle x \rangle$ of the transmitted field when the incident field is changed stepwise from zero to the value 21.04: a, deterministic theory, $q = 0$; b, $q = 10^{-3}$; c, $q = 0.05$; d, $q = 0.02$; e, $q = 0.1$.

evolution of the mean value $\langle x \rangle(\tau)$ for different values of the noise level q. Even for $q = 10^{-3}$ the length of the horizontal critical slowing-down plateau is dramatically smaller than the one of the deterministic curve. These results suggest that the transient bistability phenomenon, as well as the presence of a broad switching time distribution, persist for $q = 10^{-4}$ and even smaller.

Our analysis shows that even very little noise produces dramatic fluctuations on the behaviour of the system, when critical slowing-down is involved. The noise effects described here seem accessible to experimental observation. Finally, since the critical slowing-down condition is so sensitive to noise, it can even be used to estimate the noise level of the system. A more detailed description of this analysis is given in Broggi & Lugiato (1984).

We are grateful Professor G. Caglioti, who triggered our collaboration, and to Professor C. E. Bottani and Professor V. Balakrishnan for helpful discussions. We thank Dr M. Beghi, who wrote the three-dimensional plotting routine for us. This work has been done in the framework of the European Joint Optical Bistability Project (EJOB) of the Commission of the European Communities.

References

Baras, F., Nicolis, G., Malek-Mansour, M. & Turner, J. W. 1983 *J. statist. Phys.* **32**, 1–23.
Barbarino, S., Gozzini, A., Maccarrone, F., Longo, I. & Stampacchia, R. 1982 *Nuovo Cim.* **71B**, 183–191.
Benza, V. & Lugiato, L. A. 1979 *Lett. Nuovo Cim.* **26**, 405–410.
Bonifacio, R., Gronchi, M. & Lugiato, L. A. 1978 *Phys. Rev.* A **18**, 2266–2277.
Bonifacio, R. & Lugiato, L. A. 1976 *Optics Commun.* **19**, 172–178.
Bonifacio, R. & Meystre, P. 1978 *Optics Commun.* **21**, 131–135.
Broggi, G. & Lugiato, L. A. 1984 *Phys. Rev.* A **29**, 2949–2952.
Frankowitz, M. & Nicolis, G. 1984 *J. statist. Phys.* (In the press.)
Garmire, E., Marburger, J. H., Allen, S. D. & Winful, H. G. 1979 *Appl. Phys. Lett.* **34**, 374–378.
Grant, D. E. & Kimble, H. J. 1983 *Optics Commun.* **44**, 415–420.
Mitschke, F., Deserno, R., Mlynek, J. & Lange, W. 1983 *Optics Commun.* **46**, 135–140.
Schenzle, A. & Brand, H. 1978 *Optics Commun.* **27**, 85–90.

Phil. Trans. R. Soc. Lond. A **313**, 429–431 (1984)

Printed in Great Britain

Solitons in optical bistability

By J. V. Moloney

Optical Sciences Center, University of Arizona, Tucson, Arizona 85721, U.S.A.

Concentric spatial rings (saturable media) or symmetrically disposed filament-like structures (Kerr media) are predicted to slowly evolve across the turned-on cylindrical spot of the two-dimensional beam profile in a bistable optical resonator.

Introduction

In this paper we report on a novel nonlinear dynamical phenomenon involving the switching of an optical beam between low and high transmission states of a bistable optical ring resonator. While the non-linear intensity-dependent phase change encoded across the beam profile may be small in a single pass, a significant cumulative phase change may occur over many resonator passes due, in particular, to a large build-up of the intracavity field. If diffractive coupling is weak, only the more intense central portion of the beam exceeds threshold for switching to the high transmission branch. The end result is an intense 'on' spot with a sharp gradient at its outer edge. Furthermore, if the driving laser frequency is tuned to the self-focusing side of the atomic transition, strong diffraction will cause spatial rings to appear on the outer edge of the 'on' spot and slowly evolve towards the centre of the beam (Moloney & Gibbs 1982). The transverse spatial rings may reach an asymptotically stable state or undergo slow periodic oscillations. At higher input driving-field amplitudes the rings may undergo a bifurcation to periodic and chaotic temporal motion while retaining spatial coherence. The evolution of spatial rings is fundamentally different for saturable and Kerr media, and if a single transverse spatial dimension is assumed, the rings can be identified with transverse solitary waves and solitons, respectively (McLaughlin *et al.* 1983). Recently, we have extended our studies to the full two-dimensional transverse-beam profile and preliminary numerical results will be reported here. Unlike conventional self-focusing experiments, the shapes (heights and widths) of the transverse spatial rings are not dependent on initial conditions, but are determined as fixed points of an infinite-dimensional map.

Theory

In the good-cavity limit, the atomic medium (two-level atom) variables may be adiabatically eliminated from the Maxwell–Bloch equations to lead to the following nonlinear evolution equation for propagation of the electromagnetic field through the medium:

$$2i\frac{\partial G_n}{\partial \zeta}+\left(\frac{\partial^2}{\partial x^2}+\frac{\partial^2}{\partial y^2}\right)G_n-\frac{G_n}{1+2|G_n|^2}=0, \tag{1}$$

and the ring resonator boundary conditions become

$$G_n(x,y,0)=a(x,y)+Re^{ikL}G_{n-1}(x,y,p), \quad G_0=0. \tag{2}$$

28-2

These equations together constitute an infinite-dimensional map in the discrete time variable n, where n counts the number of circuits of the field around the resonator. G_n is the normalized intracavity field amplitude; (x, y) and ζ refer to suitably normalized coordinates in the transverse and propagation directions, respectively. These equations are written in non-dimensional form and are discussed by Moloney & Gibbs (1982) and by McLaughlin *et al.* (1983).

Equations (1) and (2) are solved as follows. The initial input beam profile $a(x, y)$, which we assume to be gaussian, acts as initial data for the nonlinear evolution equation (1). This equation is solved over the effective nonlinear medium length p and the result substituted into (2) determines the new initial data for (1). This procedure is repeated until the system reaches an asymptotic state, which may be stable or unstable.

The fact that solitons may arise as asymptotic states of (1) is most easily seen if we assume that the laser is tuned far from any medium resonance (Kerr limit, $|G_n| \ll 1$) and if we drop one transverse dimension. In this limit (1) simplifies to

$$2\mathrm{i}\frac{\partial G_n}{\partial \zeta} + \frac{\partial^2 G_n}{\partial y^2} - (1 - 2|G_n|^2) G_n = 0, \tag{3}$$

which is the well-known nonlinear Schrödinger equation with soliton solutions

$$G_n(y, \zeta) = \lambda \operatorname{sech}(\lambda y) \exp\{\tfrac{1}{2}\mathrm{i}(\lambda^2 - 1)\zeta + \mathrm{i}\gamma\}.$$

Here λ specifies the amplitude and width $(1/\lambda)$ of the soliton and γ is its phase. The analogue of (3) for a saturable medium is non-integrable and admits solitary wave solutions.

The three-dimensional problem

We now present some preliminary results of a numerical study of (1) and (2) for a two-dimensional transverse gaussian input profile $a(x, y)$. Figure 1 shows the dynamical evolution of one quadrant of the two-dimensional beam profile $|G_n(x, y, p)|$ at every 20th circuit of the ring resonator (i.e. $n = 20, 40$, etc.). On the 20th circuit, the sharp gradient is evident at the outer edge of the cylindrical 'on' spot. By the 40th circuit, the transverse solitary waves are already well developed as outer concentric rings and are slowly evolving towards the centre of the beam. The two outer rings appear to quickly stabilize while the centre keeps oscillating. In fact, at the parameter values specified in this figure the asymptotic state appears to be a slow recurrent periodic oscillation. We predict for this saturable case that an amplitude threshold exists below which the rings are unstable and break into filaments while they remain as concentric rings at large amplitudes. Physically, this is reasonable, as large-amplitude rings saturate the nonlinearity and the system is quasi-linear, while at low amplitude the saturable nonlinearity is 'Kerr-like' and filamentation is expected to occur. Unlike the Kerr case, however, the filaments cannot critically focus because of saturation. Details of the analysis and further results will be presented by McLaughlin *et al.* (1984).

References

McLaughlin, D. W., Moloney, J. V. & Newell, A. C. 1983 *Phys. Rev. Lett.* **51**, 75.
McLaughlin, D. W., Moloney, J. V. & Newell, A. C. 1984 (In the press.)
Moloney, J. V. & Gibbs, H. M. 1982 *Phys. Rev. Lett.* **48**, 1607.

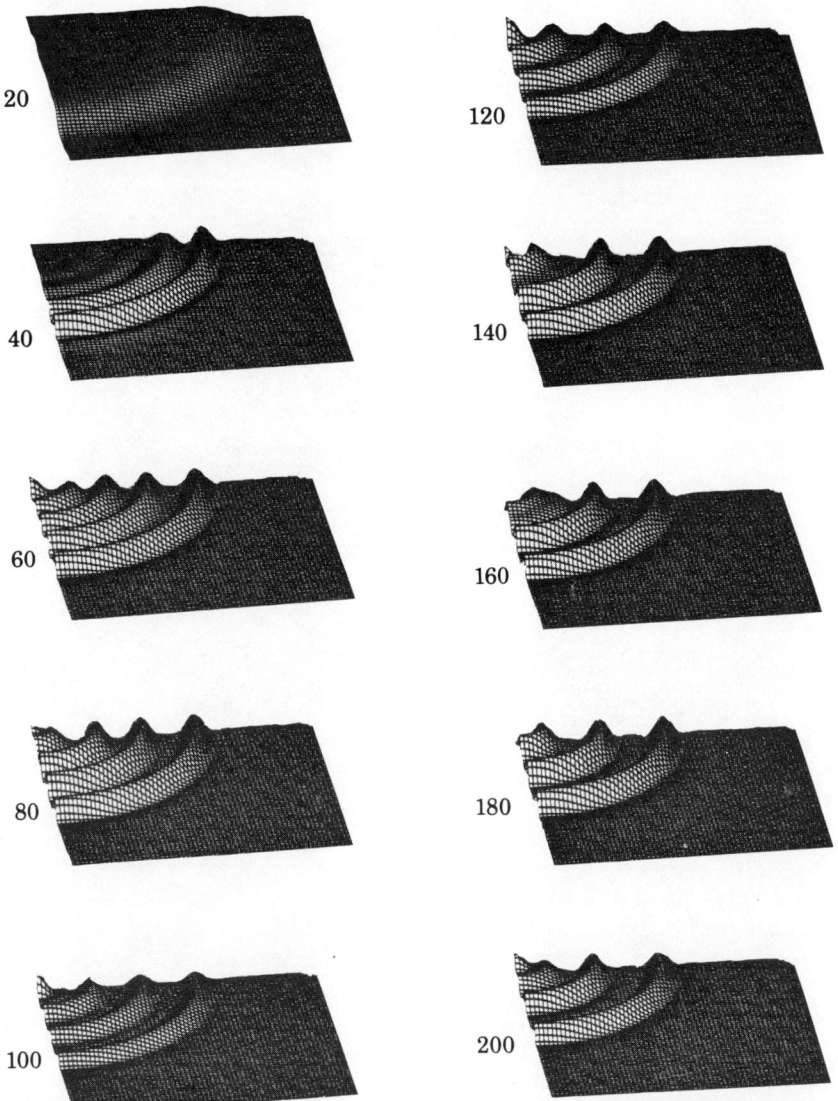

FIGURE 1. Time evolution of one quadrant of the two-dimensional intensity profile as the beam switches from a low to a high transmission state of the bistable system. The numbers indicate the resonator pass. The nonlinearity is saturable as in (1) with kL = 0.2, F = 100 and input peak intensity $|a(0,0)|^2 = 0.0375$. These parameters are explicitly defined by McLaughlin *et al.* (1983).

Phil. Trans. R. Soc. Lond. A **313**, 433–437 (1984)

Printed in Great Britain

Self-oscillation in a detuned cavity

By H. J. Carmichael

Department of Physics, University of Arkansas, Fayetteville, Arkansas 72701, *U.S.A.*

A simple symmetry relates optical bistability in a ring cavity tuned near resonance and multimode instability and self-oscillation in a cavity excited midway between resonances.

It has been predicted that the continuous wave (c.w.) output of a bistable cavity may become unstable to multimode self-oscillation. Such multimode instabilities were first discussed by Bonifacio & Lugiato (1978) for absorptive bistability in a ring cavity. Their analysis was later extended by Lugiato (1980) to dispersive bistability. Recently, much attention has been given to this subject following Ikeda's identification of an instability leading to period-doubling and chaos in dispersive bistability (Ikeda 1979; Ikeda *et al.*1980).

A detailed study of the work done by Ikeda led myself and co-workers to discover a second multimode instability for a saturable absorber in a ring cavity (Carmichael *et al.* 1982). A telling distinction exists between this and the instability studied by Bonifacio & Lugiato. They studied absorptive bistability in a resonant cavity. The instability that we have discovered occurs with the injected laser and resonant absorber tuned midway between cavity resonances. In a high-finesse cavity the Ikeda instability behaves similarly (Firth 1981; Carmichael *et al.* 1982; Bar-Joseph & Silberberg 1983). These observations provide the clue to the central result of this paper; the stability analysis for a nonlinear ring cavity exhibits a symmetry that establishes a one-to-one correspondence between optical bistability in a cavity tuned near resonance and multimode self-oscillation in a cavity tuned between resonances. The theory of absorptive and dispersive bistability can then be transferred as a whole to the description of corresponding multimode instabilities in a cavity tuned between resonances.

For simplicity I consider the plane-wave theory of absorptive bistability for a two-level homogeneously broadened medium in a ring cavity, and give detailed results only for the mean-field limit. My central conclusions are, however, quite general. They hold for dispersive bistability, for a gaussian-mode theory, and beyond the mean-field limit. On the other hand, they do not hold (at least not without qualification) in a standing-wave cavity, although it must be recognized that multimode instabilities have been predicted there also (Casagrande *et al.* 1980; Firth 1981).

The general stability analysis for a ring cavity containing a two-level homogeneously broadened absorber gives the following characteristic equation for eigenvalues λ governing the linearized dynamics (Carmichael 1983):

$$1 + R^2 e^{-2\lambda\tau} \left[\frac{E(L)}{E(0)}\right]^{2/(1+\lambda T_2)} \left[\frac{(1+\lambda T_1)(1+\lambda T_2) + E(0)^2}{(1+\lambda T_1)(1+\lambda T_2) + E(L)^2}\right]^{\frac{1}{2}(2+\lambda T_2)/(1+\lambda T_2)}$$
$$-R e^{-\lambda\tau} \left[\frac{E(L)}{E(0)}\right]^{1/(1+\lambda T_2)} \left\{1 + \left[\frac{(1+\lambda T_1)(1+\lambda T_2) + E(0)^2}{(1+\lambda T_1)(1+\lambda T_2) + E(L)^2}\right]^{\frac{1}{2}(2+\lambda T_2)/(1+\lambda T_2)}\right\} \cos\theta = 0. \quad (1)$$

Here R is the mirror reflection coefficient, T_1 and T_2 are atomic relaxation times, τ is the cavity round-trip time, θ is the cavity detuning $(-\pi \leqslant \theta < \pi)$ and $E(0)$ and $E(L)$ are dimensionless field amplitudes at either end of the medium. Solutions to (1) with $\mathrm{Re}\,(\lambda) = 0$ define instability boundaries where a stable mode, $\mathrm{Re}\,(\lambda) < 0$, becomes unstable, $\mathrm{Re}\,(\lambda) > 0$, as system parameters are varied. Switching points in absorptive bistability $(\theta = 0)$ are defined by the requirement

$$1 + R^2 \frac{E(L)^2}{E(0)^2} \frac{1 + E(0)^2}{1 + E(L)^2} - R \frac{E(L)}{E(0)} \left[1 + \frac{1 + E(0)^2}{1 + E(L)^2} \right] = 0, \tag{2}$$

where, if (2) is satisfied, (1) has a solution $\lambda = 0$, indicating marginal stability for the resonant cavity mode. In the limit $\lambda T_1 \to 0$, $\lambda T_2 \to 0$, (1) is a function of $\exp(\lambda\tau)$ alone and all the cavity modes become unstable at the bistable switching points, i.e. when (2) is satisfied, (1) has solutions $\lambda_n = in\pi/\tau$, $n = 0, \pm 2, \pm 4, \ldots$, where $\mathrm{Re}\,(\lambda_n) = 0$ for every n and $\mathrm{Im}\,(\lambda_n)$ identifies the cavity mode frequencies measured with respect to the *resonant* laser frequency (see figure 1). Observe now, that, with $\lambda T_1 = \lambda T_2 = 0$, (1) is invariant under the transformation $\lambda \to \lambda + i\pi/\tau$, $\theta \to \theta + \pi$. It follows that all cavity modes become unstable at the same instability boundaries (defined by (2)) in a cavity tuned midway between resonances $(\theta = -\pi)$. Now (1) has solutions $\lambda_n = in\pi$, $n = \pm 1, 3, \ldots$, where $\mathrm{Im}\,(\lambda_n)$ identifies cavity mode frequencies measured with respect to the *detuned* laser frequency (see figure 1).

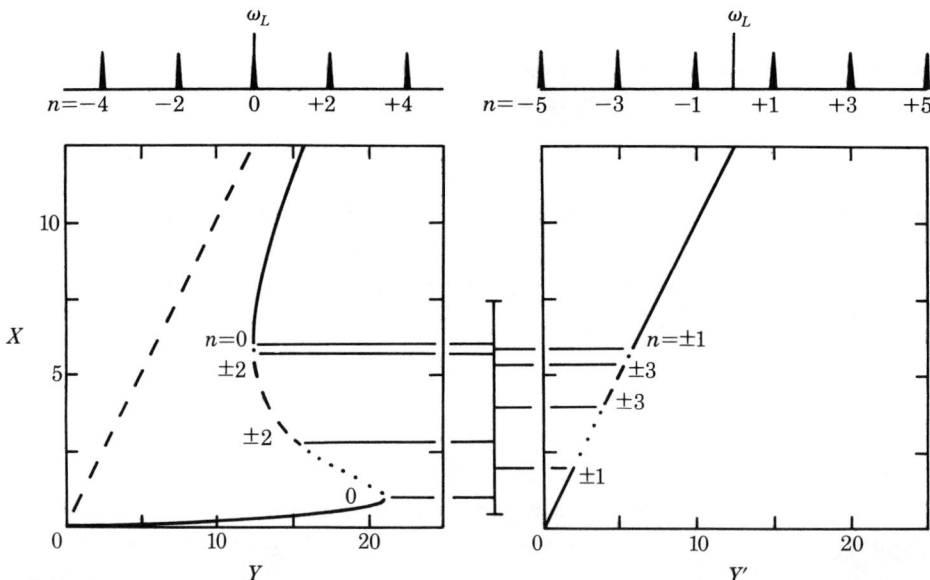

Figure 1. Multimode instabilities for $C = 20$, $T_1/\tau = 1.0$ and $T_1/T_2 \gg 1$. To the left is a plot of the state equation for absorptive bistability $(\theta = 0)$ and to the right the state equation for the corresponding detuned system $(\theta = -\pi)$. In dotted regions one mode is unstable and in regions of broken lines two modes are unstable.

An exact correspondence between the range of bistability $(\theta = 0)$ and the range of instability in a detuned cavity $(\theta = -\pi)$ exists only for $T_1/\tau \to 0$, $T_2/\tau \to 0$. I will illustrate the situation for finite T_1/τ and T_2/τ with explicit results for the mean-field limit $(1 - R) \ll 1$, $\alpha L \ll 1$, with $C = \alpha L/4(1 - R)$, where α is the resonant absorption coefficient. With $E(L) = X$ and $E(0) = X[1 + (1 - R)\,2C/(1 + X^2)]$, perturbative solutions to (1) yield

$$\mathrm{Re}\,(\lambda_n) = -\tau^{-1}(1 - R)\,\mathrm{Re}\left[1 + \frac{2C}{1 + X^2} \frac{1 - X^2 + in\pi\,T_1/\tau}{(1 + in\pi\,T_1/\tau)\,(1 + in\pi\,T_2/\tau) + X^2} \right], \tag{3}$$

with $n = 0, \pm 2, \pm 4, \ldots$ for $\theta = 0$, and $n = \pm 1, 3, \ldots$ for $\theta = -\pi$. In figure 2 the boundaries of instability $\mathrm{Re}\,(\lambda_n) = 0$ are plotted as a function of T_1/τ and X for $n = 0, \pm 2, \pm 4$ in absorptive bistability, and $n = \pm 1, \pm 3, \pm 5$ in the corresponding detuned cavity. The range of X between the vertical lines labelled $n = 0$ is the range of the negative slope branch in the bistable system. Here the resonant mode is unstable for all T_1/τ. The non-resonant modes are unstable whenever T_1/τ and X define a point lying under, or inside, the plotted curves. In figure 2 (a) the horizontal bar follows successive changes of stability as a function of X in a system with $T_1/\tau = 1.0$, $T_1/T_2 \gg 1$. In figure 1 these changes of stability are displayed on the respective steady-state curves:

$$Y = X[1 + 2C/(1 + X^2)] \tag{4}$$

for $\theta = 0$, and

$$Y' = X \tag{5}$$

for $\theta = -\pi$. Here Y and Y' are dimensionless input field amplitudes, with $Y = (1 - R)^{-1}\,(2\mu/\hbar)\,(T_1\,T_2)^{\frac{1}{2}} E_i$ and $Y' = \frac{1}{2}(2\mu/\hbar)\,(T_1\,T_2)^{\frac{1}{2}} E_i$, where μ is the atomic dipole moment.

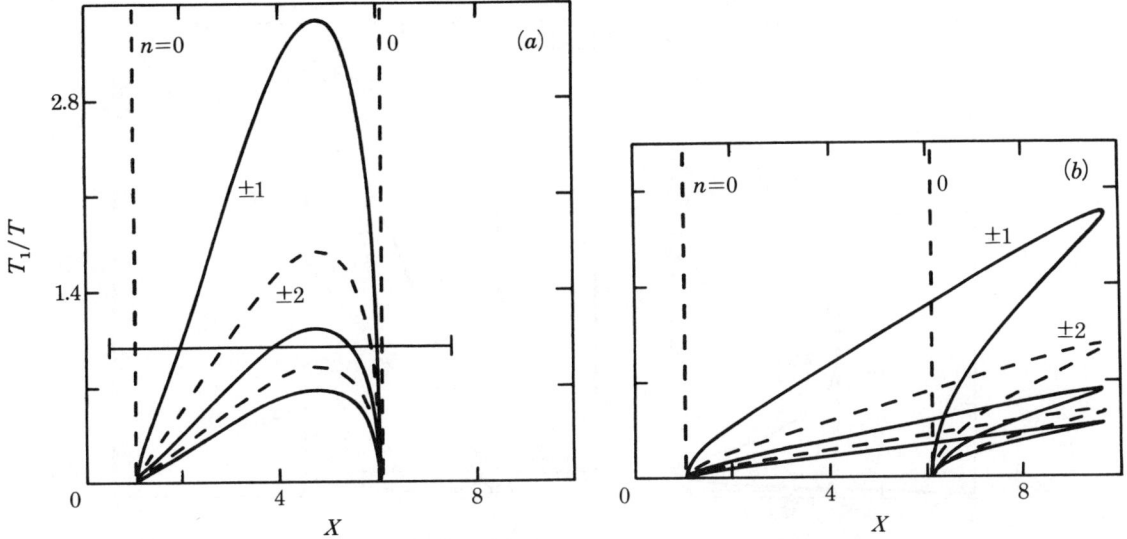

FIGURE 2. Instability boundaries for $C = 20$ and (a) $T_1/T_2 \gg 1$; (b) $T_1/T_2 = 0.5$. Broken curves are for $\theta = 0$ and solid curves are for $\theta = -\pi$.

In figure 2b the stability boundaries are distorted so that they extend outside the range of the resonant mode instability. This is the basis for self-pulsing instability along the upper branch in absorptive bistability, as studied by Bonifacio & Lugiato (1978). The only significance it holds in the corresponding detuned cavity is that the $n = \pm 3$ modes, for example, might be unstable, while the $n = \pm 1$ modes remain stable. This crossing of the instability boundaries does not occur if $T_1/T_2 \gtrsim C$ ($C \gg 1$), as in figure 2a. It is also eliminated in a gaussian-mode theory (Lugiato & Milani 1983).

If we return to the limit $T_1/\tau \to 0$, $T_2/\tau \to 0$, the relation between bistability and self-oscillation in a detuned cavity becomes even closer when we consider the form that this oscillation takes. In this limit cavity dynamics can be modelled by a nonlinear map (as in Ikeda (1979)). For the mean-field limit and $\theta = -\pi$.

$$X_{n+1} = 2Y' - X_n\{1 - (1 - R)\,[1 + 2C/(1 + X_n^2)]\}. \tag{6}$$

The fixed point $X = Y'$ is unstable for $|dX_{n+1}/dX_n| > 1$. This is equivalent to the condition $dY/dX < 0$ obtained from (4), from which my central conclusion again follows. Now if we look for a two-cycle, an oscillation between X_1 and X_2, to replace the unstable fixed point, this requires

$$X_1 + X_2 = 2Y',$$

$$X_1\left(1 + \frac{2C}{1 + X_1^2}\right) = X_2\left(1 + \frac{2C}{1 + X_2^2}\right). \tag{7}$$

It follows that X_1 and X_2 are a pair of states satisfying the state equation (4) for some Y, a function of Y'. Figure 3 shows these oscillatory states in detail. For $X_a \leqslant Y' \leqslant Y'_b = \frac{1}{2}(X_b + \check{X}_b)$ there is a stable oscillation between the middle and lower branches of the corresponding bistability curve. For $Y'_b \leqslant Y' < Y'_a \simeq \frac{1}{2}(X_a + \check{X}_a)$ there is a stable oscillation between the upper and lower branches of the corresponding bistability curve. For $X_b < Y' < Y'_a$ there is an unstable oscillation between the upper and middle branches of the corresponding bistability curve.

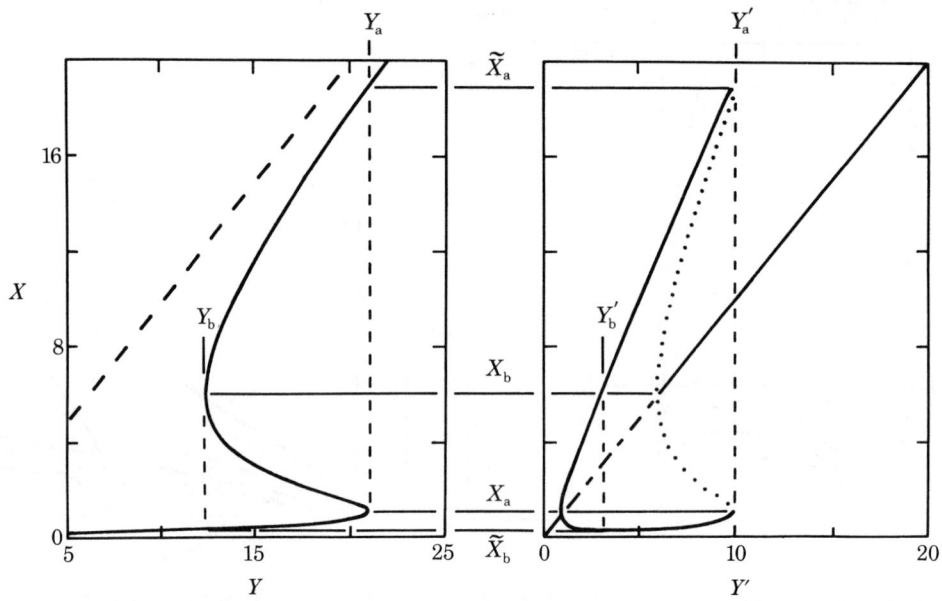

FIGURE 3. The relation between oscillation in a detuned cavity ($\theta = -\pi$), to the right, and the steady states of the corresponding bistable system ($\theta = 0$), to the left, for $C = 20$. Fixed points along the broken portion of $X = Y'$ are unstable. The solid branches bifurcating from $Y' = X_a$ are the states X_1 and X_2 of stable two-cycles. The dotted branches bifurcating from $Y' = X_b$ correspond to unstable two-cycles.

To summarize, for every example of bistability in a nonlinear ring cavity tuned near resonance, there exists a corresponding multimode instability leading to self-oscillation between states of the bistability curve, in a cavity tuned between resonances. This correspondence is one-to-one in the limit $T_1/\tau \to 0$, $T_2/\tau \to 0$, and identifies a range of multimode instability that extends to finite T_1/τ, T_2/τ.

References

Bar-Joseph, I. & Silberberg, Y. 1983 *Optics Commun.* **48**, 53–56.
Bonifacio, R. & Lugiato, L. A. 1978 *Lett. nuovo Cim.* **21**, 510–516.
Carmichael, H. J. 1983 *Lecture Notes Phys.* **182**, 64–87.
Carmichael, H. J., Snapp, R. R. & Schieve, W. C. 1982 *Phys. Rev.* A **26**, 3408–3422.
Casagrande, F., Lugiato, L. A. & Asquini, M. L. 1980 *Optics Commun.* **32**, 492–496.
Firth, W. J. 1981 *Optics Commun.* **39**, 343–346.
Ikeda, K. 1979 *Optics Commun.* **30**, 257–261.
Ikeda, K., Daido, H. & Akimoto, O. 1980 *Phys. Rev. Lett.* **45**, 709–712.
Lugiato, L. A. 1980 *Optics Commun.* **33**, 108–112.
Lugiato, L. A., Asquini, M. L. & Narducci, L. M. 1982 *Optics Commun.* **41**, 450–454.
Lugiato, L. A. & Milani, M. 1983 *Z. Phys.* B **50**, 171–179.

Phil. Trans. R. Soc. Lond. A **313**, 439–443 (1984)

Printed in Great Britain

Quantum statistical properties of a quantum theory of optical bistability

By G. P. Hildred, S. S. Hassan†, R. R. Puri and R. K. Bullough

*Department of Mathematics, University of Manchester Institute of Science and Technology,
P.O. Box 88, Manchester M60 1QD, U.K.*

The strictly quantum driven Dicke model of N two-level atoms on the same site shows conventional optical bistability if and only if cavity feedback is included. In this model we find that for $N \to \infty$ the photon statistics of the transmitted field are Poisson on the lower branch of the output–input curve, but on the upper branch approach Bose–Einstein in the hysteresis region, with reversion to Poisson only for very strong input fields.

The device potential of optical bistability (o.b.) is now so plain that it may be salutary to recall that the phenomenon also has fundamental interest. Within a semiclassical (decorrelation) approximation it can be exhibited as a fine example of a phase transition far from equilibrium (see, for example, Agarwal *et al.* 1978). So studies of any quantum features are important and one way of studying them is through the photon statistics of the different branches of the transmitted field. Unfortunately even the model problem of the o.b. of two-level atoms in an extended cavity is not solved as a quantum theory. So nothing can be said about the photon statistics.

However we have solved the quantum problem of the steady state of N two-level atoms on a single site (a Dicke model) driven by a coherent state continuous wave (c.w.) laser field completely (see, for example, Hassan *et al.* 1984). We have calculated the intensity correlations of the fluorescent radiation $g^{(n)}(0) \equiv \langle S_+^n S_-^n \rangle / \{\langle S_+ S_- \rangle\}^n$, where $S_\pm$ are collective atomic operators: $g^{(n)}(0) = 1$ below a certain threshold but, for example, $g^{(2)}(0) \to 1.2$ (when $N \to \infty$) above it. So there is a second-order type phase transition from coherence to partial coherence at the threshold (cf. Hassan *et al.* 1980, 1984 and references therein). Although cavity feedback is no longer viewed as essential for o.b. (see, for example, the papers by Garmire *et al.* (this symposium) this quantum model shows conventional o.b. if, but only if, the model is coupled to a cavity (Puri *et al.* 1984). Moreover the quantum character is essential; the o.b. disappears if the theory is decorrelated. Figure 1 illustrates the results: as $N \to \infty$ a cusp develops at the switch-up point ($x = 1$) and the second-order type phase transition occurs there.

Although the Dicke model seems unphysical, results for the model coupled to a single cavity mode of black body radiation are in remarkable agreement with recent observations on high rydberg atoms (Raimond *et al.* 1982*a, b*; Hildred *et al.* 1984). This suggests similar agreement will be possible for coherent light. A way of investigating the o.b. (rather than the second-order phase transition) is through the statistics of the transmitted light. This paper sketches the theory. The main result is that for $N \to \infty$ the photon statistics are rigorously Poisson on the lower branch, but on the upper branch quantum fluctuations persist and in the hysteresis region the statistics approach Bose–Einstein.

† On leave from Department of Applied Mathematics, Ain Shams University, Cairo, Egypt.

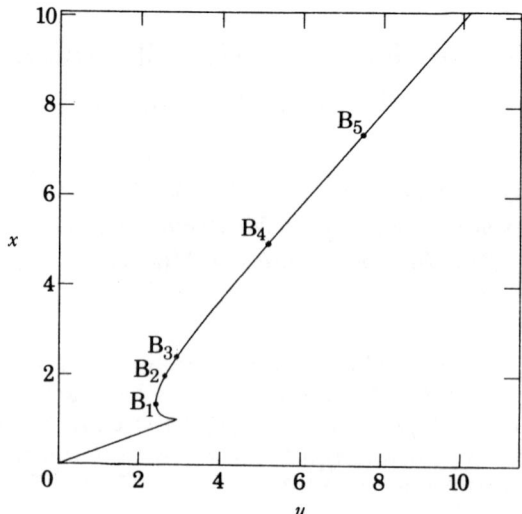

FIGURE 1. Plot of the transmitted field x against the incident field y. The points B_1 to B_5 on the upper branch refer to figure 2.

The density operator ρ for the whole system satisfies

$$d\rho/dt = -i\{(H_0 + H_1 + H_2), \rho(t)\} + \Lambda_a \rho(t) + \Lambda_f \rho(t), \tag{1}$$

$$H_0 = \omega_s S_z + \omega_s a^\dagger a,$$

$$H_1 = -iga S_+ + iga^\dagger S_-,$$

$$H_2 = i\kappa(a^\dagger E(t) - aE^*(t))$$

(Puri *et al.* 1984). The collective N atom operators $S_\pm$, S_z satisfy angular momentum commutation relations, $E(t)$ is the external field, $a(a^\dagger)$ are single-resonant-mode cavity field-operators, the atomic resonance is ω_s, g is a coupling constant, $\Lambda_a \rho$ and $\Lambda_f \rho$ describe collective spontaneous emission and cavity damping respectively. Details are in Puri *et al.* (1984). For large enough N we find $y \equiv 2|E|g/\gamma_0 N$ (γ_0 is the A-coefficient) and $x \equiv 2|\alpha|g/\gamma_0 N$ ($\alpha \equiv \langle a \rangle$) satisfy the input–output relation

$$y = (1+C)x + O(N^{-1}), \qquad x \leqslant 1$$

$$= x[1 + C\{1 - (x^2-1)^{\frac{1}{2}}/x^2 \arcsin(x^{-1})\}] + O(N^{-1}), \quad x \geqslant 1. \tag{2}$$

Figure 1 is a plot of output x against input y for $N \to \infty$ and $C = 2$; $C \equiv g^2 \kappa^{-1}/\gamma_0$ is a co-operation number and, since $C = Ng^2\kappa^{-1}/N\gamma_0$, collective $N\gamma_0$ replaces γ_0. Note the cusp at switch-up ($x = 1$) and that there is true bistability since the negative slope region is unstable. To reach (2) we 'decorrelate' matter and field operators in the equation of motion for atomic operators. This should be good for $N \to \infty$ (see Puri *et al.* 1984 and references therein).

Despite the interest of super-radiant emission in the approach to the steady state (Raimond *et al.* 1982 *a*, *b*; Hildred *et al.* 1984), here we calculate only $P_n(\infty)$, the probability of finding n photons in the cavity mode (output x) in the steady state $t \to \infty$. Since

$$P_n(\infty) = \mathrm{Tr}\, \rho(\infty)|n\rangle\langle n| \equiv \langle |n\rangle\langle n|\rangle$$

and

$$|n\rangle\langle n| = (n!)^{-1} \sum_{r=0}^{\infty} (-1)^r (r!)^{-1} (a^{\dagger})^{r+n} (a)^{r+n}$$

(compare with Sarkar & Elgin 1984)

$$P_n(\infty) = (n!)^{-1} \sum_{r=0}^{\infty} (-1)^r (r!)^{-1} \langle (a^{\dagger})^{r+n} (a)^{r+n} \rangle. \tag{3}$$

Since, in the steady state, $a = E + g\kappa^{-1}S_-$, $P_n(\infty)$ can be expressed in terms of $(|E|^2)^r$ and $\langle (S_+)^j (S_-)^l \rangle$. The latter have been calculated in the steady state already (Hassan et al. 1980 and references therein). So after some manipulation we can reach a form

$$P_n(\infty) = (|E|^2/n!)^n \sum_{r=0}^{\infty} \{(-1)^r/r!\}|E|^{2r} \sum_{v,w=0}^{r+n} \binom{r+n}{v}\binom{r+n}{w}\{\dots\}_{v,w}, \tag{4a}$$

where

$$\binom{r+n}{v} \equiv (r+n)!/v!(r+n-v)!$$

and

$$\{\dots\}_{v,w} \equiv \left[\frac{C}{\{1+C(1-(N+1)/D)\}}\right]^{v+w} D^{-1} \sum_{m=\max(v,w)}^{N} \left|\frac{g\alpha}{\gamma_0}\right|^{2(N-m)} H_{N,m}. \tag{4b}$$

This form is exact for $N \to \infty$ (given the decorrelation mentioned). Definitions and manipulative details can be deduced from Puri et al. (1984) and they will be published at length elsewhere. Here we shall use the results

$$(N+1)/D = \begin{cases} 0, & x \leqslant 1, \\ (x^2-1)^{\frac{1}{2}}/x^2 \arcsin(x^{-1}), & x \geqslant 1; \end{cases}$$

$$D^{-1} \sum_{m=\max(v,w)}^{N} (\tfrac{1}{2}Nx)^{2(N-m)} H_{N,m} = \begin{cases} 1, & x \leqslant 1 \\ \{1-(x^2-1)^{\frac{1}{2}}/x^2 \arcsin(x^{-1})\}S, & x \geqslant 1, \end{cases}$$

with

$$S = \sum_{m=0}^{\max(v-1,w-1)} (2/x)^{2m}\{(m!)^2/(2m+1)!\}, \quad x \geqslant 1.$$

The latter form comes from Drummond (1980) and is strictly valid only for $v, w \ll \infty$.

This way we find on the lower branch, $x \leqslant 1$, that

$$P_n(\infty) = (n!)^{-1}(|E|/(1+C))^{2n} \exp[-\{|E|/(1+C)\}^2]. \tag{5}$$

This is a Poisson distribution with mean $\bar{n} = \{|E|/(1+C)\}^2$, and the output x is coherent. Note that $|E| = (N\gamma_0 g^{-1})(\tfrac{1}{2}y)$: formally, as $N \to \infty$, $|E| \to \infty$ so that y remains finite; this is a 'thermodynamic limit' needed to establish an actual phase transition (characterized by (2) for $N \to \infty$) at $x = 1$ (Hassan et al. 1980). Alternatively we can consider the limit $N \to \infty$ so that $Ng^2\kappa^{-1}$, $N\gamma_0$, C and $|E|$ all stay finite. In practice we can work with N small (say $N \approx 50$) without changing the o.b. curve (2) very much (cf. Puri et al. 1984). In the experiments on Rydberg atoms (Raimond et al. 1982a) $g \approx 3 \times 10^5$ Hz, $\gamma_0 \approx 50$ Hz, so $N\gamma_0 g^{-1} \approx 1$ even for $N \approx 10^4$.

On the upper branch

$$P_n(\infty) = |E|^{2n}(n!)^{-1} \sum_{r=0}^{\infty} (-1)^r (r!)^{-1} |E|^{2r} \sum_{v,w=0}^{r+n} \binom{r+n}{v}\binom{r+n}{w}\{\dots\}, \tag{6}$$

[251]

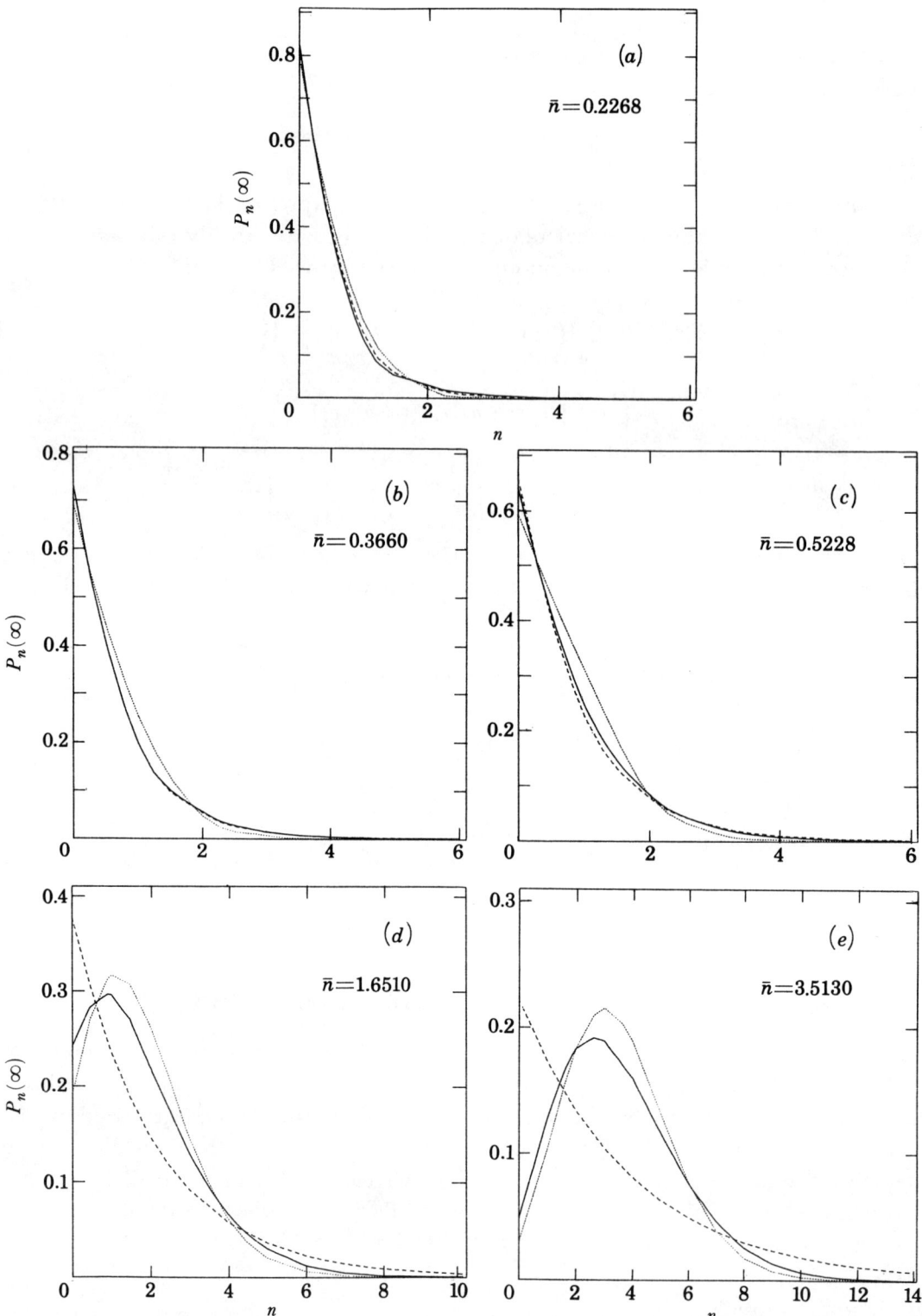

FIGURE 2. Plots of $P_n(\infty)$ (solid lines) against n for $C = 2$ corresponding, (a)–(e), to the points B_1 to B_5 on the upper branch of figure 1, respectively. The broken and dotted lines represent the Bose–Einstein and Poisson distributions for the same $\bar{n}$, respectively.

where
$$\{\ldots\} \equiv [\![-C/[1 + C\{1 - \sqrt{x^2 - 1}/x^2 \arcsin(x^{-1})\}]]\!]^{v+w}[\ldots] \tag{7a}$$

and
$$[\ldots] \equiv [1 - \{(x^2 - 1)^{\frac{1}{2}}/x^2 \arcsin(x^{-1})\} \sum_{m=0}^{\max(v-1,\, w-1)} (2/x)^{2m} m!/(2m+1)!]. \tag{7b}$$

This has no simple analytical form, so it is plotted for $C = 2$ and $N\gamma_0 g^{-1} = 0.5$ in figure $2a$–e for the points $B_1(y = 2.439)$, $B_2(y = 2.632)$, $B_3(y = 3.0)$, $B_4(y = 5.15)$ and $B_5(y = 7.503)$ on the upper branch of the o.b. curve (figure 1). The solid lines are $P_n(\infty)$ (from (6)) against n; the Poisson (dotted lines) and Bose–Einstein (broken lines) distributions for the same $\bar{n}$ are also plotted. At B_1 (switch-down) $P_n(\infty)$ is evidently near Bose–Einstein, remains so beyond B_3 (switch-up), but tends to Poisson for large input fields. The calculated $g^{(2)}(0)$ are 2.508, 1.869, 1.598, 1.196 and 1.093, respectively: the first is particularly interesting because it exceeds 2.0 (the Bose–Einstein value). For $C = 0.1$ and $N\gamma_0 g^{-1} = 1.0$ we find $g^{(2)}(0) = 1.0012$, 1.0018, 1.0023, 1.0002 and 1.0001, so the low-cooperation number means the curves coincide with a Poisson distribution everywhere. Increasing C has an opposite effect.

These results give a measure of quantum fluctuations in the optically bistable region. The recent experiments on Rydberg atoms suggest they might be observable. It is open what significance such quantum fluctuations will have to real o.b. devices. In particular, their role in bifurcation to turbulence has also still be be determined, but they would presumably induce chaos earlier in any period multiplication sequence.

REFERENCES

Agarwal, G. S., Narducci, L. M., Feng, D. M. & Gilmore, R. 1978 In *Coherence and quantum optics IV* (ed. L. Mandel & E. Wolf), pp. 281–292. New York: Plenum.

Drummond, P. D. 1980 *Phys. Rev.* A **22**, 1179.

Hassan, S. S., Bullough, R. K., Puri, R. R. & Lawande, S. V. 1980 *Physica* **103**A, 213.

Hassan, S. S., Hildred, G. P., Puri, R. R. & Bullough, R. K. 1984 In *Coherence and quantum optics V* (ed. L. Mandel & E. Wolf). New York: Plenum. (In the press.)

Hildred, G. P., Puri, R. R., Hassan, S. S. & Bullough, R. K. 1984 Submitted to *J. Phys.* B. *Lett.*

Puri, R. R., Hildred, G. P., Hassan, S. S. & Bullough, R. K. 1984 In *Optical bistability 2* (ed. H. M. Gibbs & C. M. Bowden). New York: Plenum. (In the press.)

Raimond, J. M., Goy, P., Gross, M., Fabre, C. & Haroche, S. 1982a *Phys. Rev. Lett.* **49**, 117.

Raimond, J. M., Goy, P., Gross, M., Fabre, C. & Haroche, S. 1982b *Phys. Rev. Lett.* **49**, 1924.

Sarkar, S. & Elgin, J. N. 1984 In *Coherence and quantum optics V* (ed. L. Mandel & E. Wolf). New York: Plenum. (In the press).

Phil. Trans. R. Soc. Lond. A **313**, 445–448 (1984)

Printed in Great Britain

Optical multistability and Zeeman degenerate transitions

By R. J. Ballagh and V. Jain

Physics Department, University of Otago, Dunedin, New Zealand

A variety of system behaviours appears possible when $J_1 = 1 \leftrightarrow J_u = 0$ atoms interact with σ^+ and σ^- radiation modes in a cavity. Certain choices of the two ground-level collision rates allow asymmetry to develop between the output σ^+ and σ^- fields, as is also found for $J = \frac{1}{2} \leftrightarrow J = \frac{1}{2}$ atoms. However, this possibility is excluded when physical requirements are made on the rates.

Introduction

One expects that increasing the complexity of the model atom in an optical cavity will result in system behaviour more complex than the simple bistable behaviour associated with a two-state atom. A three-state (lambda) model atom, for example, can mediate a competitive interaction between σ^+ and σ^- radiation modes, as shown by Kitano *et al.* (1981). Subsequent improvements in the treatment of this model (see, for example, Savage *et al.* 1982; Arecchi *et al.* 1983) have resulted in the prediction of a variety of interesting behaviours. Real atomic transitions have at least four states, which are coupled by radiation and collisions in gas phase. Thus the validity of simple atom models in optical bistability, especially in description of experiment, must be questioned. It has been shown (Ballagh *et al.* 1981) that if the cavity radiation is constrained to a pure polarization, a limited number of dipole transitions reproduce the 'two-state' form of macroscopic dipole. Saturation fields, though, are altered by the effect of optical pumping into radiatively inactive states.

When the system is free to determine its own polarization, its behaviour becomes more dependent on the choice of atom. The $J = \frac{1}{2} \leftrightarrow J = \frac{1}{2}$ atom in a ring cavity has been studied in detail by Hamilton *et al.* (1982), who confirm that an input field with equal amplitudes of σ^+ and σ^- radiation may produce an output in which one of the modes is dominant. The $J = \frac{1}{2}$ atom, however, does not develop ground level coherence and thus it is not an accurate realization of the lambda model atom. The simplest candidate for 'lambda' behaviour is an atom with lower level angular momentum $J_1 = 1$ and upper level angular momentum $J_u = 0$. In this paper we examine the system behaviour of such atoms interacting resonantly (absorptively) in a ring cavity with σ^+ and σ^- radiation modes.

Mean field solution for $J_1 = 1 \leftrightarrow J_u = 0$ atoms in a ring cavity

The real amplitudes $X_\pm$ of the $\sigma^\pm$ plane wave cavity fields are written in units of a saturation field $\hbar\{3\gamma\Gamma_1(\mathrm{lu})\}^{\frac{1}{2}}/d$. Here γ is the spontaneous decay rate, $\Gamma_1(\mathrm{lu})$ is the decay rate of the optical dipole coherence and d is the reduced dipole matrix element. In an irreducible representation (Omont 1977) the relevant atomic density matrix elements evolve according to a set of ten

29-2

coupled equations and give rise to an absorption $\frac{1}{2}\alpha\eta_{\pm}$ for $X_{\pm}$. Here α is the weak field absorption coefficient,

$$\eta_{\pm} = \{1+\beta_1(X_+^2+X_-^2)\pm(\beta_2-\beta_1)(X_+^2-X_-^2)\}/D, \tag{1}$$

$$D = \{1+\beta_1(X_+^2+X_-^2)\}\{1+\tfrac{4}{3}(2+\beta_2)(X_+^2+X_-^2)\}+\tfrac{1}{3}(8+\beta_2)(\beta_2-\beta_1)(X_+^2-X_-^2)^2, \tag{2}$$

and the ground level collisional relaxation rates, Γ_1 for orientation ($K=1$ multipole) and Γ_2 for alignment ($K=2$), appear in β_1 (equals γ/Γ_1) and β_2 (equals γ/Γ_2). The steady-state mean field solutions for the ring cavity system obey the coupled equations

$$Y_{\pm} = X_{\pm}(1+2C\eta_{\pm}), \tag{3}$$

where $Y_{\pm}$ are the fields incident upon the cavity (scaled by mirror transmittance) and C is the usual cooperativity parameter. By specifying the input polarization $\xi = Y_+/Y_-$ and eliminating Y_+, Y_- from (3), a fifth-order polynomial in X_+ (with coefficients in X_-) is obtained. The solution of the polynomial leads to system curves (Y_+, X_+, X_-), which can be mapped as a projection on the (Y_+, X_+) (or X_+, X_-) plane.

LINEAR INPUT POLARIZATION

The most interesting behaviour occurs when $\xi = 1$, so that neither input mode is initially favoured. Subtracting the equations in (3) leads, with the transformation $u = X_+^2 + X_-^2$ and $v = 2X_+X_-$, to the state equation

$$(X_- - X_+)\,[\beta_2\{\beta_1 + \tfrac{1}{3}(8+\beta_2)\}\,u^2 + \{\beta_1 + \tfrac{1}{3}(8+4\beta_2) + 2C\}\,u + (1+2C)$$
$$+ 2C(\beta_2-\beta_1)\,v - \tfrac{1}{3}(8+\beta_2)(\beta_2-\beta_1)\,v^2] = 0. \tag{4}$$

A symmetric output ($X_+ = X_-$) is always present, which gives the familiar optical bistability state equation $2^{\frac{1}{2}}Y_+ = u^{\frac{1}{2}}[1+2C\{1+\tfrac{4}{3}(2+\beta_2)\,u\}^{-1}]$, where the factor $\tfrac{4}{3}(2+\beta_2)$ shows the effect of optical pumping into the $m=0$ ground state.

An asymmetric branch ($X_+ \neq X_-$) may also appear, and is described by the u, v polynomial in (4). The coefficients of u and u^2 are positive, hence only one real solution for u and only one asymmetric branch (but with degeneracy $X_+ \leftrightarrow X_-$) exist. This branch forms a simple closed loop in (Y_+, X_+, X_-) space and crosses the symmetric branch at two bifurcation points (B_1 and B_2 in figure 1) found by setting $u=v$. The existence condition for the asymmetric branch is that these bifurcation points be real and positive and from the quadratic in u((4) with $v=u$) we get the requirement

$$\{\beta_1(1-2C) + 4\beta_2(\tfrac{1}{3}+C) + \tfrac{8}{3}\}^2 > \tfrac{16}{3}\beta_1(2+\beta_2)(1+2C) \tag{5}$$

or less strictly, but necessarily, $C > \tfrac{1}{2}$ and $\beta_1 > 2\beta_2$. Unstable parts of the curve are indicated by broken lines, and on the symmetric branch occur between turning points and between bifurcation points. In the régime $\beta_2 \ll 1 \ll \beta_1$ (figure 1a) the system's initially symmetric output switches abruptly at B_1 to asymmetric output ($X_+ > X_-$) at E, and reverts to symmetric output at F. The similarity to the $J = \tfrac{1}{2}$ behaviour can be understood from the similarity in this régime of the (u, v) polynomial (4) to the polynomial describing the $J = \tfrac{1}{2}$ asymmetric branch (Hamilton *et al.* 1982). In figure 1b ($\beta_1 = 100$, $\beta_2 = 10$) a type of behaviour that does not occur for $J = \tfrac{1}{2}$ appears in the region near B_2, where no stable outputs exist, indicating the possibility of oscillation.

[256]

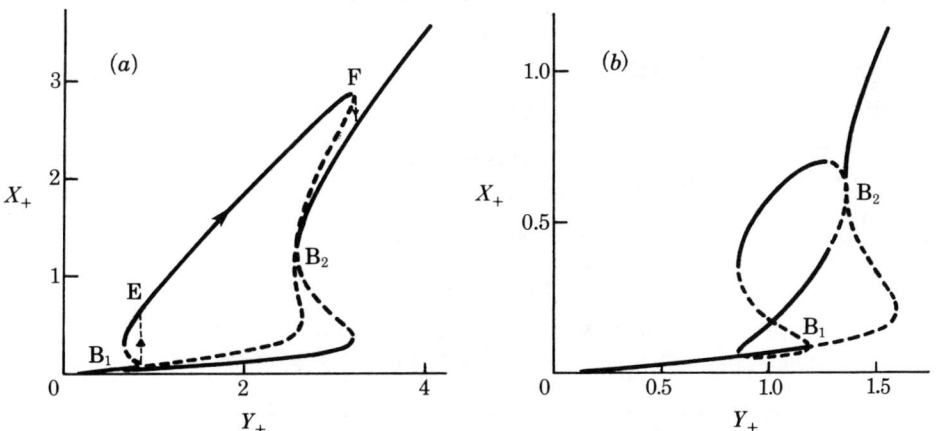

FIGURE 1. Input–output curves for $\xi = 1$ and (a) $C = 5$, $\beta_1 = 100$, $\beta_2 = 0.01$; (b) $C = 8$, $\beta_1 = 100$, $\beta_2 = 10$.

INTERPRETATION

Asymmetric output is produced when a (fluctuation generated) difference between the σ^+ and σ^- absorptions can be sustained, thus altering the relative intensities of the two cavity fields. Absorption, given (for σ^-) by

$$\eta_- = \rho_{1,1}(l) - \rho_{0,0}(u) + \frac{X_+}{X_-}\rho^2_{-2}(l), \tag{6}$$

depends on the population difference between the states of the transition $(\rho_{1,1}(l) - \rho_{0,0}(u))$ and on the ground level coherence between the $m = \pm 1$ states (the alignment $\rho^2_{-2}(l) = \rho_{1,-1}(l)$ in Zeeman representation). This coherence depends most strongly on the collisional decay rate Γ_2, since

$$\rho^2_{-2}(l) = -2\beta_2 X_- X_+ \{1 + \beta_1(X_+^2 + X_-^2)\}/3D. \tag{7}$$

Thus if β_2 is small, the difference between σ^- and σ^+ absorption is given by the difference between $m = 1$ and $m = -1$ populations, i.e. the orientation

$$\rho_{1,1}(l) - \rho_{-1,-1}(l) \equiv 2^{\frac{1}{2}}\rho^1_0(l) = 2\beta_1(X_+^2 - X_-^2)\{1 + \beta_2(X_+^2 + X_-^2)\}/3D. \tag{8}$$

This equation shows that when $X_+ > X_-$, population is pumped from $m = -1$ into $m = 1$, and is not rapidly equilibrated if Γ_1 is small (β_1 large). Thus the weaker X_- radiation is more strongly absorbed and, given sufficient cavity feedback, the X_+ radiation becomes dominant.

In the régime $\beta_2 \gg \beta_1$, the coherence (7) is important and the more intense radiation is more strongly absorbed, so that the atoms provide negative feedback, which tends to equalize the fields. This has been confirmed numerically and, for example, with initial polarization $\xi = 2$, we find $X_+ \approx X_-$ until the atoms become thoroughly saturated.

PHYSICAL RESTRICTIONS

The collision rates Γ_1 and Γ_2 are related by physical considerations. In terms of Zeeman rates $K_{1,0}$ (for transfer between $m = 0$ and $|m| = 1$) and $K_{1,-1}$ we have $\Gamma_2 = 3K_{1,0}$ and $\Gamma_1 = K_{1,0} + 2K_{1,-1}$, so that immediately $\beta_1 < 3\beta_2$. It can be shown that under the latter condition, the bifurcations occur in the physically inaccessible region between the symmetric branch turning points. Nature is even more restrictive however: experiment shows that $\beta_2 \approx 1.1\beta_1$, and this is supported by detailed collision calculations (Berman & Lamb 1969). This means that a $J_l = 1$, $J_u = 0$, atom in an isotropic collision environment will *never* allow asymmetric output. It is interesting that for $\beta_2 = \beta_1$, then $\eta_+ = \eta_-$, and the system becomes extremely stable in that the input polarization is exactly preserved in the output, i.e. $Y_+/Y_- = X_+/X_-$.

REFERENCES

Arecchi, F., Kurmann, T. & Politi, A. 1983 *Optics Commun.* **44**, 421.
Ballagh, R. J., Cooper, J. & Sandle, W. J. 1981 *J. Phys.* B **14**, 3881.
Berman, P. & Lamb, W. 1969 *Phys. Rev.* **187**, 221.
Hamilton, M. W., Ballagh, R. J. & Sandle, W. J. 1982 *Z. Phys.* B **49**, 263.
Kitano, M., Yabuzaki, T. & Ogawa, T. 1981 *Phys. Rev. Lett.* **46**, 926.
Omont, A. 1977 *Prog. quant. Electronics* **5**, 69.
Savage, C. M., Carmichael, H. J. & Walls, D. F. 1982 *Optics Commun.* **42**, 211.

Phil. Trans. R. Soc. Lond. A **313**, 449–451 (1984)

Printed in Great Britain

Closed equations of motion
for expectation values of collective operators for spontaneous emission in the presence of external driving fields

By J. Seke

Institut für Theoretische Physik, Technische Universität Wien,
Karlsplatz 13, *A*-1040 *Vienna, Austria*

By using a modified Robertson projection technique, developed by the author in previous work, three exact closed equations of motion for expectation values of collective atomic operators for spontaneous emission in the presence of a classical driving field are derived.

In this paper I shall derive exact closed equations of motion for spontaneous emission from a system of N two-level atoms in the presence of a strong classical driving field.

The modified Robertson projection-operator technique, developed in previous work (Seke 1980; Adam & Seke 1981), is based on the application of a special time-dependent projection-operator $P(t)$, which directly picks out the expectation values that are of interest. This projection operator transforms, by definition, the time derivative of the statistical density operator $\rho(t)$ of the total system S + R (S – system of N atoms with classical driving field and R – quantized electromagnetic field into which the atoms radiate) into the time derivative of $\sigma_S(t) \otimes \rho_R(0)$, where $\sigma_S(t)$ is the generalized canonical density operator (Robertson 1966; Seke 1980) of system S and $\rho_R(0)$ is the initial density operator of system R. It is assumed that systems S and R are statistically independent at $t = 0$: $\rho(0) = \rho_S(0) \otimes \rho_R(0)$.

The Hamiltonian of the total system $H(t) = H_0(t) + H_{SR}$ in the dipole and rotating-wave approximation can be written as

$$H_0(t) = H_S + H_R + H_{ext}(t)$$
$$= \omega S^z + \sum_{k,s} \omega_k a_{ks}^+ a_{ks}^- + (g S^+ e^{-i\omega_0 t} + g^* S^- e^{i\omega_0 t}), \tag{1}$$

$$H_{SR} = \sum_{i,ks} (g_{iks} S_i^+ \otimes a_{ks}^- + g_{iks}^* S_i^- \otimes a_{ks}^+), \tag{2}$$

where ω is the energy separation of the two atomic levels, ω_0 is the frequency of the harmonically varying external classical driving field (Mollow 1969; Agarwal 1974), $\omega_k = kc$ is the frequency of the kth mode of the quantized electromagnetic field, S_i^z and $S_i^{\pm}$ are the population inversion and dipole moment operators of the ith atom, $S^z \equiv \sum_i S_i^z$ and $S^{\pm} \equiv \sum_i S_i^{\pm} e^{\pm ik_0 r_i}$ (r_i is position vector of the ith atom) are the collective atomic operators, a_{ks}^+ and a_{ks}^- are the photon creation and annihilation operators for the mode ks (s is the polarization index) and g and $g_{iks} = g_{ks} e^{ikr_i}$ are the coupling coefficients.

We now introduce a generalized canonical density operator for our collective operators

$$\sigma_S(t) = \prod_{i=1}^{N} \left[\tfrac{1}{2} I_i + \sum_{\beta = (+,-,z)} \frac{d^\beta}{N} (S_{i,k_0}^\beta)^+ \langle S^\beta \rangle_t \right] = \prod_{i=1}^{N} \sigma_S^{(i)}(t) \tag{3}$$

and corresponding time-dependent projection operator

$$P(t)\ldots = \rho_{\mathrm{R}}(0) \otimes \sum_{\substack{j=1 \\ i \neq j}}^{N} \prod_{\substack{i=1 \\ i \neq j}}^{N} \sigma_{\mathrm{S}}^{(i)}(t) \sum_{\beta=(+,-,z)} \frac{d^{\beta}}{N} (S_{j,k_0}^{\beta})^{+} \mathrm{Tr}_{\mathrm{SR}}(S^{\beta}\ldots), \tag{4}$$

where I_i is the unit operator in the ith factor space, $d^{\pm} = 1$, $d^z = 2$ and $S_{i,k_0}^{\pm} = S_i^{\pm} e^{\pm ik_0 r_i}$, $S_{i,k_0}^{z} = S_i^{z}$.

Our set of operators $\{S^{\beta}\}$ is chosen so that the commutator $[H_0(t), S^{\beta}]$ can be expressed as a linear combination of the operators of the set and so that $\sigma_{\mathrm{S}}(t)$ can describe the initial conditions with uncorrelated and permutationally symmetric atoms. In the following calculations I will treat only the special initial conditions

$$\rho(0) = \sigma_{\mathrm{S}}(0) \otimes \rho_{\mathrm{R}}(0) \tag{5}$$

with $\rho_{\mathrm{R}}(0) = |\{0\}\rangle \langle\{0\}|$ being the vacuum radiation state.

By using the Robertson projection-operator formalism (Robertson 1966) from the equation

$$[d\sigma_{\mathrm{S}}(t)/dt] \otimes \rho_{\mathrm{R}}(0) = (-i) P(t) L(t) \rho(t), \tag{6}$$

we obtain the connecting equation between $\rho(t)$ and $\sigma_{\mathrm{S}}(t) \otimes \rho_{\mathrm{R}}(0)$:

$$\rho(t) - \sigma_{\mathrm{S}}(t) \otimes \rho_{\mathrm{R}}(0) = -i \int_0^t dt' U_{\mathrm{I}}(t,t') U_0(t,t') [I - P(t')] L(t') \sigma_{\mathrm{S}}(t') \otimes \rho_{\mathrm{R}}(0), \tag{7}$$

where

$$U_{\mathrm{I}}(t,t') = T \exp\left\{ -i \int_{t'}^t dt_1 U_0(t,t_1) [I - P(t_1)] L_{\mathrm{SR}} U_0^{-1}(t,t_1) \right\}, \tag{8}$$

$$U_0(t,t') = T \exp\left[-i \int_{t'}^t dt_1 L_0(t_1) \right], \tag{9}$$

with T as the Dyson time-ordering operator and $L_0(t)$, L_{S}, L_{R}, $L_{\mathrm{ext}}(t)$, L_{SR} as the Liouvillians corresponding to the Hamiltonians in (1) and (2).

If we let the operator $i[(L_{\mathrm{S}} + L_{\mathrm{ext}}(t)) S^{\beta}]$ act on (7) and afterwards take the trace over it, we obtain a set of *exact closed* equations of motion for expectation values of collective atomic operators S^{β}:

$$d\langle S^{\beta} \rangle_t/dt = i \mathrm{Tr}_{\mathrm{S}}\{[(L_{\mathrm{S}} + L_{\mathrm{ext}}(t)) S^{\beta}] \sigma_{\mathrm{S}}(t)\}$$

$$+ \int_0^t dt' \mathrm{Tr}_{\mathrm{SR}} [(L_{\mathrm{SR}} S^{\beta}) U_{\mathrm{I}}(t,t') U_0(t,t') L_{\mathrm{SR}} \sigma_{\mathrm{S}}(t') \otimes \rho_{\mathrm{R}}(0)], \quad \beta = (+,-,z). \tag{10}$$

The equations of motion obtained are closed since $U_{\mathrm{I}}(t,t')$, $P(t)$ and $\sigma_{\mathrm{S}}(t)$ are functions of the expectation values $\langle S^{\beta} \rangle_t$, $\beta = (+,-,z)$. These equations can easily be transformed to the rotating frame because the operators $\sigma_{\mathrm{S}}(t)$ and $P(t)$ are invariant under such a transformation.

For a single atom

$$\sigma_{\mathrm{S}}(t) = \tfrac{1}{2}I + \sum_{\beta} d^{\beta}(S^{\beta})^{+} \langle S^{\beta} \rangle_t,$$

$$P(t) = P = \rho_{\mathrm{R}}(0) \otimes \sum_{\beta} d^{\beta}(S^{\beta})^{+} \mathrm{Tr}_{\mathrm{SR}}(S^{\beta}\ldots),$$

and the equations of motion become much simpler. In the Markov and long-time approximations (but without making the Born approximation), these equations reduce to three linear differential equations.

Here we have shown that it is possible to derive three *exact closed* equations of motion for collective expectation values without making any restriction as to the dimensions of the atomic system. The derivation of such equations of motion is only possible by using our modified Robertson projection technique, because this method takes into account the special initial density operator of uncorrelated and permutationally symmetric atoms during the whole calculation. These special initial conditions have the consequence that all the multi-atom correlations can be expressed as functionals of the expectation values $\langle S^\beta \rangle_t$:

$$\langle S^{\beta_1}_{i_1, k_0} S^{\beta_2}_{i_2, k_0} \dots S^{\beta_n}_{i_n, k_0} \rangle_t - \left(\frac{1}{N}\right)^n \langle S^{\beta_1} \rangle_t \langle S^{\beta_2} \rangle_t \dots \langle S^{\beta_n} \rangle_t$$

$$= -\mathrm{i} \int_0^t \mathrm{d}t' \, \mathrm{Tr}_{SR} \, [S^{\beta_1}_{i_1, k_0} S^{\beta_2}_{i_2, k_0} \dots S^{\beta_n}_{i_n, k_0} U_I(t, t') \, U_0(t, t') \, L_{SR} \, \sigma_S(t') \otimes \rho_R(0)],$$

$$n \leqslant N, i_r \neq i_s \, \forall \, r \neq s, \quad (11)$$

as can easily be seen from (7).

Contrary to this, other methods that use the Zwanzig time-independent projection-operator technique (Zwanzig 1960; Emch & Sewell 1968; Agarwal 1974) cannot take into account the special initial conditions with regard to the system S and the derived master equations hold for quite general initial density operators $\rho_S(0)$. This is the reason why, by using these methods, it is not possible to derive closed equations of motion, but a hierarchy of equations containing all the expectation values $\langle S^{\beta_1} S^{\beta_2} \dots S^{\beta_n} \rangle_t$, $\beta_1, \dots, \beta_n = (+, -, z)$, $(n \leqslant N)$. But this can only be achieved for geometrically small atomic systems (Agarwal 1974) and not for arbitrary large systems like those discussed here.

REFERENCES

Adam, G. & Seke, J. 1981 *Phys. Rev.* A **23**, 3118.
Agarwal, G. S. 1974 *Springer Tracts mod. Phys.* **70**.
Emch, G. G. & Sewell, G. L. 1968 *J. math. Phys.* **9**, 946.
Mollow, B. R. 1969 *Phys. Rev.* **188**, 1969.
Robertson, B. 1966 *Phys. Rev.* **144**, 151.
Seke, J. 1980 *Phys. Rev.* A **21**, 2156.
Zwanzig, R. 1960 *J. chem. Phys.* **33**, 1338.